Instrumental Analysis
of Cotton Cellulose
and Modified Cotton Cellulose

FIBER SCIENCE SERIES

Series Editor

L. REBENFELD

Textile Research Institute
Princeton, New Jersey

Volume 1   CLOTHING: COMFORT AND FUNCTION
by Lyman Fourt and Norman R. S. Hollies

Volume 2   ESSENTIAL FIBER CHEMISTRY
by M. Carter

Volume 3   INSTRUMENTAL ANALYSIS OF COTTON CELLULOSE AND
MODIFIED COTTON CELLULOSE
Edited by Robert T. O'Connor

Volume 4   CHEMICAL MODIFICATION OF PAPERMAKING FIBERS
by Kyle Ward, Jr.

Other Volumes in Preparation

Instrumental Analysis
of Cotton Cellulose
and Modified Cotton Cellulose

EDITED BY

Robert T. O'Connor

Southern Regional Research Laboratory
New Orleans, Louisiana

MARCEL DEKKER, INC., New York 1972

MARCEL DEKKER, INC.
95 Madison Avenue, New York, New York 10016

LIBRARY OF CONGRESS CATALOG CARD NUMBER 72-78243

ISBN 0-8247-1500-4

PRINTED IN THE UNITED STATES OF AMERICA

PREFACE

During the past two or three decades the analytical laboratory has been undergoing dramatic changes in methods of operation and in general appearance. Gone are the pipet, the burette, the volumetric flask, and the beaker, replaced by the ultraviolet or infrared spectrometer, the x-ray fluorescence or diffraction unit, the mass spectrometer, the emission spectrograph, electron and scanning-electron microscopes, differential thermal analysis instruments, gas and liquid chromatographs, and more recently, gamma-ray spectrometers, neutron-activation instruments, electron emission spectrometers, Auger spectrometers, and many others.

In more recent years these instruments have been modified from manual tools to replace the manual techniques of the classical analytical chemist--modified to provide automated operation and to permit computerized simultaneous control of multiple units both to obtain data and to provide ready interpretation. Computerized automated analytical instrumentation has appeared in the research laboratories of an increasing number of industries as the cost of these sophisticated instruments has proven to be far less than the cost of the team of sophisticated analytical chemists which must be employed to compete with the output of the efficient, accurate, and versatile instruments.

These changes in the analytical laboratory have permeated the laboratories of the textile chemist and of the chemical finishing technician. Recent visits to several of these laboratories have revealed the common use of instrumental types of analysis, including even the more or less highly sophisticated instruments just recently introduced as tools of analytical chemistry.

This instrumentation and, with it, the ability to accomplish more analytical control in quality, in quantity, and in versatility, came to the textile and chemical finishing industry just when it was most urgently needed. Today, weaving or knitting of fibers to a fabric is not merely the use of a specific natural or synthetic fiber. Modern fabrics are combinations, or blends, selected and designed to provide a host of specific properties in the final fabric. Nor are these new fabrics merely combinations of two or more specific fibers. They include one, or, a combination of, several of the hundreds of chemical finishing agents available to provide precise properties for the specific end-use of a particular fabric.

iii

Today, the analytical chemist is not concerned solely with "What is the material (qualitative identification)?" or even, "How much is present in the analytical sample (quantitative determination)?" He is now being called upon to answer questions such as: "In what form (structure) is the material present? How is it bound (valence state)? Where is it spatially (location?" or, "How is it distributed through the analytical sample (homogeneity)?" The appearance of blends of several fibers into a single fabric and the treatment of fibers or fabrics with any of hundreds of chemical finishing agents has created the need for large numbers of these types of analyses in research laboratories, within textile mills, and in chemical finishing plants.

The objective of this volume is to present descriptions of the instrumental methods of analytical chemistry that have found widespread use in attempts to provide this information. Included in these pages are descriptions of instrumental techniques which have already found frequent use within the research laboratories of the textile and chemical finishing industries. Descriptions of techniques and examples of their applications to the analysis of textile materials are included for ultraviolet and infrared spectrometry, x-ray diffraction and fluorescence, nuclear magnetic resonance spectroscopy, optical, electron, and scanning-electron microscopy, mass spectrometry, gas, paper, and thin-layer chromatography, differential thermal analysis, and thermogravimetric analysis--all techniques the reader will readily find in common use in the research laboratory and the control laboratory of the textile and chemical finishing industries. Included also are descriptions of the more recent additions to this ever-growing list of automated instruments for chemical analysis, which, while they have not, as far as has been ascertained, as yet been used by the textile chemists or the chemical finishing technicians, are obviously of great potential value. This potential promise is indicated with descriptions of the newer instruments and by examples of their potential use, as obtained from surveys of their uses in other areas. Included are electron emission spectrometers, gamma-ray spectrographs, neutron-activation techniques, Auger spectrometers, among others.

Probably, no control of blends of natural and synthetic fibers into a single fabric, nor of the multiple use of the vast numbers of available chemical finishes, can be maintained without a laboratory equipped with several of these instruments for either qualitative identification or quantitative determinations of the modifications, or for periodical inspection of specific properties relating to homogeneity of treatment, nature of chemical modification, etc. Hopefully, this text will be of assistance to organizations building or modifying analytical resources, and to others using instrumental techniques of analysis in continual control of processes or in continual evaluation of product.

New Orleans, Louisiana   Robert T. O'Connor
September, 1972

CONTRIBUTORS

Anna M. Cannizzaro, Southern Regional Research Laboratory, New Orleans, Louisiana

Pronoy K. Chatterjee, Personal Products Company, Research and Engineering Division, Milltown, New Jersey

Carl M. Conrad, Southern Regional Research Laboratory, New Orleans, Louisiana

Ives V. deGruy, Southern Regional Research Laboratory, New Orleans, Louisiana

Wilton R. Goynes, Southern Regional Research Laboratory, New Orleans, Louisiana

C. Y. Liang, Department of Physics, San Fernando Valley College, Northridge, California

Robert T. O'Connor, Southern Regional Research Laboratory, New Orleans, Louisiana

Robert A. Pittman, Southern Regional Research Laboratory, New Orleans, Louisiana

Mary L. Rollins, Southern Regional Research Laboratory, New Orleans, Louisiana

Robert F. Schwenker, Jr., Personal Products Company, Milltown, New Jersey

Verne W. Tripp, Southern Regional Research Laboratory, New Orleans, Louisiana

CONTENTS

Instrumental Analysis
of Cotton Cellulose
and Modified Cotton Cellulose

Chapter 1

ELEMENTAL ANALYSIS
Detection, Identification, and Quantitative Determination of
Metals and Nonmetallic Elements

Robert T. O'Connor

Southern Regional Research Laboratory
New Orleans, Louisiana

I. INTRODUCTION

During the past two or three decades there has been an ever-increasing growth in the use of instrumental methods of analysis. The wet-chemical laboratory has almost disappeared, replaced by modern instrumental methods, most, but not all of them, spectroscopic methods.

We cannot talk of spectrochemical analysis as a substitute for volumetric analysis or gravimetric analysis, as spectroscopic analysis covers a myriad of methods and techniques, each designed to obtain specifically desired data. If we define chemical or analytical spectroscopy as the use of any portion of the electromagnetic spectrum, in any manner whatsoever, for the purpose of qualitative detection and identification, or for quantitative measurement of atomic or molecular constituents of any analytical sample, we can, theoretically at least, readily classify spectroscopic analysis into about 100 specific techniques. As shown in Table 1, the electromagnetic spectrum is, for convenience, divided into seven major spectral regions on a basis of wavelength, frequency, or energy of radiation. Radiations in each of these divisions can be employed in seven different ways for purposes of chemical analysis. Thus, as indicated in the table, forty-nine divisions of chemical spectroscopy can be classified. However, this is not all, for atomic spectroscopy can be quite different both in type of instrument employed, in techniques devised to obtain spectral data, and in the interpretation and ultimate use to which data obtained can be used. Thus, theoretically, at least, the forty-nine methods tabulated can be doubled, arriving at a total of about 100 different spectroscopic methods of analysis. (We can easily find the missing two to round the figure to 100 by consideration of a couple of spectroscopic techniques which do not strictly fit the definition of analytical spectroscopy but are well-recognized spectroscopic methods of analysis. For example, mass spectroscopy and electron emission spectroscopy, which, as they do not involve the use of the radiation of photons, do not fit into the classification scheme given in Table 1.)

The naming of any specific method of spectrochemical analysis follows logically from this classification merely by indicating the region of the electromagnetic spectrum employed, the manner in which the radiation is used, and the adjective "atomic" or "molecular." Thus we refer to "x-ray atomic fluorescence spectroscopy" or to "infrared molecular absorption spectroscopy." As will be illustrated, often the adjectives "atomic" and "molecular" are not required to eliminate ambiguity, and the two examples given above are more often referred to merely as "x-ray fluorescence spectroscopy" and "infrared absorption spectroscopy." On other occasions, however, the adjectives are essential to adequately define the method of analysis under discussion. Thus "ultraviolet atomic absorption" and "ultraviolet molecular absorption" require instruments of entirely different design, employ techniques with little or no similarity, and yield data for entirely dissimilar objectives.

Often, too, the name commonly used to indicate a specific method of spectroscopic analysis is not derived simply from a combination of the region of the electromagnetic spectrum used, the manner in which the radiation is employed, with, if required, the adjective "atomic" or "molecular." Thus "ultraviolet (or visible) atomic emission spectroscopy" is well-known as "spectrochemical analysis"; "ultraviolet atomic (Raman) scattering" simply as "Raman spectroscopy."

The number of different spectroscopic methods that can be considered experimentally practical is, however, considerably less than this

TABLE 1

Electromagnetic Spectrum for Analytical Spectroscopy

Region of the electromagnetic spectrum - type of radiation		Mode in which radiation is used		Species
I	Gamma rays	A.	Emission	
II	X rays	B.	Fluorescence	
III	Ultraviolet	C.	Absorption	Molecular
IV	Visible X		Scattering X	
V	Infrared	D.	Raman	
VI	Microwaves	E.	Reflectance	Atomic
VII	Radiowaves	F.	Tyndall scattering	
		G.	Diffraction	
Totals 7	X	7	X	2 = 98

theoretical 100. First, several techniques resulting from such a simplified classification are, obviously, nonexistent. For example, x-ray emission (or absorption) is a process independent of the combining properties of the elements, and hence these techniques are restricted to the realm of atomic spectroscopy. The same conclusions can be reached regarding nuclear emission and absorption. On the other hand, infrared spectroscopy is vibration-rotation spectroscopy and microwave spectroscopy is pure rotation spectroscopy. As only molecules can vibrate and rotate, atomic techniques in these regions need not be considered. The forty-nine techniques resulting from combination of the wavelength or frequency of the radiation together with the manner in which these radiations are employed cannot be doubled, as often either atomic or molecular radiations are nonexistent. A more realistic number would be to increase this forty-nine by about one-half, resulting in a total of 75, rather than 100, completely independent methods of analytical spectroscopy.

The seventy-five methods of spectroscopic analysis can be reduced a bit further by consideration of the fact that, even although both theoretically and experimentally possible, some of the specific methods fall victim to the fact that the data obtained cannot compete with similar data obtained by an alternate technique. However, elimination on such a basis is dangerous as it may, even at this stage in the development of spectroscopic methods of analysis, be based on inadequate investigation of all possible combinations of electromagnetic radiation available. For example, electron emission spectroscopy, which was discarded for forty years in favor of the more promising x-ray fluorescence, is now showing very definite evidence of becoming an even more useful technique than the very popular x-ray fluorescence method for qualitative identifications and particularly quantitative determinations. However, competing techniques will limit the number of spectroscopic methods which can be developed for chemical analysis. A logical consideration indicates that about forty entirely independent techniques are now available for use in analytical chemistry. Even this number might be greeted with a bit of skepticism on the part of many analytical chemists. However, if not restricted to the analysis of specific commodities, a search of the literature will reveal at least isolated descriptions of the use of at least forty independent methods of analysis. As an illustration, ultraviolet and visible (electronic) atomic emission spectroscopy is a well-known technique for the qualitative identification and/or quantitative determination of metals in all sorts of commodities. Probably thousands of such analyses (under the more common term "spectrochemical analysis") are being made every day. Spectrochemical analysis is well-known to all analytical chemists. How many, however, could offhand cite a single example of ultraviolet (or visible) molecular emission for chemical analysis? A review of the literature reveals a technique for the identification of finishes on cotton cellulose [1] by means of a determination of their molecular emission. The molecular emission spectra are illustrated in Fig. 1. With the spectra depictured in this figure as reference spectra it becomes readily apparent that finishing agents can be identified, and from a consideration of band intensities, again by comparisons with quantitative measurements of standards, be used to obtain a reliable quantitative estimation of the amount of the identified finish. The technique is not, as might be concluded from the date of this reference [1], a discarded one. A recent issue of the Varian Associates publication "Resonance Lines" [2] describes a method for the detection and quantitative measurement of the elements sulfur and phosphorus. These two elements cannot be analyzed by the method of spectrochemical analysis or atomic absorption with conventional instrumentation as the principal resonance lines in their atomic spectra lie in the far-ultraviolet region where vacuum techniques, not readily amenable to convenient routine analyses, are required. The molecular emission technique consists of burning the analytical sample in a hydrogen flame, resulting in the formation of molecular species with band heads in the visible region of the spectrum for phosphorus at 5262 Å, and in the near-ultraviolet region at 3838, 3740, and 3940 Å for sulfur.

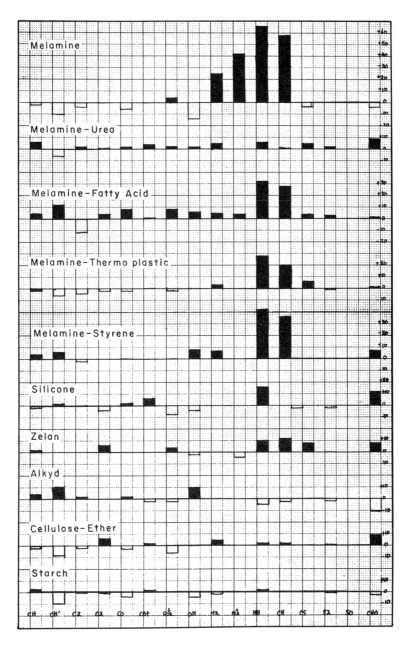

Fig. 1. Ultraviolet molecular emission spectra of various finishing agents on cotton. (Courtesy Am. Dyestuff Reptr.)

The objection might be raised that these different combinations of the wavelength of the photons employed and the manner in which they are used do not constitute different spectroscopic methods. The electromagnetic spectrum is really a manifestation of the same phenomenon, the differences between the various spectral regions being merely one of energy, wavelength, or frequency of the radiating photons. The various terms for spectroscopic methods are merely semantics--terminology introduced to differentiate between inconsequential differences in a single technique. A consideration of the experimentation involved reveals that this argument is not valid. The variations in the energy of the radiation in the different spectral regions require instrumentation to detect and measure them which varies widely. There is little resemblance, for example, between an x-ray diffractometer and an ultraviolet absorption spectrophotometer. The techniques to obtain useful spectral data also vary considerably. A specialist, or recognized expert in one area of chemical spectroscopy, for example, high resolution nuclear magnetic spectroscopy, may have only a casual knowledge of another area, such as neutron-activated gamma-ray spectroscopy. The information to be obtained from the raw data requires a considerably different training for adequate interpretation. There is little, if any, similarity in the techniques for obtaining or in the interpretation of the atomic lines in spectrochemical analysis with a high resolution spectrograph and the technique of measuring and interpreting molecular absorption bands in infrared absorption spectrophotometry. Finally the ultimate objective varies with measurements in one spectral region compared to another, i.e., crystal properties by means of x-ray diffraction, organic structural analysis by means of infrared absorption spectroscopy, determination of fractions of a part per million of a specific chemical element by means of x-ray fluorescence, etc. Each of the individual forty spectroscopic methods has its own instrumentation and techniques; each produces its own characteristic raw data requiring different methods of interpretation; and each reveals different specific facts regarding an analytical sample. Each is indeed a specific, entirely independent method of chemical analysis.

In this volume, the various chapters dealing with spectroscopic techniques will describe how a selection of these forty methods has been and is being applied to problems arising in connection with research investigations of cotton cellulose and of chemically modified cotton cellulose, resin treated cottons, and the identification and quantitative measurements of chemical finishing agents on cotton fabrics.

This chapter is concerned with the use of spectroscopic methods of analysis for the detection, identification, and quantitative measurement of the chemical elements, both metallic and nonmetallic, which may be native to cotton cellulose and, of more practical importance, of such elements introduced into the cellulose molecule incident to its chemical modification, its resin treatment, or any of the many chemical finishing processes.

Of the forty methods of spectroscopic analysis some twelve have been
introduced for qualitative identification and/or quantitative estimation of
the various chemical elements. These are listed in Table 2. Of these
twelve techniques only three (underscored in Table 2) appear to have been
employed in investigations of cotton cellulose.

In this chapter a description of the three methods which have been
adopted in analytical laboratories of several textile research organizations-
electronic atomic emission spectroscopy, x-ray fluorescence spectroscopy,
and atomic absorption spectroscopy-will be given in some detail. These
descriptions of the use of elemental analysis into research investigations
of cotton and modified cotton celluloses will be followed by an attempted
evaluation of the advantages and disadvantages of each technique. Following
this comparison, an attempt will be made to describe, somewhat more
briefly, two or three of the more powerful tools for elemental analysis
which, although not as yet employed by the cellulose chemist, are poten-
tially available to him and which will, undoubtedly, eventually find applications to investigations of cotton and modified cottons.

TABLE 2

Spectroscopic Methods for Elemental Analysis

Nuclear	I	Gamma-ray spectroscopy
	II	Neutron absorption
X-ray	III	X-ray emission
	IV	X-ray fluorescence
	V	X-ray absorption
Electronic	VI	Ultraviolet atomic emission
	VII	Ultraviolet atomic absorption
	VIII	Visible atomic emission
	IX	Visible atomic absorption
Radio	X	Wide-line nuclear magnetic resonance spectroscopy
	XI	Quadrupole moment nuclear magnetic resonance spectroscopy
	XII	Electron emission spectroscopy (electron spectroscopy for chemical analysis (ESCA))

II. X-RAY FLUORESCENCE

The spectroscopic methods listed in Table 2 are all practical methods for elemental analyses, although magnetic resonance techniques and x-ray emission have not proven to be very convenient and have found no particular applications. Only three of these methods have been applied to the analysis of cotton cellulose or to modified cotton cellulose, ultraviolet and visible (or electronic) atomic emission, ultraviolet and visible (or electronic) atomic absorption, and x-ray fluorescence (or secondary) x-ray emission. Ultraviolet and visible are listed as a single method as they are really not distinctly different techniques. The only difference between ultraviolet and visible radiation is the very important one that the eye is a natural detector of visible radiation only. Both are electronic radiations; i.e., they arise from transitions between electronic levels, the valence levels, of atoms, and both can be detected with similar instrumentation and measured with similar detectors.

Electronic transitions between the inner shells of molecules or atoms give rise to x-rays. Those arising from transitions between the levels closest to the nucleus, of greatest energy and penetrating power, and involving the K or L levels, are known as "hard" x rays; those from levels further removed from the nucleus, of less energy and penetrating power and involving the L, M, and N, etc. levels, are known as "soft x rays." When one considers x rays he is most likely to think of their power to penetrate matter. However, these rays are absorbed to some extent in passing through any material. Their absorption follows the Beer-Lambert laws of absorption, and absorption spectra of x rays can be used, just as absorption of ultraviolet, visible, or infrared radiation, for purposes of chemical analysis. However absorption, or emission, of x rays is an atomic process, independent of the chemical combination in which the particular element occurs. Another remarkable property of x-ray absorption is that if the absorptivity of a given mass of matter (rather than the linear absorptivity) through a specified path length of sample of specified concentration is considered, the mass absorption is independent of the state of matter. These two unusual properties can be beautifully illustrated with examples from Klug and Alexander [3] : the absorption of x rays by diamond, graphite, and mercury. Diamonds are very transparent to light, whereas carbon in the form of graphite is a strong absorber; both, however, have the same mass absorptivity for x rays. Liquid and solid mercury are opaque to light, but mercury vapor is almost perfectly transparent; all three states of mercury, however, have the same mass absorptivity to x rays.

If, with polychromatic x radiation, the mass absorptivity is plotted against wavelength, the mass absorptivity decreases smoothly as wavelength decreases, except for a series of sharp breaks where steep increases are observed (Fig. 2). These steep, sharp increases are called

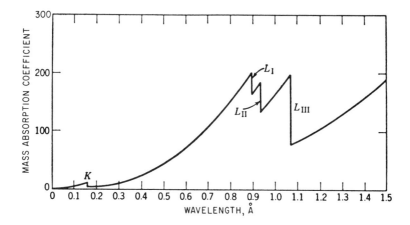

Fig. 2. X-ray absorption spectrum. (Courtesy Am. Chem. Soc.)

absorption edges. They mark points on the wavelength scale where the x rays possess sufficient energy to eject an electron from one of the shells. The positions of the discontinuities identify the element causing the absorption of the x rays, and the magnitude of the absorption edge on the mass absorptivity ordinate is a measure of its concentration, exactly analogous to a visible absorption curve of intensity vs wavelength.

X-ray absorption thus appears to be a convenient method for identifying qualitatively and measuring quantitatively. However, outside of the area of hydrocarbon chemistry it does not appear to have been widely used. It has been a useful method for the determination of lead content, to measure the tetraethyl lead in gasoline, and has been cited in examples to determine sulfur and halogens in hydrocarbon mixtures, additives in lubricating oils, etc. It has not, as far as has been ascertained, ever been used in textile or cellulose chemistry.

Theoretically there is no reason why x-ray emission cannot be used to identify elements in a manner exactly analogous to electronic emission, the so-called spectrochemical analysis in the ultraviolet or visible regions of the electromagnetic spectrum. When a beam of electrons impinges on a metal target, x rays characteristic of the particular metal will be expelled. Such an apparatus is an x-ray tube, which may be constructed with Cu, W, Mo, etc. , targets to obtain x rays of different characteristics. For use as an analytical tool it would be necessary to construct an x-ray tube for each analysis. Actually, attempts to perform analyses with x-ray emission have been described with devices which permit the changing of the target, in this case the analytical sample, within the x-ray tube. However, this technique has not proven to be particularly convenient, rapid, or simple.

A simpler and very commonly used technique which employs x radiation
consists of simply allowing the x rays from an x-ray tube to strike the
analytical sample. Fluorescent rays, which appear as secondary x-ray
emission, are dispersed by utilizing a crystal with a known lattice constant,
frequently a bent mica crystal, to act as a diffraction grating. The
diffracted fluorescent x rays are detected by a Geiger counter goniometer,
and their intensities are automatically recorded as a function of the
goniometer angle, which is proportional to the wavelength or frequency of
the x rays. Figure 3 illustrates the x-ray flourescence spectra of cotton
cellulose and modified cotton cellulose. As shown in this figure, the
position on the abscissa identifies individual elements, both metallic and
nonmetallic, and the magnitude of each is obtained from the ordinate by
calibration with known standards.

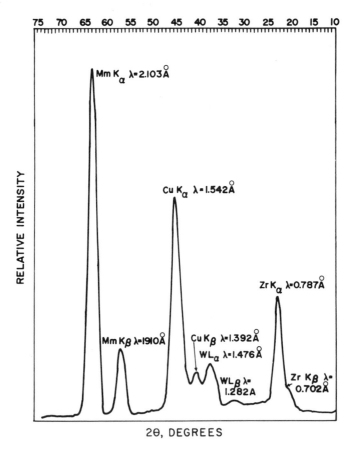

Fig. 3. X-ray fluorescent spectrum of a chemically modified cotton.

X-ray fluorescent techniques are undoubtedly being used in some textile research laboratories, although little has appeared regarding their use in the literature. In the author's laboratory the x-ray instrumentation is a General Electric XRD-6 designed especially for x-ray fluorescence measurements. Radiation is obtained from tungsten, molybdenum, or chromium tubes, operated at 50 k. For detection or determination of the heavier elements, emitted radiation of short wavelengths is used with a LiF crystal and xenon-filled proportional counter. For lighter elements, long wavelengths were used with an ethylenediamine tartrate (EDT) or with a potassium acid phthalate (KAP) analyzing crystal. Both of these crystals were used with gas-flow (90% argon, 10% methane) proportional counters and with a helium path.

X-ray fluorescent spectra to permit qualitative identification of any element can be obtained from the cotton cellulose in almost any physical form. Fiber can be measured as a fiber bundle, fabric as a small square of the cloth. This is one of the major advantages of the x-ray fluorescent method of spectroscopic analysis. No prepreparation of the sample, or at least only very minimal preparation, is required. No time is required for dissolution or ashing, and thus the x-ray fluorescent method is the most rapid. For best quantitative results, standardization of the particle size of the textile materials is highly desirable. The textile material in the form of fiber, yarn, or fabric is ground to pass a 20-mesh screen in an intermediate Wiley mill. A disk of the ground material, about 300 mg in weight, is pressed in a 1-in. die under pressure of 25 tons for 3 min. The resulting disks have good mechanical stability, giving reproducible counts over a period of several months. Working, or calibration, curves are prepared from disks made by mixing solutions containing approximately known quantities of the desired element with the ground cotton, drying, agitating any loose material to ensure homogenous distribution, and pressing in the die. If the sample material cannot be readily dissolved, it is finely ground and intimately mixed with the ground cotton.

X-ray fluorescence can be used to detect, identify, or quantitatively determine almost all elements in the periodic table. This is the second of its major advantages--it can be used for the detection or determination of nonmetals as well as metallic elements. Spectrochemical analysis and atomic absorption techniques are limited, as will be shown, to the more metallic elements, the alkali metals, alkaline earths, true metals, and, with some loss in sensitivity, to some of the amphoteric elements. The practical limitation in the scope of analyses by means of x-ray fluorescence is the detection and determination of the lighter elements. Magnesium is the lightest element which can be detected and determined at concentrations less than 0.1% without a vacuum spectrometer and a special analyzing crystal such as lead stearate, involving techniques which make the x-ray fluorescent method somewhat awkward for rapid analyses. With elements lighter than atomic number 12, Auger emission (see section below under

the heading "Electron Emission Spectroscopy") becomes very prominent at
the expense of x-ray fluorescence.

Count rates of equivalent disks, prepared as described, show a maximum
difference of 2%. The relationship of percentage composition to counting
rate was found to be linear over a range of 0.01 to 3.0% for most elements
examined. The lower limit for the three lightest elements determined,
aluminum, magnesium, and silicon, is 0.1%. Above concentration of about
3% a loss in counting rate was noted, indicating self-absorption. This loss
becomes very marked at a concentration of about 10%, and dilution with
pure cellulose, free of the element to be determined, must be made.

In Table 3 conditions for x-ray fluorescence measurements of twenty-
five elements found in native cotton or added to the cotton to impart specific
properties are listed. Seventeen of the heavier elements are detected and
measured with a LiF crystal detector and eight of the lighter elements with
an EDT crystal, including one use of the KAP crystal (for the detection of
aluminum). The purpose for which the specific elements have been added
in the chemical modification of the cotton cellulose, usually as an organometallic compound or as a simple inorganic salt or oxide, is given for
each element. The analytical x-ray line used to detect and for the
determination of each element is listed, and the concentration range over
which such measurements can be made is included.

Examples of the use of x-ray fluorescence measurements of elemental
content in investigations of the chemical modification of cotton fabric are
illustrated by the data in Tables 4 and 5. In investigations to determine
the effectiveness and permanency of cotton fabric treated with cadmium
selenide (CdSe) and cadmium sulfoselenide (CdSSe) to weather exposure,
the fate of the selenide pigments was followed by x-ray fluorescent
measurements of the cadmium and selenium content. Fabrics treated with
these two reagents were exposed to weather for a period of two years. At
four-month intervals samples were removed from the exposure racks.
The breaking strength of the fabric was measured and the Cd and Se content obtained by x-ray fluorescent measurements. The percent retention
of the Cd and Se in the original samples before exposure is reported in
Table 4. Both Cd and Se contents show excellent correlation with the
breaking strength of the fabric during the entire exposure. These data
demonstrate that either Cd or Se content, as obtained by x-ray fluorescence
measurements, can be used as a criterion of the duration of effectiveness
of the chemical treatment. The excellent correlation between the gradual,
progressive loss of breaking strength with time and the corresponding
gradual and progressive loss of Cd and Se indicates that deterioration of
the fabric is caused by the gradual loss of the protective selenide pigments
and that for longer useful life of the fabric a more permanent treatment is
required.

TABLE 3

Summary of Conditions of X-Ray Fluorescence Analysis of Elements in Modified Cottons

Chemical element	Function in modified cotton	Line	Analytical Wavelength		Range of conc. (%)	Typical net count rate over conc. range (counts/sec)
			2θ (deg)	Crystal type		
Ag	Fungicide	K	16.1	LiF	0.03–3	0.34 – 5.81
Al	Water repellency	K	36.8	KAP	0.1 –3	Variable
Br	Flame resistance	K	30.0	LiF	0.03–3	0.29 – 24.5
Ca	Native cotton	K	44.9	EDT	0.03–3	0.29 – 22.8
Cd	Fungicide	K	15.3	LiF	0.03–3	0.07 – 4.87
Cl	Flame resistance	K	65.0	EDT	0.03–3	0.26 – 13.2
Co	Fungicide	K	52.8	LiF	0.03–3	1.20 – 61.3
Cr	Mildew proof	K	69.4	LiF	0.03–3	1.37 – 55.6
Cu	Fungicide	K	45.0	LiF	0.03–3	1.06 – 61.3
Fe	Pigment	K	57.5	LiF	0.03–3	2.40 – 48.3
Hg	Fungicide	L	35.9	LiF	0.03–3	0.16 – 27.4
K	Native cotton	K	50.3	EDT	0.03–3	1.26 – 55.2

aMo target.
bCr target.

TABLE 3 (continued)

Summary of Conditions of X-Ray Fluorescence Analysis of Elements in Modified Cottons

Chemical element	Function in modified cotton	Analytical Wavelength			Range of conc. (%)	Typical net count rate over conc. range (counts/sec)
		Line	2θ (deg)	Crystal type		
Mg	Native cotton	K	43.9	KAP	0.1 –3	Variable
P	Flame resistance	K	88.7	EDT	0.03–3	0.06 – 2.94
Pb	Outdoor protection	L	28.2	LiF	0.02–3	0.73 – 41.1
S	Crosslink	K	75.2	EDT	0.03–3	0.12 – 4.18
Sb	Flame resistance	K	13.5	LiF	0.05–3	0.61 – 34.4
Se	Fungicide	K(Mo)[a]	31.9	LiF	0.1–10	1.51 – 19.2
Si	Water repellency	K	108.1	EDT	0.1– 3	0.08 – 3.48
Sn	Flame resistance	K	14.0	LiF	0.05–3	0.90 – 50.7
Te	Fungicide	L	109.5 (He path)	LiF	0.05–3	0.14 – 3.08
Ti	Pigment	K	86.1	LiF	0.05–3	1.01 – 41.7
W	Fungicide	L(Cr)[b]	43.0	LiF	0.03–3	0.24 – 13.8
Zn	Catalyst	K	41.8	LiF	0.03–3	1.20 – 29.5
Zr	Water repellency	K	22.6	LiF	0.05–3	0.25 – 11.4

aMo target.
bCr target.

TABLE 4

Relative Cadmium and Selenium Contents of
Weather-Exposed Treated Cotton Fabrics

	CdSe[a]			CdSSe[b]		
Months of exposure	Cd	Se	Breaking strength	Cd	Se	Breaking strength
	% of original			% of original		
0	100	100	100	100	100	100
4	83	88	87	87	86	93
8	64	80	77	74	73	84
12	39	66	54	21	58	67
16	30	52	38	8	40	50
20	23	49	32	--	31	45
24	17	47	23	--	24	32

[a]Original cadmium and selenium contents: 0.63% and 0.37%,
respectively.

[b]Original cadmium and selenium contents: 0.50% and 0.25%,
respectively.

Table 5 shows some relationship between the phosphorus content of
fabrics treated to impart flame resistance and a standard flame retardency
test. Several durable flame-retardent finishes have been developed over
the past few years. Most are based on treatment with a phosphorus-
containing resin, tetrakis (hydroxymethyl) phosphonium chloride (THPC)
having attained some commercial acceptance. A recently developed flame-
retardent treatment is based on the reaction product of methanolic sodium
hydroxide and THPC. The product is tetrakis (hydroxymethyl) phosphonium
hydroxide (THPOH).

The data in Table 5 are selected from hundreds of analyses of phosphorus in cotton fabrics treated with THPOH resin. All samples, various
types of cotton fabrics, were padded with solutions of the THPOH resin of
varying concentrations, all dried to 10% moisture and cured for varying
times with ammonia vapor. The phosphorus content was obtained by x-ray
fluorescence and correlated with the vertical flame test for flame resistance. Char length is defined by ASTM specification D 626-55T as the

TABLE 5

(a) Relationship of % Concentration and Curing Time
on Phosphorus Content and Char Length

Sample number	Resin solution	Curing time in NH_3	% Add-on	P %	Char length
7-1	15%	2	8.2	3.31	4.70
7-3	15%	4	8.3	3.96	3.75
7-5	15%	6	8.4	3.91	3.50
7-7	15%	8	8.6	4.04	4.38
7-9	15%	10	8.7	4.17	3.70
7-2	20%	2	9.5	4.22	3.05
7-4	20%	4	10.2	4.23	3.53
7-6	20%	6	10.1	4.39	2.65
7-8	20%	8	11.3	4.67	3.63
7-10	20%	10	12.0	5.19	3.45

(b) Relationship Between % Concentration of Resin and
% Phosphorus and Char Length

Sample number	Resin solution	P %	% Add-on	Char length
127-1	10% THPOH	0.95	8.4	BEL
127-2	15% THPOH	2.91	8.9	BEL
127-3	20% THPOH	4.16	12.9	3.3
127-4	25% THPOH	3.45	12.7	4.2
127-5	30% THPOH	4.07	12.8	6.1

length of tear of a fabric after it has been subjected to the open flame of a
Bunsen burner for 12 sec.

Table 5 (a) shows the relationship between the concentration of the resin
solution and the curing time with the % add-on, the phosphorus content, and
the char length. At the low resin concentration there is no significant
change in % add-on or in phosphorus content with curing time in ammonia
and only a relatively poor negative correlation between add-on or phosphorus content and char length. At the higher resin solution concentration
there is good correlation between curing time and % add-on and % phosphorus content and some correlation with char length.

Table 5 (b) illustrates the relationship between the % concentration of
the resin solution and the % add-on, % phosphorus content, and the char
length. Above 20% THPOH, it will be noted, there is no appreciable change
in % add-on or % phosphorus content.

These data illustrate how determination of elemental concentration by
x-ray fluorescence can be used to investigate and control the chemical
modification of cotton fabric to obtain predetermined specifications.

III. ELECTRONIC EMISSION - SPECTROCHEMICAL ANALYSIS

Spectrochemical analysis, the identification and determination of metallic elements by identification and intensity measurements of atomic lines
in the electronic, that is, ultraviolet and visible, regions of the electromagnetic spectrum, emitted when the analytical sample is excited as by
a flame, arc, or spark, is the oldest of the spectroscopic methods of
elemental analysis. While there are now available to the analytical spectroscopist several techniques for qualitative and quantitative determinations
(Table 2) and while some of the newer methods offer special advantages,
in actual numbers of analyses made in research and production control
laboratories spectrochemical analyses probably exceed, by a considerable
factor, all other types added together. This technique has replaced
chemical methods of analysis in every type of application. In laboratories
dealing with a wide range of commodities, where elemental analyses of
any considerable number are required, they are now being made by scores
and hundreds per day by means of spectrochemical analysis.

There are two almost entirely different techniques of spectrochemical
analysis, depending upon the type of detector used. The first and oldest
is the photographic technique, where the raw data are obtained on a photographic plate and analysis is made by means of wavelength comparators
and intensity measurements with comparative densitometers. The second
and newer type of detecter is employed in the direct-reading spectrometers,
the quantometers and atom counters, employing photoelectric cells or
Geiger counters. The second technique has become very popular, in spite
of considerably increased cost for initial equipment and instrumentation,
as it is more rapid and superior in precision and accuracy. The earlier,

photographic plate technique, is, however, of considerable advantage if
the problem is one mainly of qualitative identification of the elements present in a specific analytical sample, as a record of all possible atomic
lines is obtained and can be used for any qualitative search at any subsequent time. Choice then of the technique to be selected will depend
largely on the nature of the specific analytical problem. The advantages
of the direct reading instruments in convenience, speed, and accuracy of
quantitative determinations have recently, in more sophisticated laboratories, been largely reduced by the complete digital and computer handling
of the raw data as obtained by either technique.

Regardless of the detector used for the spectrochemical analysis, if
quantitative values are needed, the most rigorous requirement is that of a
uniform matrix. This requirement for a uniform matrix is absolute.
Regardless of the control of conditions for the production of atomic emission
lines, the presence of one metal in the analytical sample will affect the
intensities of the atomic spectral lines of other metals. This effect can be
readily illustrated by a simple experiment. Prepare a sample of pure,
spectrochemically metal free, silicon dioxide to which exactly one part per
million of copper is added. Similarly add to a corresponding pure sample
of sodium chloride exactly the same quantity of copper. Obtain the emission
spectrum of each of these samples in the region where the strong copper
arc lines appear, 2961.16 and 3273.96 Å, under identical, very carefully
controlled conditions, as for example, from an arc source operated at
identical voltage, of exposure times of identical duration, and photographed
on the same portion of a single photographic plate, or measured with the
same photoelectric cell or Geiger counter detector. Despite all precautions to
avoid differences in intensities, the copper line from the first sample the silicon
dioxide, will appear relatively strong, while from the second sample, the
sodium chloride, they will appear as weak lines. This observed difference
is readily explained by the fact that the energy available, in this case as
electrical energy from the arc or spark, is not equally distributed between
the two elements in the analytical sample. Sodium will "rob" more than its
share of the energy from the copper, and the copper lines will appear weak in a
sodium chloride matrix. Copper, however, will take more than a proportional
share of the electrical energy at the expense of the silicon atoms and produce
emission lines of copper with considerably enhanced intensity.

From this discussion it will be obvious that, if an analytical sample of
unknown composition (but actually sodium chloride) is to be analyzed by
spectrochemical methods for its copper content, and if this analysis is
conducted against standards prepared by careful and precise addition of
known amounts of copper to pure silicon dioxide (or more exactly by means
of working or calibration curves prepared for such a series of standards),
the intensities of the copper lines will be weak relative to the standard and
the resulting calculated concentrations of the copper in the analytical sample
will be too low. The requirement that the matrix for both standards and

analytical samples be kept constant is thus of primary importance in precise quantitative spectrochemical analyses.

In the earliest popular applications of spectrochemical analysis the
requirement for a constant matrix was not of such paramount importance.
These early uses of spectrochemical analyses were confined, to a considerable degree, to analysis within the metallurgical industries. In the
scores and hundreds of analyses in a steel mill the matrix is always iron,
the analyses being conducted to follow the concentration of alloying elements
in minor quantities in this constant matrix. Similar conditions prevailed
in the determination of specific metals in brass, bronze, and aluminum
samples. (It has been demonstrated that various metals present only in
trace quantities will not produce undesirable matrix effects [4].)

While of paramount importance, the requirement for a uniform matrix
is not the only requirement for high precision and accuracy in quantitative
spectrochemical analyses. Obviously, even in a constant matrix, the
intensities of the emission lines will depend upon the ability of the technician to maintain his arc or spark source at exactly predetermined voltages,
to keep the image of the source exactly aligned with the fine entrance slit
of the spectrograph, to overcome very small differences in the speed of
emulsion and differences in development conditions from plate to plate, if
photographic techniques are used, or to repeat exactly intensity measurements in densitometers, or maintain constant counting times or rates with
direct reading detectors. Careful analysis has shown that the overall precision of spectrochemical analysis is more a matter of algebraic addition
of several of these factors rather than the magnitude of any one of them
[5]. In many of these cases the individual errors in each specific step of
the procedure are so small that attempts to reduce them further are usually
futile. As these errors are random with ± signs, it should be obvious that
analyses made in duplicate or triplicate will result in more precise values.

Spectroscopists have worked out methods to overcome their inability to
obtain absolute control over all parameters, when producing spectral lines
and when measuring intensities, by the technique known as "internal standard method." The usual technique is to prepare "standards" containing a
series of concentrations of the element(s) to be determined in the preselected constant matrix. The intensities of the selected atomic lines of
each element to be determined are measured and these values plotted
against the known concentrations to obtain a calibration or "working" curve.
From measured intensities of the same line in the spectrum of each
unknown analytical sample, the concentration can be read from such calibration curves. The internal standard method for control of instrument
parameters is a simple extension of this procedure. To every analytical
sample and to each calibration standard a constant amount of a rare element,
known to be absent from the samples to be analyzed, is added. Consider,
for example, two samples of cellulose fiber containing two different amounts
of copper. These samples are to be analyzed by means of previously

prepared "working curves" for copper to which a constant amount of, let us say the element beryllium, has been added. The same concentration of beryllium is added to each of the two analytical samples. An atomic line of beryllium is selected at a wavelength close to the previously selected copper line (so the two lines will appear on the same portion of the photographic plate and intensity differences arising from emulsion differences will be minimized). The intensity of the beryllium line cannot change because of changes in concentration (as beryllium content is the same in all standards and samples). Any change that does occur in the intensity of the beryllium line must, therefore, be a result of change in conditions incident to some step in either the production or the measurement of the intensity of the beryllium line, for example, uncontrollable fluctuations in the dc arc source voltage, unknown variations in the emulsion on the photographic plate, etc. However, these factors will affect the intensities of the copper line in a similar manner, and if the two lines are close together on the photographic plate and they have been properly selected (they should, for example, both be lines arising from the same spectrum, i.e., both arc lines or both spark I and spark II, etc., lines which would be expected to react to changes in instrument parameters in an identical manner), the effect on the intensity of one line will be, within limits of measurement, identical to the effect on the second line. Factors that tend to increase or decrease the intensity of the beryllium line will tend to increase or decrease the intensity of the copper line by the same increment if these factors are not too large. But the constant intensity of the beryllium line will, in multiple analyses, tend to become known (as remember the beryllium content is constant in all analytical samples and in all standards). Any gross variation from this more or less established constant intensity value should be looked upon with suspicion, and very probably a specific analysis where such a gross departure from the established constant intensity of the beryllium line occurs should be rejected. The working curves are prepared by plotting the known copper concentration of the standards, not against the intensities of the copper lines, but against the ratio of the intensities of these lines to those of the beryllium line. This intensity ratio of copper/beryllium is much less likely to be affected by uncontrolled or uncontrollable factors when obtaining or measuring the intensities of the spectral lines. Combination of the internal standard technique with properly prepared working or calibration curves and critical control of the matrix will permit very accurate analyses by either photographic or direct reading techniques.

While the nature of some analytical problems more or less results in a constant matrix, as illustrated in the discussion of the use of spectrochemical analysis in the field of metallurgy, in an agricultural research laboratory successive samples may involve appreciably different matrices, necessitating different standards and numerous calibrations. A method to overcome the problems of continually preparing calibration curves for varying matrices was solved by O'Connor and Heinzelman [6] some years ago. They described a method applicable to all agricultural materials, of which cotton cellulose is a good example. The method is based on the fact that agricultural samples are organic materials and must be ashed, whether analyzed by photographic or direct reading techniques, otherwise they would burn too rapidly to permit recording of atomic emission line intensities. The method devised permits the problem of constant matrix and of suitable internal standards to be adequately satisfied during the ashing with very little extra effort in the preparation of the ash for the spectrochemical analysis of native cotton cellulose or of any of the several chemically modified cottons or blends with other natural or synthetic fibers.

Cotton cellulose, in the form of fiber, yarn, or fabric, is an excellent example of an agricultural product which can be handled by this general technique for agricultural commodities or constituents [7]. In principle, the method involves the ashing of the cellulose sample in the presence of magnesium nitrate and the addition, if desirable, of an "internal standard" during this ashing. The magnesium nitrate serves four distinct purposes: (a) it acts as an ashing aid, reducing the temperature during ashing of the sample, thereby avoiding loss by volatilization; (b) it acts as a carrier for the small quantity of ash, which often results from the ashing, and permits analysis of very small samples or samples with very low ash content; (c) it acts as a spectroscopic buffer, it becomes the constant matrix and the analytical problem is reduced to a determination of the selected elements in a matrix of magnesium oxide-magnesium carbonate (the ashing need not, and probably should not, be carried to a white ash); and (d) the magnesium provides the required constant internal standard line, or, if desirable for specific analyses where magnesium lines would be somewhat far removed from the preselected analytical lines, an internal standard can be added in known and constant amount to each sample and standard during preparation. For complete reproducibility the standards are prepared by being subjected to the same ashing technique with magnesium nitrate.

In the earlier report, describing the general method for the spectrochemical analysis of plant products [6], two procedures were recommended, one for samples with an ash content of less than 1% (where there could be no matrix effect from the composition of the ash [4]) and another for samples with ash content over 1% (where the composition of the ash might effect the condition for constant matrix). In a paper which appeared some years later [8] the second of these procedures was used, mainly because the cotton samples being analyzed were fabrics chemically modified by the introduction of metal-containing reagents to produce specifically desired effects. Both procedures are described here as originally published:

Procedure A: Ash Content of Sample Less than 1%. Exactly 16.67 grams of the sample are weighed into a Vycor dish (90-mm. diameter), and 0.50 gram of magnesium nitrate in ethanol (2 ml. of a solution of 250 grams of magnesium nitrate hexahydrate per liter of 95% ethanol) is added. The reagents used must be free of the trace metals for which samples are to be analyzed. Obviously the ratio of sample weight to the magnesium nitrate buffer is arbitrary and can be varied for different samples depending upon the sensitivity required, if the ratio actually used is considered in obtaining and using the working curves. The ratio suggested here will permit analysis with a sensitivity of about 1 part of metal in 10,000,000 parts of sample for the more sensitive metals, copper and iron, for most types of samples.

Ashing with magnesium nitrate is, of course, well known. The procedure was adapted from that used in a micromethod for phosphorus using ethanol instead of hydrochloric acid as a solvent. The dish is covered with an inverted short-stemmed borosilicate glass funnel, with a maximum diameter less than the maximum diameter of the dish. The sample is heated on a hot plate, and the temperature is gradually and cautiously raised until a temperature of approximately 300° C. is attained. The charred sample is then ashed in a muffle furnace, with an initial temperature of 225° C. increasing in increments of 25° C. at 30-minute intervals until a temperature of 450° C. is reached. The samples are held at this temperature in the furnace overnight. They are then removed, cooled, and quantitatively transferred to a small mullite mortar with aid of a camel's-hair brush. The ash is finely ground and thoroughly mixed in final preparation for analysis.

Working standards are obtained by ashing, in an identical manner, portions of the magnesium nitrate solution containing graduated amounts of the various elements to be determined.

Procedure B: Ash Content of Sample over 1%. Exactly 5 grams of sample are ashed in a tared Vycor dish as described in Procedure A, but without addition of the magnesium nitrate solution. After removal from the furnace the dish is cooled and reweighed to determine the ash content of the sample. A volume of a very dilute aqueous solution of germanium dioxide is added so as to introduce a quantity of the dioxide equal to 4.14% of the ash. The ash when dried, ground, and thoroughly mixed is ready for analysis.

Working curves are obtained by preparing a mixture of salts representing the major constituents of a typical ash. The particular mixture selected contains 65.18% potassium carbonate, 0.96% sodium chloride, 11.02% calcium carbonate, 18.70% magnesium carbonate, and 4.14% germanium dioxide. Such a mixture is believed to be sufficiently representative of the ash of most plants so as to avoid any significant error due to extraneous ion effects. The amount of germanium dioxide, selected as an internal standard, was obtained from study of photographic densities. Under the conditions of photographing, this quantity gave ideal densities for use with line-width measurements.

Working curves are prepared by adding graduated amounts of elements to portions of the salt mixture.

Tests of analyses of cotton cellulose samples made by these procedures are shown in Tables 6, 7, and 8. In Table 6 the nickel content of a chemically modified cotton fabric has been repeatedly determined to establish the reproducibility of the method. A coefficient of variation of 7.9% of the amount of zinc present at the concentration range of about 0.1% was obtained. In Table 7 the usual analytical recovery tests are shown. Three metals, copper, iron, and manganese, were first determined and then known amounts of each of these elements added. The total amount recovered is compared with the total amount present. The % deviation in these tests is the sum of the deviations in the determination of the amount originally present and in the determination of the amount present after this analytically determined amount has been augmented by a known concentration. The deviation varies from -26.7 to +26.3% of the amount present at the range from about 3 to 12 ppm. Finally, in Table 8 accuracy tests are presented. The amount of copper, iron, and manganese added to cotton fiber entirely free of these elements was determined by the spectrochemical method and compared to the amount known to be present with deviations ranging from -14.9 to +16.3% of the amount present, which was varied from 40 to 2000 ppm.

Analyses of native cotton fiber are illustrated in Table 9. In this table results of the analysis for copper, iron, and maganese in eight varieties of cotton grown at a single location under identical conditions and for a single variety grown at thirteen different locations are tabulated. The results show that, even when grown under identical conditions, genetic character has a marked effect on the metal content, but a single genetic variety, when grown at different locations, will exhibit substantial differences, which (although not evident from the data shown) reflect the concentrations of these three metals in the soil in which the cotton fiber was raised.

The types of analyses described in Table 9, while they illustrate the application of the spectrochemical analysis procedure to determination of elements found in native cotton cellulose, do not, of course, constitute any considerable activity in textile research laboratories. The chemical modification of cotton, its treatment with various reagents to impart properties not native to cellulose has, however, created an additional need, or opportunity, for applications of spectroscopic techniques for elemental analyses. As will be described in a later chapter (Chapter 8) infrared absorption spectroscopy has been of considerable assistance to the cotton

TABLE 6

Precision of Spectrochemical Method
Determination of Nickel in a Modified Cotton Fabric[a]

Spectrum	% Metal		% Dev.
1	0.12		5.26
2	0.11		3.51
3	0.11		3.51
4	0.11		3.51
5	0.13		14.04
6	0.12		5.26
7	0.11		3.51
8	0.10		12.28
Ave.	0.114	Ave. deviation	6.36

[a]Standard deviation = + 0.009. Coefficient of variation (%) = 7.9.

cellulose chemist in attempts either to identify the manner in which a specific sample of cotton fiber or fabric has been modified or to determine the extent of the chemical modification.

The sensitivity of infrared absorption spectroscopy is, however, often not sufficient for adequate quantitative determination of the small amounts of reagents used in many of these chemical modifications to produce cotton resistant to mildew, rot, soiling, actinic degradation, and flame, or to impart, by crosslinking the cellulose molecules, such properties as permanent creasing and wrinkle resistance. Many of these reagents contain elements which lend themselves to spectrochemical analysis. Several of them are organometallic compounds and many others are inorganic salts and oxides. Once a modification of this type has been identified, as by infrared absorption spectroscopy, it is a relatively simple procedure to obtain a reasonably accurate determination of the elemental content, and to calculate the concentration of the organometallic compound, or the extent of modification with the inorganic salt or oxide.

In Table 10, eleven elements commonly found in chemically modified cottons are listed, together with the spectral lines used for their determination and the concentration range in which they are most commonly

TABLE 7

Spectrochemical Analyses for Cu, Fe, and Mn
Recovery Tests

Sample, cotton and added metal	Amount originally present (ppm)	Total amount present (ppm)	Amount recovered (ppm)	Deviation (%)
Cu	0.96	2.96	2.87	- 3.0
Cu	1.44	3.44	2.52	-26.7
Cu	0.99	2.99	2.75	- 8.0
Fe	5.60	9.60	9.30	- 3.1
Fe	7.47	11.47	12.52	+ 9.2
Fe	6.87	10.87	13.73	+26.3
Mn	6.82	10.82	13.03	+20.4
Mn	3.02	7.02	7.91	+12.7
Mn	8.55	12.55	12.61	+ 0.5

TABLE 8

Spectrochemical Analysis for Cu, Fe, and Mn
Analysis of Known Standards

Synthetic ash	Amount in standard (ppm)	Amount determined (ppm)	Deviation (%)
Cu	40	42.7	+ 6.8
Cu	100	85.1	-14.9
Cu	200	214	+ 7.0
Fe	400	468	+17.0
Fe	1000	977	- 2.3
Fe	1600	1860	+16.3
Mn	210	214	+ 1.9
Mn	841	871	+ 3.6
Mn	2000	1740	-13.0

TABLE 9

Spectrochemical Analyses for Cu, Fe, and Mn (a) of Varieties
of Cotton Grown at a Single Location (Stoneville, Miss.),
and (b) of a Single Variety of Cotton (Coker 100 Wilt)
Grown in Various Locations

(a)

Cotton variety	Ash (%)	Copper	Iron	Manganese
		Parts per million		
Acala 4-42	1.42	1.99	11.55	7.22
Acala 1517	1.41	1.60	12.19	7.21
Coker 100 Wilt	1.13	1.59	11.14	7.46
Deltapine 15	1.09	1.40	7.90	5.90
Mebane (Watson)	1.36	1.01	5.91	3.18
Rowden 41B	1.28	1.74	10.84	5.36
Stoneville 2B	1.19	1.53	12.55	6.28
Wilds	1.26	1.00	7.97	4.91
Average	1.27	1.48	10.01	5.94

(b)

Location of growth	Ash (%)	Copper	Iron	Manganese
		Parts per million		
Auburn, Ala.	1.56	0.78	9.03	8.90
Sacaton, Ariz.	1.49	2.72	12.89	12.39
Shafter, Calif.	1.35	0.93	5.60	5.81
Tifton, Ga.	1.26	1.67	7.91	9.30
St. Joseph, La.	1.33	0.68	3.74	2.79
Stoneville, Miss.	1.13	1.59	11.14	7.46
Statesville, N.C.	1.53	1.55	6.88	7.44
State College, N.M.	1.56	1.05	4.85	5.82
Chickasha, Okla.	1.37	1.13	7.25	7.87
Florence, S.C.	1.26	2.10	7.29	9.08
Jackson, Tenn.	1.22	1.45	7.25	9.20
College Station, Tex.	1.26	4.01	13.77	8.17
Greenville, Tex.	1.22	1.95	7.03	8.42
Average	1.37	1.66	8.05	7.90

TABLE 10

Element Determined by Spectrochemical Analysis of Chemically Modified Cotton

Element	Analytical line wavelength (Å)	Magnesium control line wavelength (Å)	Conc. from typical analyses (%)	Used as
Al	3082.16	2942.11	0.018 – 0.038	Oxide: soil resistance. Hydroxide: water repellent. 8-Quinolinolate: resistance to microorganisms.
Bi	3067.72	3073.99	0.51 – 0.56	Oxide: flameproofing. Hydroxide: water repellent, weatherproofing.
Cd	3261.06 3261.06	2915.52 2942.11	0.037 – 0.36	Chloride: resistance to termites and microorganisms. Pentachlorophenate: rotproofing. Hydroxide: flame and weatherproofing.
Cr	2835.63 3197.08	2942.11 2942.11	0.003 – 2.61	Cr complex perfluoroctanoic acid: water and oil repellency. Oxide: rot- and weatherproofing, fungicide and bactericide, protection against light.

TABLE 10 (continued)

Element Determined by Spectrochemical Analysis of Chemically Modified Cotton

Element	Analytical line wavelength (Å)	Magnesium control line wavelength (Å)	Conc. from typical analyses (%)	Used as
Cr (cont)				Lead chromate: weatherproofing, flame retardant. Potassium dichromate: resistance to termites and microorganisms.
Co	3044.00	2942.11	0.10 – 0.17	Hydroxide: mildew-, rot-, and general weatherproofing. Metaborate: mildew-, rot-, and general weatherproofing.
Cu	2961.16 3273.96 3273.96	2942.11 2915.52 2942.11	0.00005 – 2.02	Naphthenate: rotproofing, fungicide, mildew-proofing. 8-Quinolinolate: water-, rot-, mildew-, and weatherproofing. Phosphate: waterproofing. Carbonate: fungicide and bactericide.
Fe	3017.63 3020.64 3057.45	2942.11 2942.11 2942.11	0.17 – 4.34	Hydroxide: fungicide and bactericide. Oxide: rot- and weatherproofing, protection against light. Phosphate: waterproofing.

Mn	2576.10 3054.36	0.030 – 8.04	Oxide: protection against actinic degradation, waterproofing. Flurosilicate: catalyst in stabilization against shrinkage.
Ni	2943.91 3002.49	0.13 – 3.16	Acetate: water repellency. Oxide: water repellency. Phosphate: weatherproofing and protection against actinic degradation.
Sn	2571.59 3175.02	0.002 – 2.14	Oxide: weatherproofing and protection against action of light; flameproofing and mildewproofing. Phosphate: weatherproofing, protection against actinic degradation, fire resistance. Bis(tri-n-butyl tin) oxide: antibacterial agent.
Zn	3345.02 4722.16	0.030 – 0.24	Oxide: flame retardant. Chloride: flame retardant. Phosphate: fire and weather resistance. Nitrate: crease resistance, wrinkle resistance (catalyst).

found in chemically modified cotton. The particular form in which these elements are used and the property imparted to the cellulose by their presence are included. The data in Table 10 illustrate the type of modifications that have created a considerable interest in spectroscopic methods of elemental analysis by the cellulose chemist. They have created the need for spectroscopic methods, of one type or another, by the textile research laboratories which deal with cotton fabrics either as pure cotton fibers or in blends with the numerous synthetic fibers.

IV. ATOMIC ABSORPTION SPECTROSCOPY

A third method which has been used by the cotton cellulose chemist and by the textile industry for the determination of both metallic and nonmetallic elements is atomic absorption spectroscopy. Each of the procedures used for elemental analyses has its individual advantages and disadvantages, as will be discussed in the following section. The major advantage of atomic absorption which has created a considerable interest in this technique is its simplicity and the relatively lower cost of initial equipment. These factors have led several groups to initiate investigations of atomic absorption as a potential substitute for better known methods of spectrochemical analysis and x-ray fluorescence. Committees within the American Society for Testing and Materials, the American Association of Textile Chemists and Colorists, and The Technical Association of the Pulp and Paper Industry have, among others, initiated collaborative investigations of the method for the detection and determination of elements in textile materials and long-chain polymers.

The atomic absorption process is essentially the reverse of atomic emission (spectrochemical analysis). In the latter the atoms are excited by introduction of energy in the form of electricity or heat as with a flame, arc, or spark source. The energy is absorbed promoting valence electrons to excited states, to orbits of the atoms of higher energy levels. When the electrons fall back to lower energy states, the energy reappears as radiation in the electronic portion of the electromagnetic spectrum, that is, in the visible or ultraviolet division. It is well known that the vapor of an element, usually with excited atoms, is an excellent absorber of radiation from that specific element. Thus sodium vapor is a very good absorber of the D lines in the emission spectrum of sodium. This principle was well understood by the earliest spectroscopists. It was used to explain the dark lines in the solar spectrum, the Fraunhofer lines, appearing as the emission lines of elements in the sun, are absorbed in passage through a layer of the vapor of the specific element. While this principle of atomic absorption appears to have been well understood, its use as an analytical tool is relatively new [9, 10]. Its applications to the textile industry and to cotton cellulose have only begun to be investigated.

In the experiments to be discussed here to illustrate the application of atomic absorption to cellulose chemistry, a Perkin Elmer atomic absorption instrument, model 303, equipped with a Digital Concentration Readout Accessory (DCR-I) was used. In atomic absorption, the emission lines to be absorbed are excited in a hollow-cathode lamp. A specific lamp is selected for the analysis of a specific element (although in some cases hollow-cathode lamps designed to provide radiation from more than a single element have been proposed which can be used, in turn, for the analysis of two or more elements). The analytical sample is atomized into a Boling burner head, using air-acetylene oxidizer fuel system. The raw absorption data are printed out on tape by the DCR-I.

For atomic absorption techniques the sample must be in true solution otherwise difficulties will be encountered during attempts to aspirate the analytical sample into the oxidizer-fuel system. As cotton is insoluble in any suitable solvent, recourse must be had to an ashing technique. This necessity for ashing is one of the major disadvantages of atomic absorption techniques for a material which cannot be readily prepared as a suitable solution. A sulfate ashing is used. The following procedure is recommended:

Procedure: A typical procedure for the atomic absorption analysis of a cellulose sample involves a sulfate ashing. Five grams of the analytical sample are accurately weighed into a platinum dish and wetted with 1 M sulfuric acid and ashed in a muffle furnace at 575°C until all carbonaceous material is destroyed. The ashing time is usually about 3 hr. The analytical sample is removed from the furnace, cooled, and dissolved in 10 ml of 2 M hydrochloric acid, warmed slightly if necessary and brought to volume in a 50-ml volumetric flask with distilled water. If undissolved matter is present this solution is filtered before analysis.

The analytical sample is atomized into the flame and burned as in a Boling burner head operating with an air-acetylene oxidizer fuel system. The line sources of the emission spectrum are hollow-cathode lamps of the specific elements being analyzed, operating at the specific current for which each lamp is rated. The reduced intensity is measured directly by a counter appropriately arranged to detect and measure only the atomic line selected for the specific analysis. If the atomic absorption instrument is equipped with a digital concentration readout, an accessory now commonly included, the results obtained are printed out on tape directly in terms of percent concentration. If this accessory is not available, intensity measurements must be converted to percent concentration by means of previously prepared working or "calibration curves."

Typical data obtained for nine metals commonly found in chemically modified cotton are given in Table 11. Data of these types are compared with similar data obtained by other techniques in the following section.

Ant-Wuorinen and Vaspapää [11] have recently published a description of the use of atomic absorption spectrometry for the determination of inorganic impurities in cellulose. They describe a dry ashing technique followed by solution in 6 N HCl. Data are presented for the determination of Na, K, Mg, Ca, Fe, Cu, and Zn in bleached cotton linters, bleached sulfite pulp, and in viscose stable fiber before and after wet-disintegration in distilled water and after Wiley-milling followed by washing in distilled water containing dissolved carbon dioxide and by washing in dilute hydrochloric acid. X-ray fluorescence analysis revealed that the material not dissolved in the 6 N HCl was mainly SiO_2 but that is was not pure, containing small amounts of Ca, Fe, S, Ti, Mn, Cu, and Zn, in concentrations, however, which probably did not cause appreciable error in the results obtained by atomic absorption spectrometry. However, the insolubility of the ash and resulting removal of certain traces of metals by adsorption must be considered in the use of atomic absorption techniques.

Ant-Wuorinen and Vaspapää concluded that atomic absorption spectrometry offered several advantages over chemical methods and other spectroscopic techniques: (a) The method is specific, only in a very few cases are interferences possible; (b) only a single portion (about 10 g) of cellulose is required for determination of all the elements; (c) concentrations down to fractions of a part per million are easily determinable; (d) a single determination of any given element contained in the ash solution is effected in a few seconds (not including, however, the lengthy ashing and solution procedures); (e) the instrument is exceptionally simple to use, relatively little experience is required for the attainment of reliable results; and (f) the reproducibility and accuracy of atomic absorption spectrometry are superior to those obtainable by flame photometry.

V. COMPARISON OF THREE SPECTROSCOPIC TECHNIQUES

X-ray fluorescence, spectrochemical analysis, and atomic absorption appear to be the only three spectroscopic methods which have been used in reported analyses of cotton cellulose or in textiles. A consideration of the merits and disadvantages of each should, therefore, be considered.

Results from repeated determination of zinc content by the three methods are compared in Table 12. The sample was a cotton fabric treated with dimethylol ethyleneurea and varying amounts of zinc salts as catalyst to form intermolecular crosslinks between cellulose chains to produce a wrinkle resistant, or durable-press fabric. These data, in Table 12, illustrate that each of the three spectroscopic techniques is capable of yielding, within the experimental limits of any of the three methods, the same concentration values. While a laboratory working with one specific

TABLE 11

Determination of Trace Metals in Cellulose
by Atomic Absorption

				Trace metal (ppm)					
Ca	Cu	Fe	K	Mg	Mn	Na	Pb	Zn	
310	2.35	6.5	10.5	219	34.4	27.0	11.7	1.70	
17.4	0.32	1.6	11.0	5.1	0.15	222	13.0	0.53	
413	1.11	3.6	10.5	100	0.53	59.0	12.4	0.85	
384	1.48	2.6	14.0	106	0.53	163	13.0	0.74	

TABLE 12

Zinc Content of Chemically Modified Cotton by
Three Analytical Methods

Sample No.	Zn content (%)		
	Atomic absorption	Emission	X-ray fluorescence[a]
1	0.06	0.04	0.07
2	0.06	0.06	0.08
3	0.06	0.04	0.07
4	0.18	0.08	0.22
5	0.03	0.07	0.04
6	0.09	0.08	0.10
7	0.03	0.06	0.04
8	0.58	0.88	0.72
9	0.87	0.89	0.80

[a] Zn K_α line, LiF crystal, tungsten target, 1 mA current.

method will, very probably, become more proficient with that particular technique, it is our conclusion that with sufficient experience any laboratory, working with any one of the three methods, could perfect its techniques until the same results would be obtained from any of the three methods. In other words, selection of the method to adopt need not, and probably cannot, be based on demonstrated superior results.

What criteria, then, can be used by a laboratory seeking to initiate identifications and determinations of the chemical elements by means of spectroscopic techniques. In Table 13 an attempt has been made to rate each of the three techniques in six categories with a "plus" value if a specific technique appeared to offer additional advantages over the other two, a "negative" value if it appeared to have specific disadvantages, and no rating if it appeared to offer neither advantages nor disadvantages. The tabulation reveals that each of the four techniques (dividing spectrochemical analysis, for convenience, into two subtechniques depending upon whether photographic or direct-reading detectors are used) offer about the same number of + and - values.

TABLE 13

Relative Criteria for Selection of Spectroscopic Techniques
for Elemental Analyses of Cotton Cellulose[a]

	Emission analysis, "spectrochemical analysis," photographic	Emission analysis, "spectrochemical analysis," direct reading	X-ray fluorescence	Atomic absorption
I. Initial equipment cost	Moderate	(-) Expensive	(-) Expensive	(+) Relatively low
II. Operator requirements	(-) Highly skilled	Moderately skilled	(-) Highly skilled	(+) Relatively low
III. Time – sample preparation	(-) Slow – involves ashing	(-) Slow – involves ashing	(+) Very fast – negligible	(-) Very slow – involves ashing and solution
IV. Time – instrument analysis	(+) Reasonably fast	(+) Very fast	(+) Very fast	Moderately fast

TABLE 13 (continued)

Relative Criteria for Selection of Spectroscopic Techniques
for Elemental Analyses of Cotton Cellulose[a]

	Emission analysis, "spectrochemical analysis," photographic	Emission analysis, "spectrochemical analysis," direct reading	X-ray fluorescence	Atomic absorption
V. Scope of analysis	(+) Excellent for qualitative scan (−) Metallic and amphoteric elements only	(−) Metallic and amphoteric elements only	(+) Metallic and nonmetallic elements (−) No light elements below AW 12 (Mg)	(−) Metallic and amphoteric elements only
VI. Sensitivity	(+) Very high	(+) Very high	(−) Somewhat low	Moderately high

Original equipment cost and the degree of sophisticated skill required for suitable operation are part of the cost of establishing and maintaining an atomic spectroscopic laboratory. In these two factors, atomic absorption has a clear-cut and decisive advantage. The cost of initial equipment with adequate accessories for atomic absorption operation is probably less than 10% of that required for adequate equipment for x-ray fluorescence or for direct-reading emission analysis and probably less than half the initial cost to establish a photographic emission laboratory. This advantage is coupled with the equally decisive advantage in degree of skill required for successful operation. An operator can become qualified to produce satisfactory results with atomic absorption with far less training and in a much shorter time than a sufficiently qualified spectrochemical analyst or x-ray fluorescence spectroscopist can be developed. To operate an x-ray fluorescence spectrometer efficiently and to interpret data produced requires a reasonable degree of sophisticated understanding of the principles and theory of x-ray analysis. Similarly, time and experience are required before an operator becomes very skillful in the recognition of atomic lines in atomic emission spectrograms to make reliable identifications of the elements to be measured. It should be noted, however, that these two categories, cost of initial equipment and operator requirements, are really the only advantages offered by atomic absorption. In all others it is either deficient or no more than equal to the other techniques. However, as costs are of importance, these two factors alone appear to guarantee that atomic absorption will receive considerable attention from research laboratories throughout the textile industry where the problem of elemental analysis arises.

Categories III, time for preparation of the sample for analysis, and IV, time of the instrumental analysis, can be combined. In these two categories x-ray fluorescence has a clear and decisive advantage. No ashing nor solution of the analytical sample is required and the detection and determination by modern instrumentation can be made in a few minutes. Thus, for over-all time of analysis x-ray fluorescence is, for either qualitative identification or quantitative determination, far superior to the other two techniques.

The advantages in scope of analysis and ultimate sensitivity appear to be somewhat divided between the photographic technique of spectrochemical analysis and the x-ray fluorescence method. The photographic spectrochemical technique is probably the simplest and most direct answer to the question involving more or less complete qualitative analysis which arises frequently in an analytical laboratory. Not "Does the analytical sample, i.e., a chemically treated cotton, contain copper or manganese and if so, how much?"; but, "What elements are present in this unknown sample?" For this latter task the emission spectrogram, covering a considerable portion of the visible and ultraviolet electromagnetic spectrum, is ideal.

The spectrograms can be examined and reexamined at a later date to
verify the presence or absence of any specific element. This technique
also offers very high sensitivity which, by taking advantage of integration,
that is, longer exposures with larger samples, can be made considerably
higher than either x-ray fluorescence or atomic absorption.

However, spectrochemical analysis is limited in scope to a determination of the metals and probably a few of the amphoteric elements. It is not
suitable for the identification or determination of nonmetallic elements.
The less basic an element, in general, the less chance it has of obtaining
an equitable share of the electrical energy available to produce electronic
transitions and exhibit characteristic atomic lines. The more acidic elements are represented in their spectrum by a larger number of lines than
most of the more basic elements, and the energy available is consequently
divided among more transitions, with the result that no single transition is
predominant to produce a very strong line. Finally, the strongest lines of
the acidic elements lie, usually, in the far ultraviolet region of the electromagnetic spectrum, unavailable to any convenient analytical technique, and
less sensitive lines have to be resorted to for purposes of analysis. For
these three reasons spectrochemical analysis is, for most practical purposes, limited to basic elements, the true metals and a few amphoteric
elements.

The advantage of x-ray fluorescence in scope and sensitivity lies in the
fact that this technique is very useful for the determination of nonmetallic
elements, the true acid-formers. Of particular importance to cotton cellulose research, B, P, Se, and Cl can be determined. None of the other
techniques permit satisfactory identifications and determinations of these
elements. However, as the Auger spectroscopy takes over from x-ray
fluorescence for light elements, x-ray fluorescence is not available, without extremely sophisticated accessories, for identification or determination of elements below about Mg (AW 24) in the periodic table. In scope or
sensitivity atomic absorption does not appear to offer any advantages over
the other techniques.

From these discussions the following conclusions regarding choice of
spectroscopic techniques can be made:

1. If in a specific textile laboratory all three techniques are available
the personnel will find occasions where, for a specific problem, one of
the three will be preferred over the other two, and other occasions where
another technique will be the definite choice.

2. Selection of the method to be used, particularly if the choice is to be
limited to a single spectroscopic technique, will depend upon the principal
operation of the laboratory. If cost must be a major consideration, atomic
absorption must be seriously considered; if the analytical spectroscopist
is being called upon to identify the elements present in a fiber or fabric,

photographic spectrochemical analysis will probably be most satisfactory; if rapid quantitative determinations for specific metals and nonmetals is the major requirement, x-ray fluorescence will probably be the most judicious choice.

3. In the final selection of instrumentation, the limitations as well as advantages of particular equipment must be kept in mind. If, for example, the principal problems to be encountered are the analysis for acidic elements, Se in weather-proofing, P in flameproofing, etc., neither spectrochemical analysis nor atomic absorption will be of any particular help and x-ray fluorescence is the choice by elimination. If determinations for the very light elements, for example, B in flame retardant investigations, F in mildew-proofing investigations, etc., then x-ray fluorescence cannot be used and either spectrochemical analysis or atomic absorption equipment are to be selected.

VI. A NOTE CONCERNING OTHER SPECTROSCOPIC TECHNIQUES FOR ELEMENTAL ANALYSES OF POTENTIAL INTEREST TO TEXTILE CHEMISTS

There have been to date very few published reports concerning the application of spectroscopic methods to the analysis -- qualitative identification or quantitative determination -- of the chemical elements, either metallic or nonmetallic, to textile research. Visits to textile research groups, discussions with textile chemists, and particularly activities of specific committees of such organizations as the American Society for Testing and Materials, Committee E-2 on Emission Spectroscopy, the American Chemical Society, Division of Cellulose, Wood and Fiber Chemistry, committees from the American Association of Textile Chemists and Colorists, and the Technical Association of the Pulp and Paper Institute bear testimony of the considerable interest and activity among textile workers in the use of spectroscopic methods for such analyses. To date, as far as can be ascertained, activity has been confined to the three techniques described. However, as indicated by the expanding use of infrared absorption during the past decade, and of elemental techniques during the past few years, interest in spectroscopic methods within the textile industry has been increasing rapidly. It would appear, therefore, appropriate to close these discussions with brief descriptions of additional techniques which could very well find application and probably rapid acceptance with this expanding interest in spectroscopic instrumentation.

A somewhat newer spectroscopic technique, which is finding wide acceptance as a spectroscopic means for identification and determination of the chemical elements and which will undoubtedly be suggested to the textile chemist, is the coupling of gamma-ray spectrometry with neutron activation. If an analytical sample is a naturally radioactive material, the

nuclide can be identified without difficulty by beta- or gamma-ray spectroscopy. But, obviously, such analyses would be limited in scope. Activation analysis, whereby a nonradioactive analytical sample is made radioactive by bombardment with neutrons, electrons, protons, x rays, etc.,
has no such limitation in scope. A complete analysis by this method involves three more or less independent steps. First, the analytical sample
must be activated as with a beam of neutrons (neutron activation) to produce
gamma rays for analysis. Activation analysis is not limited to the use of
neutrons, as the analytical sample target can be transmuted by many particles, such as protons, deutrons, tritons, alpha particles, and x rays.
The popular choice of neutrons as a means of activation results from the
fact that most elements have a high probability of capturing slow neutrons.
In addition, neutron sources are now both readily available and convenient
to use. The gamma rays emitted from the now radioactive analytical sample are, in the second step of the analysis, detected by means of the convenient Tl-activated NaI crystal. The gamma rays are not, usually, measured
directly, but by secondary effects, the most successful and most popular of
which seems to be with the use of NaI-type scintillation counters. A gamma-
ray spectrometer is used in the third step of the procedure to sort out the
gamma rays as a function of their energy. The usual presentation is a
chart of energy values as the "X" parameter vs intensity, usually a relative
counting rate, as the "Y" parameter. Identification of the nuclide is made
by the position of the peaks, on the energy scale, and quantitative determinations by the height of these peaks on the intensity scale. Figures 4 and 5
illustrate gamma-ray spectrometric curves of neutron-activated gamma
rays from two elements which have been shown to be of interest to textile
chemists.

   High flux thermal-neutron activation is an extremely sensitive means
of analysis and a technique of extremely wide scope. Extreme sensitivity
is achieved by irradiating elements having large cross sections with a high
neutron flux. Using a reactor as a source of slow neutrons, the sensitivity
can be increased to the extent that one part per 10^{13} of some elements can
be detected. The technique has, in addition to its extremely high sensitivity which probably exceeds any other spectroscopic technique, the advantage
of wide applicability. It is capable of detecting and quantitatively measuring 75 of the chemical elements. It has the additional advantage of being
an entirely nondestructive technique. The precision, accuracy, and a
comparison with mass spectrometric determinations are illustrated in
Tables 14, 15, and 16, from the excellent monograph entitled Applied
Gamma-Ray Spectrometry, edited by C. E. Crouthamel. The reader interested in further details of activation analysis, scintillation counting
techniques, and gamma-ray spectroscopy is referred to this Volume 2 of
the International Series of Monographs on Analytical Chemistry [12]. Disadvantages of neutron activation - gamma-ray spectroscopy are probably

TABLE 14

Precision and Accuracy of Gamma Scintillation
Spectrometric Methods

	Samples			
	1	2	3	4
	^{235}U, weight %			
	3.002	21.46	65.24	92.43
	2.986	21.71	65.79	94.17
	2.989	21.51	64.30	92.75
	2.937	21.78	65.00	94.28
	2.964	21.59	65.65	93.35
	3.001	21.41	65.86	92.10
	2.952	21.46	65.97	93.42
	2.983	21.19	65.51	94.34
	2.997	21.70	65.37	93.69
	2.933	21.82	64.61	93.29
Ave.	2.9744,	21.563,	65.330,	93.382,
	±0.0262	±0.195	±0.551	±0.734
Coeff. var.,%	0.88	0.90	0.84	0.79
^{235}U by mass spec.,%	2.9687, ±0.003	21.506, ±0.022	65.516, ±0.066	93.177, ±0.014
Error,%	0.17	0.27	0.28	0.22

only the obviously rather sophisticated instrumentation involving relatively
high-cost initial installation.

A second technique, or a series of very closely related techniques, of
potential interest to the textile chemist interested in elemental analyses, is
the nonradiative method of electron spectroscopy. It will be recalled that
at the introduction to this chapter a discussion of the factors that reduced
the number of divisions of chemical spectroscopy to less than the 98 shown
in Table 1 was balanced by techniques that did not involve applications of
the electromagnetic spectrum, and did not employ photons of any energy

TABLE 15

Accuracy of Activation Analysis

Component determined	Material	Concentration of element (%)	
		Present	Found
Cu	Aluminum	0.0055	0.0054
	Al alloy	2.48	2.65
		0.30	0.32
Fe	Al alloy	0.21	0.20
		0.90	0.85
Mn	Beryllium	0.015	0.0149
	Al alloy	0.66	0.68
		0.041	0.04
Ti	Al alloy	0.016	0.016
		0.035	0.038
Zn	Al alloy	0.019	0.017
		0.077	0.072
Si	Zirconium	0.059	0.062
U	U-Al	0.160	0.160
Co	Inconel	0.150	0.155
S (beta counted)	Al alloy	0.0110	0.0114

range or electromagnetic radiation of any frequency. Hence such techniques do not fit, strictly, our definition of chemical spectroscopy as the use of any part of the radiation from the electromagnetic spectrum for the purposes of chemical analysis.

A well-known method of analysis which does not involve radiation in the electromagnetic spectrum, but has always been recognized as a spectroscopic technique, is mass spectroscopy. In this method, the positively ionized particles resulting from bombardment of the sample are sorted as a function of their mass (or more strictly, as a function of the ratio of the

TABLE 16

Determination of ^{235}U in Uranium

Radioactivation (%)	Mass spectrometer (%)
0.391 ± 0.006	0.388 ± 0.007
0.388 ± 0.006	0.392 ± 0.007
0.385 ± 0.006	0.390 ± 0.007
0.391 ± 0.006	0.390 ± 0.007
0.391 ± 0.006	0.391 ± 0.007
0.374 ± 0.006	0.373 ± 0.007
0.389 ± 0.008	0.395 ± 0.007
0.387 ± 0.008	0.388 ± 0.007
0.398 ± 0.008	0.396 ± 0.007

mass to their charge, m/e). The raw data are obtained as a "spectrum" with the "X" axis a function of the m/e values and the "Y" parameter a function of their intensities. Applications of mass spectroscopy to elemental analyses have been known for a long time and have proven most successful in identification of isotopes and in the determination of isotope abundance.

Other nonradiative methods of spectroscopic chemical analysis are becoming known as: (1) "Electron Spectroscopy for Chemical Analysis" (ESCA), or "Induced Electron Emission Spectroscopy" (IEES); (2) "Photoelectron Spectroscopy" (PES); and (3) "Auger Electron Spectroscopy" (AES). As will be seen shortly, the three are related. They have not as yet been investigated sufficiently to permit full evaluation and have not, as far as has been ascertained, been applied to any applications in textile chemistry. The most significant trends today are the announcements by a number of instrument companies of the availability of instruments for these analyses. These announcements emphasize advantages of simplified instrumentation, permitting analytical applications of electron spectroscopy in most spectroscopic laboratories. Development of instrumentation has preceded rapid introduction and acceptance of most advances in spectroscopic methods of chemical analysis. There would appear to be very small possibility that electron spectroscopy will fail to play a major role in future applications of chemical analysis by means of spectroscopic techniques.

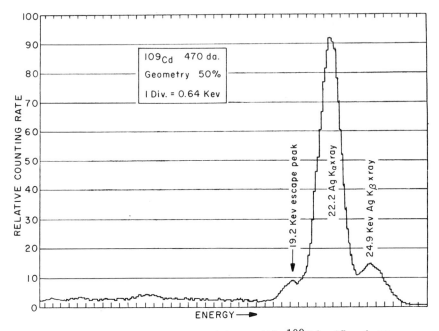

Fig. 4. Gamma-ray spectrum of the nuclide ^{109}Cd. (Courtesy Pergamon Publishing Co.)

When a beam of x rays strikes a material, such as an analytical sample, as we have seen (p. 8) the beam, after penetration, is weaker than the incident beam, the difference in intensity is a measure of the absorption of x rays. As has been shown, x-ray absorption is another spectroscopic tool for the analytical chemist.

As the x rays are absorbed by a sample, the energy gained is used to raise electrons in the inner orbits, the K, L, M, etc., shells to higher electron states. This absorbed energy can be reemitted as the electrons fall back to lower states, the energy appearing as reemitted x rays, i.e., secondary x-ray emission, or x-ray fluorescence, the x-ray fluorescence that is used for analytical purposes, as described. A certain number of the electrons in the analytical sample may receive sufficient energy, as x rays are absorbed, to overcome their binding power and obtain enough kinetic energy to escape from the atom entirely. The atom from which the electron escapes becomes ionized. Such a beam of escaping electrons always accompanies x-ray absorption and subsequent fluorescence. Until very recently, this beam of electrons was ignored by analytical spectroscopists, although it has been known for a long time that the kinetic energy of these photoelectrons is equal to the quantum energy of the incident radiation reduced by the binding energy of the electrons in the solid. It was also recognized that absorption of x rays in materials is due to this photoelectric

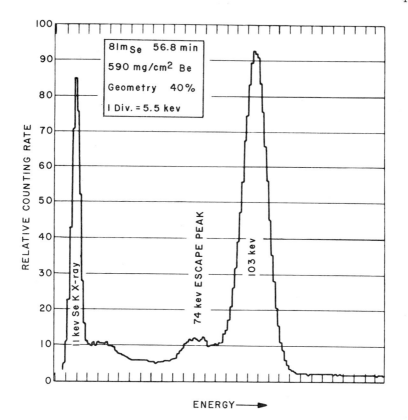

Fig. 5. Gamma-ray spectrum of the nuclide ^{81}Se. (Courtesy Pergamon Publishing Co.)

effect and that during the process of x-ray absorption some of the atoms become ionized. It should be recognized that the energy of the electrons from this electron-ejection type of spectroscopy is not, like most spectra resulting from electronic relaxation processes, a measure of the differences in energies between two electronic states, but represents the energy of the ejected electron at the time of the impact with the incident x rays. By varying the x-ray length of the incident x-ray beam it is possible to investigate the onset of ionization of deep-lying electron shells. This is x-ray absorption. Bergengren [13] showed in 1920 that the positions of the x-ray edges (Fig. 2) depend slightly on the types of bonding or valence states of the atoms of the material. This effect was explained by Pauling [14] in 1929 as being due to the charges appearing on atoms from ionic bonding. X-ray diffraction edges, however, cannot conveniently be measured with sufficient resolution to permit the accuracy required for analytical chemistry.

Most recent and detailed experiments with x-ray induced beams of emitted electrons, particularly at the University of Uppsala by Swedish investigators [15], have shown that not only do these electrons appear at frequencies characteristic of the specific energy level of the individual elements, but that the exact energy will be characteristic of the particular environment in which the electron exists at the time the sample absorbs the x radiation. The emitted energy will be that of the incident x-ray beam minus the binding energy of the electron, modified by the effect of the molecular or chemical environment from which the electron escapes.

This latter effect is reminiscent of the effect of the molecular environment on the resonant signal, at radio-frequency wavelengths, in high resolution nuclear magnetic resonance experiments which creates the useful "chemical shifts," as observed in high-resolution photon magnetic resonance where scores of signals are received from the hydrogen atom. These effects have very important implications in analytical applications. They provide a means for detecting and quantitatively measuring not merely the presence and total quantity of a specific element in an analytical sample, but they permit the differentiation of various oxidation or combining states of the elements or of the specific nature of the functional group in which the atom occurs in the analytical sample. These modifications in the observed energies of the escaping electrons are, in ESCA, being referred to as "electron chemical shifts" [16], probably in recognition of the analogy with high resolution nmr spectra.

Examples of this important effect for diagnostic purposes are given in Figs. 6, 7, and 8. Figure 6 is the carbon electron spectrum for ethyl trifluoroacetate. Four peaks are observed, one representing each of the four different carbon atoms in the compound. All four peaks appear at about equal intensities as there is one and only one carbon of each type in the compound. Figure 7, the electron spectrum of 2(4-nitrobenzene sulfonamido) pyridine is another illustration of the differentiation of a chemical element. In this spectrum the three different nitrogen atoms are represented by peaks of equal intensity, clearly differentiating the $\overset{O\diagdown}{\underset{O\diagup}{}}N$, N-H, and $\underset{C\diagdown_N\diagup C}{}$ ring nitrogen, and permitting measurement of their relative concentrations. Figure 8 illustrates, similarly, the differentiation of carbon peaks in the electron spectrum of ethyl chloroformate. The Cl-$\overset{O}{\overset{\parallel}{C}}$-O, O-$\overset{H}{\underset{H}{C}}$-C, and the C-$\overset{H}{\underset{H}{C}}$-H groups are clearly differentiated.

As electron spectroscopy is developed as an analytical tool, undoubtedly correlation charts or tables, relating the binding energies to specific forms of the different elements, similar to tables and charts used by infrared absorption and nuclear magnetic resonance spectroscopists for analytical

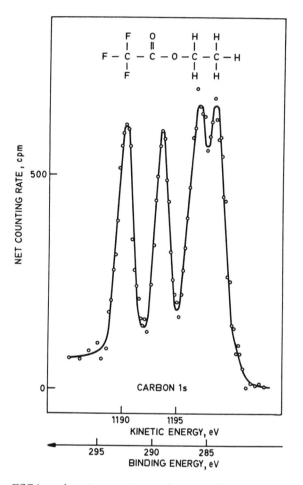

Fig. 6.   ESCA carbon 1s spectrum of ethyl trifluoroacetate.   (Courtesy Am. Chem. Soc.)

identifications, will appear. Preliminary correlation charts of this nature have been suggested, and two are reproduced in Figs. 9 and 10. In general the lighter elements with greater kinetic energy, thus less binding energy, appear first. For a given element the lower valence states generally appear at the lower energies. Thus the order for sulfur is sulfide, S^{2-}, elemental sulfur, S^0, sulfite, S^{4+}, and sulfate, S^{6+}. Similarly, in the energy range of nitrogen atoms, cyanide, CN^- will appear at lower energy, about 398 eV, followed by amide, $RCONH_2$, about 399 eV, amine, RNH, about 400 eV, nitrite, $-NO_2^-$, about 404 eV, and nitrate, $-NO_3^-$, about 407 eV.

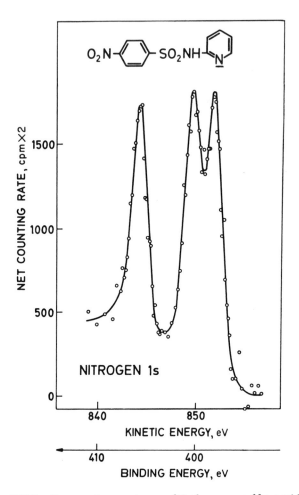

Fig. 7. ESCA nitrogen 1s spectrum of 2-(benzenesulfonamido), 2-(4-nitrobenzene sulfonamido) pyridine. (Courtesy Am. Chem. Soc.)

Most spectroscopic techniques for identifying and measuring the elements are capable of identifying and quantitatively determining only the total amount of a specific element, for example, total sulfur, total nitrogen, total carbon, etc. Mass spectroscopy has been used by analytical spectroscopists very successfully for determinations of specific isotopes. Mössbauer spectroscopy has been employed to differentiate various valence states of some elements, for example, to determine the amounts of iron as ferrous, Fe^{2+} and ferric, Fe^{3+}. X-ray absorption edges can be used, at least theoretically, to determine the nature of the type of bonding of the specific element. High resolution nuclear magnetic resonance has been thus far

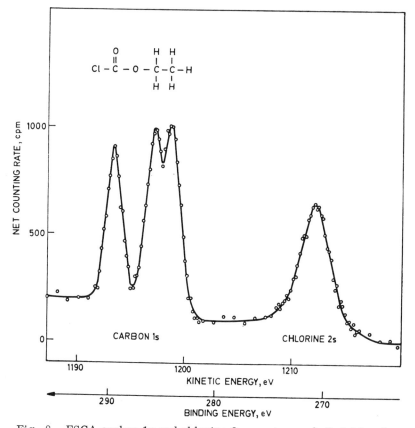

Fig. 8.   ESCA carbon 1s and chlorine 2s spectrum of ethylchloroform-
iate.

the most effective spectroscopic method for differentiating elements according to the particular functional group in which they exist in the analytical sample. This factor, the measurements of "chemical shifts," has in
a few years made this method one of the most important tools for the organic analytical chemist.

A rapid method of analysis, such as ESCA, which will differentiate
among the particular molecular or chemical environments in which specific chemical elements exist in the analytical sample, which can quickly
and accurately supply data from which relative amounts of sulfur as sulfate,
sulfide, sulfite, or nitrogen as cyanide, nitride, nitrite or nitrate is obviously a tool of exceptional value to the analytical spectroscopist and one
which will receive a deal of attention in the immediate future.

It is not essential that ionization be achieved by bombardment of the
analytical sample with x rays. The use of high energy protons from an

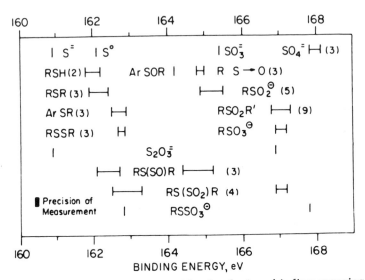

Fig. 9. Correlation chart for sulfur 2s electron binding energies and functional groups. Lines mark range of observed energies; numbers in parentheses indicate number of compounds used in correlation. (Courtesy Am. Chem. Soc.)

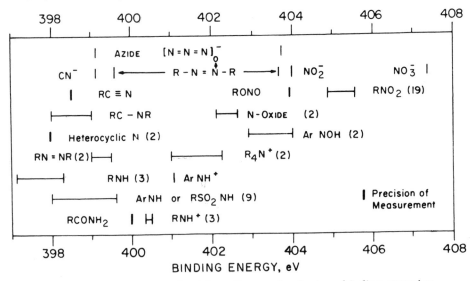

Fig. 10. Correlation chart for nitrogen 1s electron binding energies and functional groups. Lines mark range of observed energies; numbers in parentheses indicate number of compounds used in correlation. (Courtesy Am. Chem. Soc.)

ultraviolet source (such as a helium discharge lamp) has been investigated. These lower energy particles cause ejection of electrons from the valence shells of the atoms of the analytical sample. As x-ray-produced electron spectra yield the bonding energies of the ejected core electrons, ultraviolet photons yield information as to the ionization potential from ejected valence shell electrons. The use of photons to produce electron spectra is similar to the use of x rays and, in theory at least, might be considered as a variation of the same process. However the two techniques differ in the type of source used to eject the electrons, in the types of instrumentation used to identify and measure the ejected electrons, in the type of raw data obtained, and in the use to which the data can be put for analytical purposes. The technique using photons in the high energy electronic regions of the electromagnetic spectrum has been named "Photoelectron Spectroscopy" (PES). PES has been employed primarily to study ionization potentials and in investigations with gases and therefore with small molecules. ESCA is used for investigations of solids with large molecules and primarily to obtain bonding energies or to obtain information as to the types of groups in which specific chemical elements are to be found in an analytical sample.

Applications of PES to analytical spectroscopy arise from the observations that electrons ionized from valence orbitals give rise to narrow spectral bands, the positions of which are quite dependent upon the chemical environment of the element. The data are recorded in intensity units, usually expressed as counting rates as the "Y" axis and ionization potential, rather than the bonding energies of ESCA, as the "X" parameter. However, it has been found that the identification of the atoms present in an analytical sample is not as certain by PES as with ESCA. The valence shell electrons are frequently highly delocalized throughout the molecular framework. Most reported experiments with PES have been the determination of ionization potentials of small molecules with the analytical samples in the form of a gas or vapor. PES spectra of larger molecules reveal the presence of many overlapping bands, so that exact analysis is not possible. Betteridge and Baker [17], in their excellent review of the analytical potentials of photoelectron spectroscopy, suggest that "the photoelectron spectrum is a valuable 'fingerprint' of a molecule." They cite the use of PES to differentiate between the cis and trans forms of 1, 3-dichloropropene. In the ionization potential range for this molecule, the higher values between 14 and 19 eV, the spectra reveal marked and significant differences by means of which they could be readily distinguished. However, these experiments are dealing with mixtures of two pure compounds composed of only ten atoms. It would appear that any attempts to use PES with mixtures will be feasible only after careful separations, as by thin-layer or gas-liquid chromatography. Another disadvantage of PES is that, unlike ESCA where, as has been shown, the relative peak intensities in the spectrum are directly proportional to the relative number of atoms in the molecule, and quantitative analyses can be made directly from the raw data from the spectrum, peak intensities in PES spectra, while giving some information as to the relative numbers of atoms, are complicated by so many factors which determine the intensity that such a direct procedure for quantitative analysis is probably impossible.

At the present time ESCA appears to be further advanced as a tool for analytical spectroscopy although future potential applications of PES spectroscopy cannot be overlooked. PES spectroscopy has been investigated principally by Turner and his co-workers. The reader interested in more details of this division of spectroscopy is referred to his two review papers on the subject [18, 19]. Examples of applications of the technique, limited to the determination of ionization potentials, will be found in a series of papers in the Journal of the Chemical Society of London by Jobowry and Turner [20]. For a discussion of the potential applications of PES as an analytical tool for the chemical spectroscopist, the above-mentioned paper by Betteridge and Baker [17] should be consulted.

A third type of electron spectroscopy is "Auger Electron Spectroscopy" (AES). The Auger effect was described by Auger in 1925 [21], but the spectra resulting from the effect were almost completely ignored as a tool for analytical spectroscopy.

The various methods or techniques of analytical spectroscopy which have been discussed can be divided into two very general classifications, "A" on the basis of the character of the energy investigated as it reaches the detector of the spectrometer and "B" depending on the origin of this energy. Under classification "A" may be "1" radiative energy, any portion of the electromagnetic spectrum from gamma rays to radio frequencies, or "2" nonradiative energy, the positively charged particles of mass spectrometry or the electrons of ESCA or PES. Classification "B" consists of either "1" energy produced during the process of relaxation of an atom or molecule which has been excited or ionized by the absorption of energy in any form, or "2" energy released as a result from the direct impact of the incident energy.

Most analytical spectroscopic techniques involve measurements of electromagnetic radiation resulting from relaxation of excited states of ionized atoms or molecules, that is, they are A-1 - B-1 types. Most nonradiative procedures involve measurements of particles resulting from the direct impact of the incident energy, i.e., they are B-1 - B-2 types. Auger spectroscopy occupies an almost unique position of being a nonradiative technique resulting from electronic relaxation, i.e., it is an A-2 - B-1 type.

When an atom or molecule in an analytical sample is subjected to the impact of x rays, as in ESCA spectroscopy, it may be ionized and the escaping electrons are used for the chemical analysis. The atoms or molecules of the analytical sample are ionized. They will return to their normal states by relaxation processes involving transitions of electrons from higher energy states to fill vacancies in the lower energy states. Thus

when ESCA spectra are produced there may be vacancies or "holes" in the
K level, creating the ionized atom. One process of relaxation, that is,
return of the atom to its normal state, involves transitions of electrons
from one of the higher energy levels to fill this "hole" in the K shell and,
very probably, setting up a series of such transitions until all the levels
are filled by capture of free electrons from above the "free electron level."
The energy gained by the transitions of the electrons from higher levels to
fill holes in the lower levels may appear as bundles of quantized energy
h υ , that is, secondary x rays. This is the process of x-ray fluorescence.

Auger's experiments showed that there is an alternate and competing
process of electronic relaxation, involving electron ejection. The energy
released when an electron leaves a higher energy state to fill a "hole" in
the K shell is used to simultaneously eject a second electron from the atom.
This ejected electron is called the "Auger electron" and the process "Auger
Electron Emission." Like ESCA, Auger electrons will appear with the
energy characteristic of the parent atom and can be used to identify it.
However, unlike ESCA, the energy of the Auger electron, resulting from a
relaxation process, is independent of the exciting energy.

The Auger process involves three energy shells or subshells. The process is illustrated in Fig. 11. Assuming a hole in the K shell resulting
from production of ESCA with incident x rays, the left side of this figure
depicts the normal x-ray fluorescence procedure whereby this hole is filled
from the higher levels, the 2p and 3d orbitals to yield the normal Kα and
Kβ fluorescence lines. The right-hand portion of the diagram demonstrates
typical Auger electron emission. A 2p electron undergoes a transition to
fill the primary 1s vacancy and the energy released is used to eject a second 2p electron, the Auger electron. The process is referred to as KLL
Auger electron emission, the first letter indicating the shell where the
primary vacancy occurred; the second the level from which an electron
underwent transition to fill this "hole"; and the third the shell from which
a second electron was ejected by the energy created by this transition to
fill the primary vacancy. This third electron is, of course, the Auger
electron and will appear at an energy characteristic of the 2p level of the
specific atom. A second example in Fig. 11 involves transition of an electron from the 2p shell to fill the K level vacancy with a transfer of the
energy released in this process to eject an electron from the 3p level, which
becomes the Auger electron, and the process is a KLM Auger electron
emission.

Auger electron spectroscopy has not been developed fully as a tool for
the analytical spectroscopist. Most publications on this subject deal with
the theory, the potentialities, and the instrumentation, particularly newly
proposed instrumentation of considerably simplified design.

AES is not restricted to the use of x rays as the incident source. As
with ESCA, instruments are being designed using electrons for the initial
bombardment of the analytical sample. As the energy measured is in the

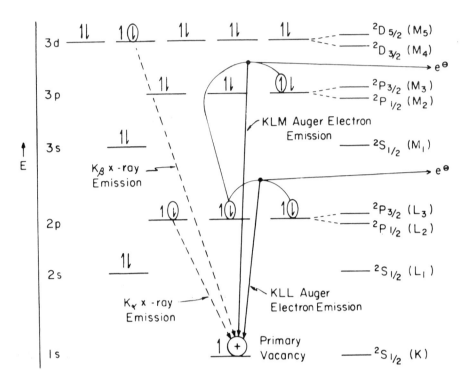

Fig. 11. Electronic relaxation processes as origin of x-ray fluorescence and Auger electron spectra.

form of electrons, it has been suggested that this process be called "secondary emission electron spectroscopy." As electrons are employed to create the beam of electrons measured, it might well be called "fluorescence electron spectroscopy." Neither AES nor "fluorescence electron spectroscopy" has been developed sufficiently to properly evaluate its potential as a tool of analytical spectroscopy.

Auger electron spectroscopy can be compared to several spectroscopic techniques. From a comparison with x-ray absorption spectroscopy it should be observed that AES peaks should occur at energy values comparable to the absorption edges (Fig. 2), the energy values where the incident source has just the energy to cause an electron to be expelled from a specific shell. Plots of x-ray absorption energy on the same scale as the energies of Auger electron spectra reveal that the peaks of the latter appear coincidental with the absorption edges of the former. Auger electron spectroscopy can be compared to x-ray fluorescence. Both are the result of an ionized atom undergoing relaxation processes to achieve its normal state.

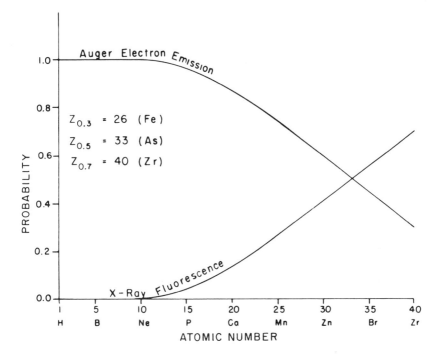

Fig. 12.   Probability of Auger electron emission and x-ray fluorescence as function of atomic number.

As Auger spectroscopy is a competing relaxation process with x-ray fluorescence process, the data obtained are of comparable utility. While theoretically Auger electron spectroscopy and x-ray fluorescence are competing processes, for purposes of analytical spectroscopy, they might be thought of as complementary. Figure 12 shows the probability of Auger electron emission and x-ray fluorescence occurring as competing processes. The Auger electron process excels for light atoms where it is almost unity and x-ray fluorescence virtually zero below atomic number 11, sodium. This advantage of Auger over fluorescence decreases until about atomic number 35, bromine, when the two become essentially equal. Considering the probability of x-ray fluorescence below atomic number 20, calcium, it is a tribute to x-ray fluorescence spectroscopists that they have been successful in extending the range of this useful technique down to approximately sodium, atomic number 11. The advantage of Auger electron spectroscopy to complement the very popular x-ray fluorescence technique for elements with atomic number less than 10 is strongly suggested. Photoejection of electrons during relaxation within atoms is not equally probable for all atoms. Generally such excitation is inversely proportional to the square of the orbital radius, thereby favoring small atoms. But for a given electron, i.e., 1s electron, it varies as the cube of the nuclear charge.

Auger electron spectroscopy should be a valuable tool for detection, identification, and quantitative measurement of the light elements, i.e., to complement x-ray fluorescence, when x rays are used as the exciting source. It might be expected to reveal chemical shifts. However whether resolution, probably with electron excitation, will permit these effects to be measured with analytically useful accuracy probably remains to be established. Its first applications will very probably be concerned with the identification and determination of the light elements beyond the reach of x-ray fluorescence techniques.

REFERENCES

[1] D. P. Norman, W.W.A. Johnson, and N.S. Johnson, Am. Dyestuff Reptr., 37, 838-848 (1948).

[2] K.G. Brodie, Resonance Lines, Vol. 1, No. 3 (G.P. Thomas, ed.), Varian Techtron, Walnut Creek, Calif.

[3] H.P. Klug and L.E. Alexander, X-Ray Diffraction Procedures, Wiley, New York, 1954.

[4] W.R. Brode, Chemical Spectroscopy, 2nd ed, Wiley, New York, 1943.

[5] R.T. O'Connor, D.C. Heinzelman, and M.E. Jefferson, J. Am. Oil Chemists' Soc., 25, 408-411 (1948).

[6] R.T. O'Connor and D.C. Heinzelman, Anal. Chem., 24., 1667-1669 (1952).

[7] D.C. Heinzelman and R.T. O'Connor, Textile Res. J., XX 805-807 (1950).

[8] R.T. O'Connor, D.C. Heinzelman, and B. Piccolo, Am. Dyestuff Reptr., 50 (No. 13 21-5) (1962).

[9] A.Walsh, Spectrochim. Acta, 7, 108 (1955).

[10] B.J. Russell, J.P. Shelton, and A.Walsh, Spectrochim. Acta., 8, 317 (1957).

[11] O. Ant-Wuorinen and A. Visapää, Paperi Puu, 48, No. 11, 649-56 (1966).

[12] C.E. Crouthamel, ed, Applied Gamma-Ray Spectrometry Pergamon, New York, 1960.

[13] J. Bergengren, Compt. Rend., 171, 624-6 (1920).
[14] L. Pauling, Phys. Rev., 34, 954-63 (1929).
[15] K. Siegbahn et al., Electron Spectroscopy for Chemical Analysis, Air Force Materials Laboratory, Air Force Systems Command, Wright-Patterson Air Force Base, Ohio, Tech. Rept. AFML-TR-68-189.
[16] D.M. Hercules, Anal. Chem., 42, 20A-40A (1970).
[17] D. Betteridge and A.D. Baker, Anal. Chem., 42, 43A-56A (1970).
[18] D.W. Turner, Physical Methods in Advanced Inorganic Chemistry (H.A. Hill, and P. Day, eds.,) Wiley-Interscience, New York, 1968.
[19] D.W. Turner, Advances in Mass Spectrometry, Mol. Spectrosc. Proc. Conf., 4th, 1968 (P. Hepple, ed.), Institute Petrol, London, England.
[20] M.I. Al. Jobowry and D.W. Turner, J. Chem. Soc., 1963, 5141; 1964, 4434; 1966A, 85; 1968B, 22.
[21] P. Auger, J. Phys. Radium, 6, 205 (1925).

Chapter 2

INFRARED SPECTROSCOPY
AND PHYSICAL PROPERTIES OF CELLULOSE

C. Y. Liang

Department of Physics and Astronomy
California State University
Northridge, California 91324

I. INTRODUCTION

In the review papers which considered the deuteration and polarization of the infrared spectra of high polymers [1] and the band assignments of the spectra of carbohydrates [2] , some limited discussions about the infrared study of cellulose have been given. In the present chapter the infrared spectra of cellulose and some wood components closely related to cellulose will be considered with particular attention to the techniques and results of deuteration and polarization measurements. The accessibility, hydrogen bonding schemes, and crystalline modifications of cellulose determined by infrared spectroscopy will be discussed, and the interpretation of the infrared absorption bands of cellulose will be given and discussed.

II. EXPERIMENTAL

In this section we do not intend to discuss the general experimental methods for the observation of the infrared spectra of high polymers, since many books and review articles have covered this part in great detail. We will consider only the methods pertinent to the investigation of the infrared spectra of cellulose and some related substances.

A. Orientation of the Samples

For the infrared polarization measurement, the samples have to be oriented. Fibers and fiber grid can be used as uniaxially oriented samples [3,4]. In order to reduce the scattering of light as well as to reduce the thickness of the fiber layer, the fibers may be swollen and pressed [5,6]. Cellophane and bacterial cellulose films can be wet stretched and dried under tension. However, the orientation of the films so obtained is usually low. Highly oriented cellulose I and cellulose II thin films can be obtained from cellulose crystallites by the following procedure [7]. The cellulose is hydrolyzed in strong sulfuric acid and a colloidal dispersion of crystallites is obtained. Ultracentrifugation of the dispersion gives a birefringent gel material which can be spread on an AgCl plate to give a thin film with double orientation (uniaxial and uniplanar with the 101 plane parallel to the film surface). Highly oriented cellulose II film can also be obtained by regeneration with caustic soda from highly oriented film of secondary cellulose acetate [8].

Thin wood sections cut in a specific direction can be used for the infrared polarization measurement [9, 10] . Some naturally grown films have uniplanar orientation such as the bacterial cellulose film (101 uniplanar orientation) and the spirogyra cell wall (101 or 002 uniplanar orientation). These films can be studied with polarized infrared radiation at inclined incidence [11].

B. Deuteration

The OH groups in cellulose can be partially replaced by OD groups by exposing the sample to D_2O vapor or by soaking in D_2O liquid. The rate of deuteration is faster when the sample is soaked in D_2O liquid than when it is exposed to D_2O vapor. It is well known that the amorphous part (disordered region) of cellulose is much more easily deuterated than the crystalline part. However, deuteration of some crystalline regions does occur when the sample is at higher temperature [12] or when it is under elastic deformation [13].

After deuteration the sample must be protected from rehydrogenation by the moisture in the laboratory. The protection may be provided by using a deuteration cell (see, for example, Ref. [1]). If the deuteration is carried out in a dry box, the sample can be protected by sealing it between two AgCl plates. To seal the edges of the AgCl plates, a hot wire may be used to fuse the edges together or a moisture-proof tape may be used for temporary protection.

Completely deuterated cellulose II film can be prepared for cellulose triacetate film with regenerating solution NaOD in D_2O [14]. Rehydrogenating such a fully deuterated sample will result in a sample deuterated only in the crystalline regions.

The exchange of the OH groups of cellulose II with the vapor of tritium oxide has also been reported [15].

C. Spectral Regions

Commercial spectrometers equipped with gratings of prisms of various materials can be used to cover quite a wide range of wavelengths. It may be noted that high resolution optics may be needed for certain regions. In the region from the near infrared to about 4μ, the use of a spectrometer equipped with a LiF prism is often satisfactory. In the region from 6 to 15μ, a spectrometer equipped with a NaCl prism or gratings is often used. The infrared spectra of cellulose materials in the region beyond 15μ to longer wavelengths have not been widely studied, and the spectra in the region from 15 to about 30μ, to be discussed later, were observed with CsBr optics.

D. Polarizers

Silver chloride polarizers may be used in the region from the near infrared to about 25μ. The polarizer can be easily constructed with six AgCl plates arranged at Brewster angle. If a double-beam spectrometer is used, two nearly identical polarizers should be used with one in each beam to insure a flat 100 percent base line.

In the longer wavelengths region (from $15\,\mu$ to the far infrared region), six or more thin polyethylene sheets arranged at Brewster angle may be easily constructed as a far-infrared polarizer.

III. BAND ASSIGNMENT

For the theoretical band assignment one has to make normal coordinate analysis and to calculate the potential energy distribution [16]. Since the cellulose chain is rather complicated, such calculations have not been carried out. At the present time, the interpretations of the spectra of various cellulose samples are primarily based upon the following experimental facts: Firstly, one may compare the spectra with those of some model compounds in order to locate some of the "group frequencies." A group frequency is the characteristic frequency associated with a vibrational mode of a particular atomic group. Such a vibrational mode is not, or only slightly, effected by its neighboring atomic groups. In cellulose, some of the vibrational modes connecting the CH, OH, or CH_2 atomic groups may be located by the group frequency method. Secondly, the bands due to OH groups can be differentiated from those due to other groups by studying the spectra of deuterated samples. Deuteration method can also be used to determine the coupled vibrational modes between the OH groups and some other groups of atoms. It is also possible to make selective deuteration of some CH groups in a model compound, and the spectrum of such a deuterated model compound may give some suggestion for the interpretation of the spectra of cellulose samples. Thirdly, polarized spectra and spectra with inclined incidence for oriented samples are certainly very helpful for band assignments.

For the analysis of the spectra of cellulose samples, a unit cell of the cellulose crystal should be considered. There are two antiparallel chains passing through the unit cell with a smallest possible chain repeating unit to be the cellobiose unit. The probable space group is $P2_1$, and each chain may possess a C_2 screw symmetry axis. Thus we expect twelve OH stretching frequencies and the same number of OH in-plane bending frequencies and of OH out-of-plane bending frequencies. These expected frequencies have not been identified in the spectra. In the following discussion, it is sometimes indicated that an observed OH frequency may be associated with a particular type of hydrogen bond.

For the analysis of the carbon-hydrogen vibrations, we assume that the interaction between the neighboring glucose units is small. Thus the carbon hydrogen vibrations are essentially originated from one CH_2 unit and five CH units. This assumption seems to be in accordance with the experimental observation.

The skeletal vibrations of cellulose chains are usually associated with quite complicated modes, and a detailed description of these modes has not been given at the present time. However, analysis has been made based on

a single glucose ring for the ring vibrations, and on a three-atom bent
COC unit for the bridge oxygen vibrations [17]. Such an analysis can only
be considered as a first step approximation.

A. Native Cellulose (Cellulose I)

The polarized infrared spectra in the region from 2.5 to 15 μ of natural flax and ramie fibers [5] and of ramie and bacterial cellulose crystallites [7, 17] have been investigated. As a representative spectrum for
cellulose I, the polarized spectrum of bacterial cellulose crystallites is
shown in Fig. 1. The spectra of various samples of cellulose I are essentially the same except for some small differences. These differences are
probably due mostly to the differences in impurity content, orientation,
and crystallinity [7, 17]. For example, the relative intensities of the two
perpendicular bands at 1317 and 1336 cm^{-1} are different in the spectra of
ramie and bacterial cellulose crystallites [17]. This is because the transition moment associated with the 1317 cm^{-1} band is nearly perpendicular
to the 101 plane, and the weaker intensity of this band in the spectrum of
bacterial cellulose crystallites indicates a highly 101 uniplanar orientation
for this sample.

The interpretation of the spectrum of cellulose I in the region from 2.5
to 15 μ is given in Table 1. Some of the assignments are discussed in the
following.

From the positions of the OH stretching frequencies, the OH groups are
all hydrogen bonded. The strong parallel band at 3350 cm^{-1} indicates that
there is a intramolecular hydrogen band, possibly of the type $O_3H...O_5'$,
where a prime indicates that the oxygen atom is in the adjacent glucose
unit of the same chain. The perpendicular bands at 3305 and 3405 cm^{-1}
indicate that there are intermolecular hydrogen bonds between chains.
(See later discussion on hydrogen bonds.)

The CH$_2$ symmetric stretching frequency at 2853 cm^{-1} and the CH$_2$ symmetric bending frequency at 1430 cm^{-1} are both parallel. This observation rules out the possibility of an intramolecular hydrogen bond involving
the C$_6$ hydroxyl group. The differentiation of the OH bending vibrations
from the CH bending vibrations and the CH$_2$ bending vibrations (including
CH$_2$ rocking, wagging, twisting, and symmetric bending) is reached by
comparing the deuterated and undeuterated spectra of amorphous cellulose,
as shown in Fig. 2. A band reduced in intensity after deuteration would
be likely to be associated with an OH vibration. For this reason, the
bands at 1205, 1336, and 1455 cm^{-1} are assigned to the OH in-plane bending vibrations. The out-of-plane OH bending frequency is known to be a
group frequency near 650 cm^{-1}. Since both of the perpendicular bands at
663 and 700 cm^{-1} are reduced in intensity after deuteration, both of the
bands are assigned to the OH out-of-plane bending vibrations.

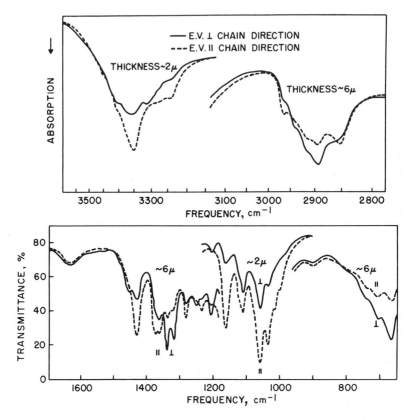

Fig. 1. Polarized infrared spectrum of doubly oriented bacterial cellulose crystallites [7, 17].

It has been suggested that, from the infrared study of C_1-deutero sugars [18], the C_1H stretching and the C_1H bending vibrations of cellulose I are identified, respectively, with the perpendicular band at 2914 cm^{-1} and the parallel band 1358 cm^{-1}.

If we assume that the bridge COC stretching modes, the C-OH stretching modes, and the glucose ring stretching modes can be approximately separated to their "group frequencies," then we would expect to observe two COC bridge stretching frequencies, five glucose ring stretching frequencies, and a number of the stretching frequencies involving the CC and CO bonds attached to the glucose rings. We thus make the following tentative assignments [17]: The stong parallel band at 1162 cm^{-1} is assigned to the antisymmetric bridge COC stretching frequency. This band is observed in the infrared spectra of a number of cellulose samples and their closely related small molecules as well as polymer derivatives, as expected. From the

TABLE 1

Infrared Spectrum of Bacterial Cellulose Crystallites [7, 4]

Frequency (cm⁻¹)	R.I.[a]	Polarization	Interpretation
663	M	⊥ ⎫	
700	M	⊥ ⎭	OH out-of-plane bending
740	Sh	⊥ ?	CH_2 rocking
800	Sh	⊥ ?	Ring breathing
895	W	‖	Antisym. out-of-phase ring stretching
985	Sh	⊥ ? ⎫	
1000	Sh	‖ ? ⎪	
1015	M	‖ ⎬	Skeletal vibrations involving C-O
1035	S	‖ ⎪	stretching
1058	S	‖ ⎭	
1110	S	‖	Antisym. in-phase ring stretching
1125	Sh	‖	
1162	S	‖	Antisym. bridge COC stretching
1205	W	⊥	OH in-plane bending
1235	W	‖	
1250	W	⊥	
1282	M	‖	CH bending
1317	M	⊥	CH_2 wagging
1336	M	⊥	OH in-plane bending
1358	M	‖ ⎫	CH bending
1374	M	‖ ⎭	
1430	M	‖	CH_2 sym. bending
1455	Sh	⊥	OH in-plane bending
1635	–		Adsorbed H_2O
2853	M	‖	CH_2 sym. stretching
2870	M	⊥ ⎫	
2897	M	⊥ ⎬	CH stretching
2914	Sh	⊥ ⎭	
2945	Sh	⊥	CH_2 antisym. stretching
2970	W	‖	CH stretching
3245	M	‖	
3275	Sh	‖	

[a]R.I. - relative intensity, S - strong, M - medium, W - weak, Sh - Shoulder, V - very (see other tables).

TABLE 1 (continued)

Frequency (cm⁻¹)	R.I.[a]	Polarization	Interpretation
3305	Sh	⊥	OH stretching (intermolecular hydrogen bonds)
3350	S	‖	OH stretching (intramolecular hydrogen bonds)
3375	Sh	‖	
3405	Sh	⊥	OH stretching (intermolecular hydrogen bond)

[a]R.I. - relative intensity, S - strong, M - medium, W - weak, Sh - Shoulder, V - very (see other tables).

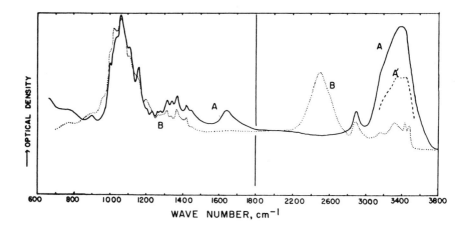

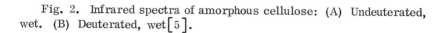

Fig. 2. Infrared spectra of amorphous cellulose: (A) Undeuterated, wet. (B) Deuterated, wet[5].

infrared polarization and the expected frequency range, we assign the band near 800 cm^{-1} to the ring breathing vibration, the band at 895 cm^{-1} to the antisymmetric out-of phase ring stretching frequency, and the band at 1110 cm^{-1} to the antisymmetric in-phase ring stretching frequency. The bands between 985 and 1058 cm^{-1} are most likely to be involved with the stretching modes of the CO and CC bonds attached to the glucose rings.

B. Mercerized Cellulose (Cellulose II)

The spectrum of cellulose II has been studied by several investigators [14, 19, 20]. As a representative spectrum of cellulose II, the polarized spectrum of mercerized ramie cellulose crystallites [20] is given in Fig. 3. The interpretation of the spectrum is given in Table 2. It may be noted that differences between the spectra of cellulose I and cellulose II are quite obvious. Especially, we note that the bands in the OH stretching region are quite different, and the polarizations of the bands near 700 and 1335 cm^{-1} are different in the two spectra. These differences exist because cellulose II has a different crystal structure and a different hydrogen bonding scheme from cellulose I. This topic will be discussed in later sections.

The perpendicular bands at 3175, 3305, and 3350 cm^{-1} are connected with the OH stretching frequencies of intermolecular hydrogen bonds, and the parallel bands at 3447 and 3488 cm^{-1} are connected with the OH stretching frequencies of intramolecular hydrogen bonds. The perpendicular bands at 2874, 2891, 2904 cm^{-1} and the parallel bands at 2955 and 2968 cm^{-1} may be assigned to the expected five CH stretching frequencies. The polarizations of the five CH$_2$ bands listed in Table 2 are approximately similar to that of the corresponding bands of cellulose I. This result indicates that the C$_6$ hydroxyl groups are not involved in intramolecular hydrogen bonding in both cellulose I and cellulose II. Most of the remaining bands in the spectrum of cellulose II can be similarly interpreted as in the case of cellulose I.

C. Near-Infrared and Far-Infrared Spectra of Cellulose

The near-infrared spectra are convenient for the study of fibrous specimens which are usually too thick for the spectra in the fundamental region. The near-infrared spectra of four cellulose materials [21] (tire cord rayon, cellophane, ramie, and bacterial cellulose) are shown in Fig. 4, and the band assignment for these spectra is given in Table 3.

TABLE 2

Infrared Spectrum of Mercerized Ramie Cellulose Crystallites [20]

Frequency (cm^{-1})	R.I.	Polarization	Interpretation
650	S	⊥	OH out-of-plane bending
700	Sh	∥	
760	Sh	⊥	CH$_2$ rocking
800	Sh	⊥	Ring breathing
892	M	∥	Antisym. out-of-phase ring stretching
965	Sh	∥	
996	S	∥	
1005	S	⊥ ?	Skeletal vibrations involving CO
1020	S	∥	stretching
1035	S	⊥	
1060	VS	∥	
1078	VS	∥	
1107	S	∥	Antisym. in-phase ring stretching
1155	S	∥	Antisym. bridge COC stretching
1200	M	⊥	OH in-plane bending
1225	M	⊥	
1257	W	⊥ ?	
1277	M	∥	CH bending
1315	W	⊥	CH$_2$ wagging
1335	W	∥	OH in-plane bending
1365	M	∥	CH bending
1375	M	∥	
1416	W	∥	CH$_2$ sym. bending
1440	Sh	∥	
1470	Sh	⊥	OH in-plane bending
1635	–		Adsorbed H$_2$O
2850	W		CH$_2$ sym. stretching
2874	Sh	⊥	
2891	S	⊥	CH stretching
2904	S	⊥	
2933	M	⊥ ?	CH$_2$ antisym. stretching
2955	W	∥	CH stretching
2968	W	∥	
2981	W	⊥ ?	

TABLE 2 (continued)

Frequency (cm^{-1})	R.I.	Polarization	Interpretation
3175	S	$\perp$	
3305	S	$\perp$	OH stretching (intermolecular
3350	S	$\perp$	hydrogen bonds)
3447	VS	$\parallel$	OH stretching (intramolecular
3448	S	$\parallel$	hydrogen bonds)

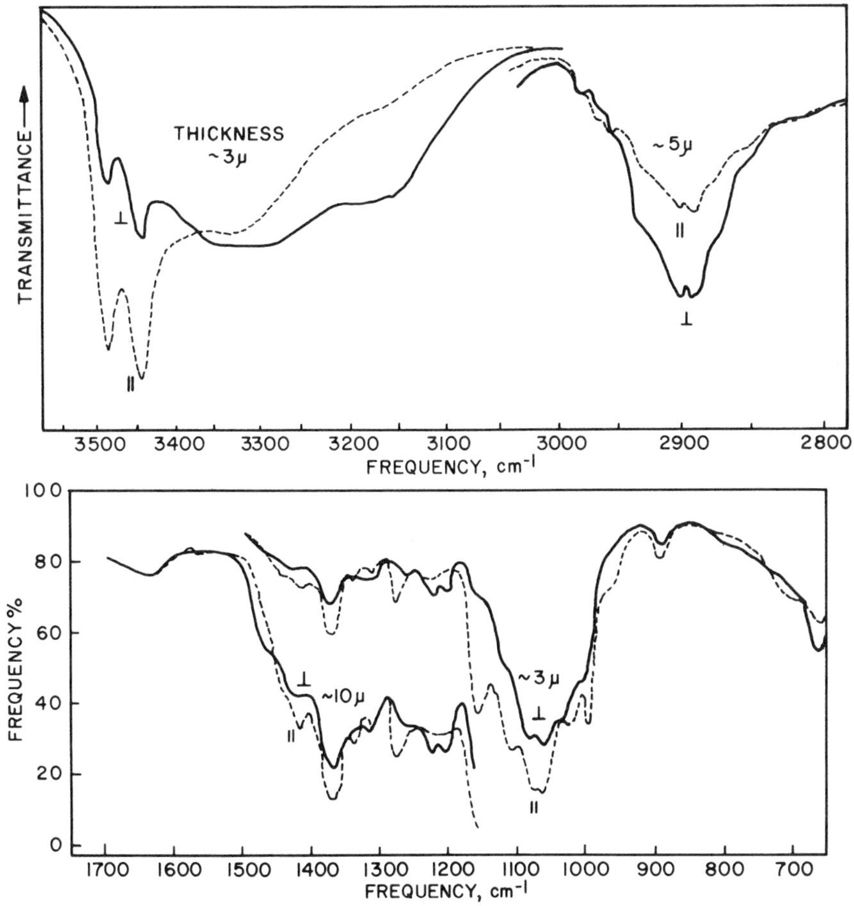

Fig. 3. Polarized infrared spectrum of doubly oriented mercerized ramie cellulose crystallites [20] .

TABLE 3

Near-Infrared Spectrum of Cellulose [21]

Frequency (cm^{-1})	Assignment
3970 } 3990 }	CO stretching + CH and CH$_2$ stretching
4235	OH and CH bending + CH and CH$_2$ stretching
4365	CO stretching + OH stretching CH bending + CH$_2$ stretching
4560	Cellophane only
4780	OH and CH bending + OH stretching
5190	Adsorbed H$_2$O
6770	OH stretching + OH stretching

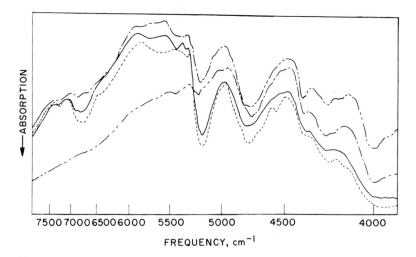

Fig. 4. Near-infrared spectra of four cellulosic materials: (——) tire cord rayon; (---) cellophane; (-·-) ramie; (-··-) bacterial cellulose [21].

Attempt has been made to observe the polarized spectra of cellulose samples in the near infrared. It has been found that no polarization could be measured for the bands in this region. This observation is probably due to overlapping, since each band in the near-infrared region could be interpreted in many ways. For example, the 6770 cm^{-1} band could be interpreted as the overlapped frequencies contributed for the overtone of a perpendicular OH stretching vibration, the overtone of a parallel OH stretching vibration, and the combination tone of one perpendicular and one parallel OH stretching vibration. One may thus expect to observe no or very weak polarization for such a band.

The near-infrared spectra of cellulose and other wood components have been studied in some detail [22]. A summary of the frequencies and possible assignments for the bands in these spectra are given in Table 4.

The polarized infrared spectra of cellulose I and cellulose II have been observed with CsBr optics and polyethylene polarizers. The observed frequencies and the dichroism are given in Table 5. A detailed interpretation of the spectra in this region is rather difficult, but a rough classification of the bands may be made as follows: The bands between 660 and 500 cm are most likely to be associated with the OH out-of-plane bending modes. The bands between 500 and 350 cm^{-1} are most likely to be associated with the in-plane skeletal bending modes. The spectra of deuterated cellulose samples in this region may serve to clarify some points of the assignments, but such data are not available at the present time.

D. Spectra of Wood Sections

Infrared spectra of wood sections have been studied by several investigators [9, 10, 23]. The polarized spectra of some wood sections are shown in Fig. 5. These polarization data clearly indicate that some wood components (cellulose, xylan, and mannan) are oriented in the wood. By comparing the spectra of wood sections with the spectra of pure cellulose, lignin, xylan, and mannan (see Fig. 6), a complete assignment of all the bands in the spectrum of wood can be made [9], and a summary of the assignments [23] is given in Table 6.

IV. HYDROGEN BONDS

The study of hydrogen bonds in cellulose by infrared spectroscopy has been reported by many investigators. In the following we will discuss the hydrogen bonding schemes for cellulose I and cellulose II suggested from the interpretation of the polarized infrared spectra of oriented cellulose samples [5, 7, 8, 14, 20, 24].

TABLE 4

Summary of Near-Infrared Spectra of Wood Components [22]

Regenerated cellulose	Bacterial cellulose	Ramie cellulose	Hardwood xylan	Ivory mannan	Wood sections	General assignments
3840	3830		3860	3820	3820	CO stretching + CH_2 symmetric stretching
3980	4000		3980	3940	3920	CO stretching + CH_2 antisymmetric stretching
	4140	4180				?
4235	4225	4265	4245	4220	4240	CH and CH_2 bending + CH_2 symmetric stretching
4365	4360	4350	4400	4335	4355	CH and CH_2 bending + CH_2 antisymmetric stretching
4780	4710	4705	4730	4725	4700	OH, CH, CH_2 deformation + OH stretching
5140	5120	5135	5155	5120	5150	Adsorbed H_2O + OH stretch
		5315		5475		?

5450	5505	5565	5610		5400	?
5635	5660	5745	5745	5400		2X CH and CH$_2$ stretching
5785	5820	5770		5900	6000	2X CH, CH$_2$, and CH$_3$ stretching
5870	5975	5900	5900	6260		CH and CH$_2$ stretching + OH stretching
6335	6290	6390	6410		6000–6500	
	6450					
	6485					
6630	6635	6720	6720	6850	6800	2X OH stretching
6715	6600					
	6695					
	6870					
7240		7245				2X ring, bridge stretching + 2X CH, CH$_2$ stretching
8180		8230	7700–8700			
		8470				3X CO stretching +
						3X CO stretching
10,060						3X OH stretching

TABLE 5

Infrared Spectra of Cellulose I (Bacterial Cellulose Crystallites) and
Cellulose II (Mercerized Ramie Cellulose Crystallites)
in the Region from 15 to 30 μ

Cellulose I			Cellulose II		
Frequency (cm^{-1})	R.I.	Polarization	Frequency (cm^{-1})	R.I.	Polarization
660	m	$\perp$	665		$\perp$
613	M	?	606	M	$\parallel$
			585		
556	M	$\perp$	552	W	? $\perp$
			535		
517	W	$\parallel$	515	M	$\parallel$
			470	W	$\perp$
455	W	$\parallel$	457	M	$\parallel$
435	W	?	438	W	$\perp$
			416	W	$\perp$
366	W	$\perp$	379	M	$\perp$
351	M	$\perp$?	352	M	$\perp$?
343			340		$\parallel$

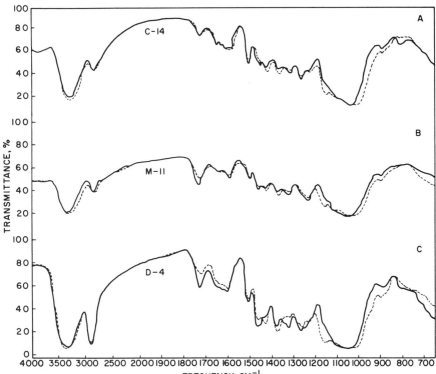

Fig. 5. Polarized infrared spectra of wood radial sections: (A) Western Red cedar in KBr disk. (B) Red maple in KBr disk. (C) Douglas fir in trace of Nujal 9 (solid line⊥ , dotted line ∥).

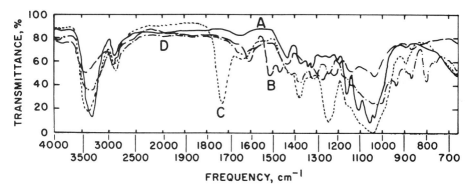

Fig. 6. Infrared spectra of (A) cellulose crystallites from prehydrolyzed cellulose wood pulp; (B) klason lignin from red cedar; (C) O-acetyl-4-O-methyl glycuronoxylan from white birch; (D) ivory nut mannan [9].

TABLE 6

Infrared Spectrum of Wood [23]

Frequency (cm^{-1})	R.I.	Polarization[a]	Assignment
3600–3200	S		Hydrogen-bonded OH stretching
2970	Sh		CH stretching
2945	Sh	⊥	CH$_2$ antisym. stretching
2914–2870	S	⊥	CH stretching
2850	M	∥	CH$_2$ sym. stretching
1730–1725	S	⊥ ?	C=O stretching of acetyl or carboxylic acid
1670	W	u	Lignin
1635	M	u	Adsorbed water
1600	S	u	COO$^-$ ion
1595	M	u	Lignin
1500	M	u	Lignin
1460	M	⊥	Lignin and CH$_2$ sym. bending on pyran ring
1455–1400	M		OH in-plane bending
1430	S	∥	CH$_2$ sym. bending of hydroxymethyl
1425	S		Carboxylic acid and COO$^-$ ion
1380	S	∥	CH bending
1335–1315	W	⊥	CH$_2$ wagging
1270	Sh		Lignin
1240	M		CO stretching of acetyl
1162	S	∥	Antisym. bridge COC stretching
1125–895	S		CO stretching and ring stretching
895	W	∥	Characteristic of β link (ring stretching)

[a] u - unpolarized.

TABLE 6 (continued)

Frequency (cm^{-1})	R.I.	Polarization[a]	Assignment
875	Sh	$\parallel$	Glucomannan
800	W	$\perp$	Glucomannan
768			Arabogalactan
700-650	M		OH out-of-plane bending

a_u - unpolarized.

A. Hydrogen Bonds in Cellulose I

As shown in Fig. 1, the parallel band at 3350 cm^{-1} should be connected with an intramolecular hydrogen bond. Since the intramolecular hydrogen bond involving the C_6 hydroxyl group can be ruled out by observing parallel dichroism for the CH_2 symmetric stretching and symmetric bending bands [7,24], the only possible intramolecular hydrogen bond is between C_3 hydroxyl group of one glucose residue and the ring oxygen of the next residue. Thus the observation of the parallel band at 3350 cm^{-1} in the spectrum of cellulose I strongly indicates that there is an intramolecular hydrogen bond of the type $O_3H...O_5'$ in the cellulose I crystal.

In order to account for the parallel dichroism of the CH_2 symmetric stretching and symmetric bending bands, the C_6 hydroxyl group of one chain may be hydrogen bonded to the bridge oxygen of the next antiparallel chain. One such set can readily be formed in the $10\bar{1}$ plane and another set can be made in the 101 plane. The bonds in the 101 plane are shorter than those in the $10\bar{1}$ plane. Thus the perpendicular band at 3305 cm^{-1} is assigned to the stretching mode of the hydrogen bond in the 101 plane and the perpendicular band at 3405 cm^{-1} to the stretching mode of the hydrogen bond in the $10\bar{1}$ plane.

Another possible set of intermolecular hydrogen bonds is in the 002 plane between the C_2 hydroxyl group and the C_6 oxygen. The identification of the bands in the spectrum associated with the hydrogen bonds in the 002 plane has not been definitely made.

According to the above discussion, a possible unit cell for cellulose I, which was based upon the Meyer and Misch cell but with bent cellobiose units, is shown in Fig. 7. Fig. 7A shows the arrangement of CH_2OH groups in the cellulose unit cell and Fig. 7B shows the hydrogen bonds in the $10\bar{1}$, 101, and 002 planes. This structure would be quite satisfactory

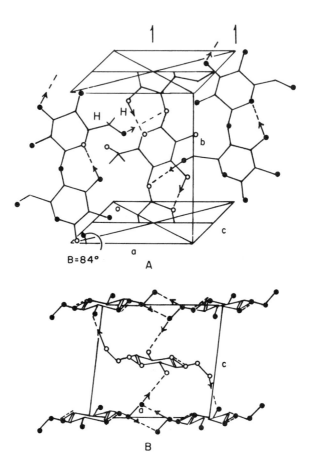

Fig. 7. Arrangement of CH_2OH group and hydrogen bonds in cellulose I unit cell [7].

for the interpretation of the polarized infrared spectrum of doubly oriented cellulose I crystallites.

It may be noted that by the variation of some chain parameters, Jones [25] proposed several possible hydrogen bonding schemes for cellulose I. Some of these proposed schemes do not seem to be in agreement with the observed infrared polarization. It is expected that the crystal structure and hydrogen bonding scheme of cellulose I would be improved in the future in order to account for both infrared and x-ray data.

B. Hydrogen Bonds in Cellulose II

Several hydrogen bonding schemes for cellulose II have been proposed by Mann and Marrinan [8] and by Marchessault and Liang [20]. A comparison of two of these proposed schemes is shown in Fig. 8. Since the CH_2 symmetric bending band at 1416 cm^{-1} is observed to be parallel (see Fig. 3), it is unlikely that there is an intramolecular hydrogen bond involving the C_6 hydroxyl group. The scheme proposed by Marchessault and Liang maintains the $O_3H...O_5'$ intramolecular hydrogen bond but the C_6 hydroxyl groups are involved in an intermolecular hydrogen bonding network. In this scheme, one-half of the CH_2 groups are oriented at a position giving unpolarized symmetric bending mode while the other half are oriented to show strong parallel polarization for this mode. The combination would give rise to a parallel polarization for the CH_2 symmetric bending mode in order to account for the infrared observation. The two parallel bands at 3447 and 3488 cm^{-1} (see Fig. 3) may be interpreted as both due to the stretching frequencies of the $O_3H...O_5'$ intramolecular hydrogen bonds. Since the O_3 atoms of alternate cellobiose units are not involved in an $O_6...O_3$ intermolecular hydrogen bonding, there are two different environments for the $O_3H...O_5'$ bonds in the crystal lattice. Thus one expects to observe two parallel OH stretching bands from the intramolecular hydrogen bonds. The higher frequency of the bands assigned to the intramolecular hydrogen bond in cellulose II compared to cellulose I indicates that a longer $O_3...O_5'$ distance is achieved for cellulose II. This longer $O_3...O_5'$ distance for cellulose II could be due to a slight twist of the bent cellobiose unit [20]. Of the other expected OH stretching bands in this scheme, one is perpendicular and one is nearly unpolarized. Both bonds are in the 10$\bar{1}$ plane. The O_2 hydroxyls are bonded in the 002 plane to the O_6 atom of the adjacent chain similar to the case in cellulose I, giving two more perpendicular bands. These predictions are essentially in agreement with the observed infrared polarization.

V. DEUTERATION AND ACCESSIBILITY

It is well known that the OH (or OD) groups in the crystalline region are not easily accessible to D_2O (or H_2O). This statement can be clearly demonstrated in Fig. 9. The solid line represents a viscose film exposed to D_2O vapor for several hours. The crystalline OH stretching bands in the region from 3000 to 3500 cm^{-1} show structure while the amorphous OD band near 2500 cm^{-1} shows no structure. When this deuterated film is exposed to the moisture air for several hours, the intensity of the OD stretching band decreases considerably. This is indicated by the broken curve in Fig. 9. The spectrum of a fully deuterated cellulose II film after several hours of rehydrogenation in the moisture air is shown by the

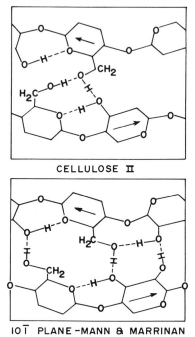

Fig. 8. Possible hydrogen bonding schemes for cellulose II [8, 20].

dash-dot curve in Fig. 9. The crystalline OD bands show quite a similarity with the crystalline OH bands.

Deuteration method has been used to determine quantitatively the accessibility of cellulose to water [15, 26, 27]. If the deuteration in the crystalline region can be avoided by using vapor phase of D_2O, an estimate of the percentage of crystallinity can be obtained. Let the optical density of a band be given by

$$\log(I_0/I) = kcL,$$

where I_0 and I are, respectively, the intensities of incident radiation and transmitted radiation, c is the concentration expressed as a mole fraction of the absorbing material, L is the thickness of the film, and k is the extinction coefficient per unit mole fraction. Thus the ratio of the optical

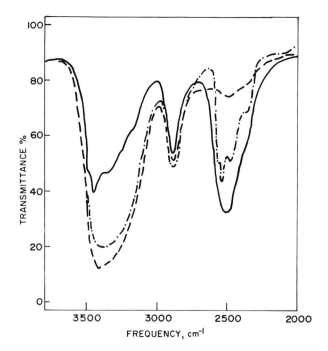

Fig. 9. Deuterated cellulose II: (——) viscose film expressed to D_2O vapor; (---) same film rehydrogenated to moisture air in the laboratory; (—·—) cellulose II film deuterated in the crystalline region.

densities of the OD and the OH bands is given by

$$\frac{\log(I_0/I)_{OD}}{\log(I_0/I)_{OH}} = \frac{k_{OD}}{k_{OH}} \frac{c_{OD}}{c_{OH}} \quad ,$$

where $c_{OD} + c_{OH} = 1$. The ratio of the optical densities of the OD and OH bands gives the relative crystallinity of the sample. Absolute values of crystallinity can be obtained [26] by extracting the deuterium from the cellulose sample with known amount of water and estimating the deuterium content of the water by measuring its refractive index. The absolute crystallinity so obtained can be used to calculate the ratio k_{OD}/k_{OH}, and thus the absolute crystallinity for some other samples can be determined by infrared measurements. The percent crystallinity of several cellulose samples so determined are given in Table 7 [26, 28].

It may be noted that the percent crystallinity of cellulose determined by the infrared method is different from that determined by x-ray diffraction measurements [15]. Using the ball-milled cellulose or amorphous

TABLE 7

Crystallinity of Cellulose [26, 28]

Cellulose samples	% Crystallinity
Saponified acetate	25
Viscose rayon	26
Precipitated cellulose	32
Viscose treated with NaOH	33.5
Mercerized bacterial cellulose	33
Bacterial cellulose	70

cellotetraose as amorphous standard, the x-ray percent crystallinity of a viscose rayon sample is determined to be about 40%, while the infrared method gives a value of only 26%. These results imply that only 26/40 or 65% of the material which is measured as being crystalline by x-ray diffraction measurement is hydrogen bonded in a regular manner. The 35% of the OH groups of the crystalline chains which are hydrogen bonded in an irregular manner may be accounted for as the surface OH groups of the crystallites, having a cross section of about 29 Å by 65 Å.

Deuteration (or rehydrogenation) at high temperature for cellulose II has been investigated by Okajima and Inoue [12], and it has been found that the accessibility of the OH (or OD) groups by D_2O (or H_2O) increases. Some of the results obtained by these authors are given in Table 8. It may be noted that crystallization occurs during the heat treatment and some of the newly formed crystallites are of the form of cellulose IV.

VI. CRYSTALLINE MODIFICATIONS OF CELLULOSE

It is known that there are other crystalline modifications of cellulose in addition to cellulose I and cellulose II. The infrared spectra of cellulose III and IV in the region from 2800 to 3600 cm^{-1} have been observed by Marrinan and Mann [8, 29]. These spectra are shown in Figs. 10 and 11. It is noted from Fig. 11 that the spectrum of cellulose IV prepared from cellulose II resembles that of cellulose II, and the spectrum of cellulose IV prepared from cellulose I resembles that of cellulose I. Thus the two materials may be indicated as IV_I and IV_{II}. Since x-ray data indicate that the unit cells of the two materials are very similar, the relative arrangements of the chains in the unit cells of these two materials are very similar.

TABLE 8

Resistant OD Groups Obtained by Room Temperature
Rehydrogenation for a Cellulose II Sample
Deuterated at Various Temperatures [12]

Condition of heat treatment with D_2O		Soaking time in H_2O at room temperature (hr)	Increase in resistant OD group (%)
(°C)	(min)		
100	30	0	5.8
		3	5.0
		25	5.5
150	30	0	11.3
		3	9.4
220	5	0	18.9
		25	18.1

However, the chain conformations and especially the intramolecular hydrogen bonds $O_3H...O_5'$ for these two materials are quite different. From
the infrared spectra it may be concluded that the intramolecular hydrogen
bond of cellulose IV_I is similar to that of cellulose I and the intramolecular
hydrogen bond of cellulose IV_{II} is similar to that of cellulose II.
    As shown in Fig. 10 the spectrum of cellulose III prepared from cellulose
I (III_I) is different from that of cellulose III prepared from cellulose II
(III_{II}). The spectrum of cellulose III_{II} is very similar to that of cellulose
II. The spectrum of cellulose III_I is not similar to that of cellulose I but it
is rather close to that of cellulose II except that there is only one absorption peak in the high frequency region. Although the unit cells of III_I and
III_{II} as determined by x-ray measurement are the same [30], the intramolecular hydrogen bonds in these two materials as revealed from the infrared data are different.
    There is another crystalline modification of cellulose indicated as cellu-
lose X [27]. The infrared spectrum of this modification in the region from
800 to 1800 cm^{-1} is rather similar to that of cellulose I. The spectrum of
this modification in the OH stretching region is not available.
    From the above consideration, we see that, by ignoring some small dif-
ferences, there are essentially only three different kinds of infrared

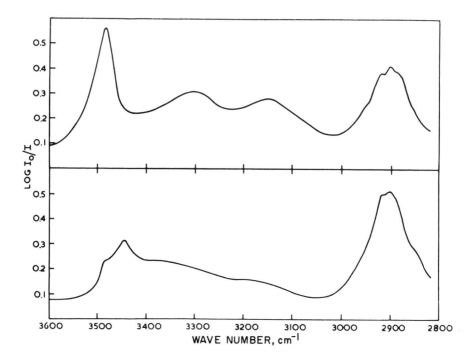

Fig. 10. Infrared spectra of cellulose III: Upper, prepared from I
(III_I); lower, prepared from II (III_{II}) [29].

spectra for the crystalline modifications of cellulose. The representative
spectra of these three kinds are the spectra of cellulose I, cellulose II,
and cellulose III_I. We have discussed the polarized spectra and the hydrogen bonding schemes for cellulose I and cellulose II in the early sections;
the only crystalline modification left for further discussion is cellulose
III_I. The polarized infrared spectra of cellulose III_I in the $3-\mu$ region
has been reported [8] and is shown in Fig. 12. Although the unit cell of
cellulose III_I determined from x-ray measurement [30,31] is somewhat
similar to that of cellulose II, the relative position of the parallel and antiparallel chains in the unit cell may shift to a certain position such that all
the intramolecular hydrogen bonds $O_3H \ldots O_5'$ have approximately the same
environment. Such a situation would account for the observation of only
one parallel band at 3484 cm^{-1}. If we assume that the CH_2 symmetric
stretching and symmetric bending bands are still parallel, then a possible
hydrogen bonding scheme for cellulose III_I may be sketched as shown in
Fig. 13 [20,32].

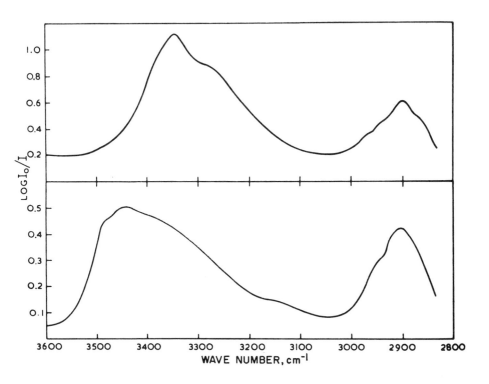

Fig. 11. Infrared spectra of cellulose IV: Upper, prepared from I
(IV$_I$); lower, prepared from II (IV$_{II}$) [29].

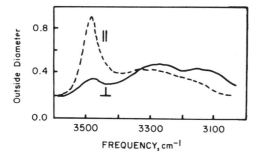

Fig. 12.   Polarized infrared spectrum of III$_I$ [8].

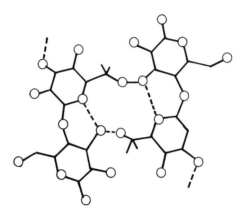

Fig. 13.   Possible hydrogen bonding scheme in III_I [20, 32].

VII. SPECTRA OF SOME OTHER CRYSTALLINE
WOOD POLYSACCHARIDES

In the last section, we are going to give a brief discussion of the spectra of xylans and mannans. The interpretation of these spectra may be considered as an application of the principles used in the study of the spectra of various crystalline modifications of cellulose.

The x-ray structure of xylans and mannans have been reviewed by Marchessault and Sarko [33]. The β-D-1,4 xylan chain forms a left-hand helical with a threefold screw axis. The fiber repeat is 14.8 Å. The mannan chain may have a β-D-1,4 linked mannose backbone. The fiber repeat of the mannan chain is about 10.3 Å, and the ratio of mannose to galactose in the chain is 2 to 1.

The polarized infrared spectrum of 4-O-methyl-D-glucuronoxylan [34] is shown in Fig. 14. Since the broad OH stretching band near 3400 cm^{-1} is unpolarized or slightly perpendicularly polarized, the question arises whether the intramolecular hydrogen bond $O_3H...O_5'$ is still present in the xylan chain. From the infrared spectrum of partially deuterated esparto xylan, two sharp OD stretching bands near 2625 and 2650 cm^{-1} were observed [34]. The pattern is somewhat similar to the OD bands of deuterated cellulose II (see Fig. 9), where two sharp bands were observed near 2540 and 2570 cm^{-1}. Since these two bands in cellulose II are related to the $O_3D...O_5'$ intramolecular hydrogen bonds, the two sharp bands in esparto xylan could be connected with similar intramolecular hydrogen bonds, but the $O_3...O_5'$ distance is somewhat longer than that in cellulose II. This is in accordance with the helical nature of the xylan chain. Such an interpretation was further confirmed by observing a parallel band near

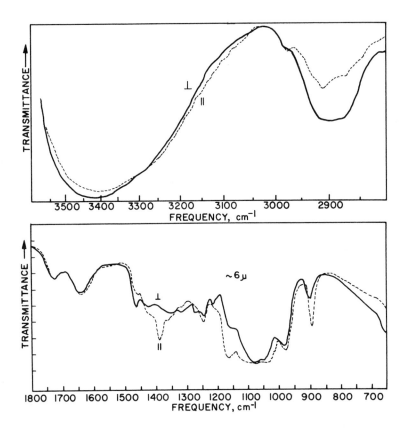

Fig. 14. Polarized infrared spectrum of oriented 4-O-methyl-D-glu-curonoxylan [34].

3500 cm^{-1} in the polarized infrared spectrum of a dry crystal form of white birch xylan [34].

It may be pointed out that there are two hydrogen atoms attached to the C_5 atom in the ring. From the helical arrangement of the chain, we expect the CH_2 symmetric and antisymmetric stretching modes, the symmetric bending mode, and the rocking mode, to be perpendicularly polarized, and the CH_2 wagging mode to be parallely polarized. These expected polarizations have been actually observed. In Table 9, the interpretation of the spectrum of 4-O-methyl-D-glucuronoxylan is given. Most of the bands have origins similar to that of cellulose. It may be noted that we prefer to interpret the band at 897 cm^{-1} as one of the ring frequencies (antisymmetric out-of-phase ring stretching) as suggested in Ref. [17].

The infrared spectrum in the fundamental region of vegetable ivory mannan has been observed and tentatively interpreted [22]. The spectrum

TABLE 9

Infrared Spectrum of 4-O-Methyl-D-Glucuronoxylan [34]

Frequency (cm^{-1})	R.I.	Polarization	Interpretation
650	M	⊥ ⎫	OH out-of-plane bending
725	Sh	⊥ ⎭	
897	M	‖	Antisym. out-of-phase ring stretching
980	S	‖ ⎫	
1045	S	‖ ⎪	
1085	S	‖ ⎬	Skeletal modes involving CO stretching
1115	S	‖ ⎭	
1130	S	‖	Antisym. in-phase ring stretching
1165	S	‖	Antisym. bridge COC stretching
1215	VW	⊥	
1245	W	‖	OH in-plane bending
1275	VW	⊥	CH bending
1317	W	⊥	OH in-plane bending
1350	W	‖ ⎫	OH in-plane bending
1305	Sh	‖ ⎭	
1390	M	‖	CH bending
1425	W	‖	COO$^-$ sym. stretching
1440	Sh		
1465	W	⊥	CH$_2$ sym. bending
1600	Sh	u	COO$^-$ antisym. stretching
1640	M	u	Water of hydration
1725	M	u	C=O stretching (acid)
2853	W	⊥	CH$_2$ sym. stretching
2873	M	⊥ ⎫	CH stretching
2914	M	⊥ ⎭	
2935	Sh	⊥	CH$_2$ antisym. stretching
2957	Sh	⊥ ⎫	CH stretching
2975	W	u ⎭	
3430	VS		OH stretching

is given in Fig. 15, and the suggested assignments are given in Table 10. The assignments are based on the deuteration effect as well as the results obtained from the infrared study of cellulose. A discussion on the hydrogen bonding scheme for this polysaccharide has to be deferred until the infrared polarization data for some mannan samples become available.

TABLE 10

Infrared Spectrum of Vegetable Ivory Mannan [22]

Frequency (cm⁻¹)	R.I.	Effect of deuteration	Assignment
650	M	Loss	OH out-of-plane bending
760	Sh		CH_2 rocking
805	M	Loss	
870	M	?	
897	Sh	None	Characteristic of β-linkage (ring st.)
934	M	None	
1013	S		
1032	S		CO stretching
1064	S		
1167	Sh		
1140	Sh		Bridge and ring stretching
1160	Sh		
1235	W	Loss	OH in-plane bending
1300	W	?	CH_2 wagging
1365	M	?	CH bending
1410	Sh	Loss	OH in-plane bending
1440	W		CH_2 sym. bending
1635	M	Loss	Adsorbed water
2300	W		
2815	W	None	CH_2 sym. stretching
2880	M	None	CH stretching
2920	Sh	None	CH_2 antisym. stretching
3340	S	Loss	OH stretching
3445	S	Loss	

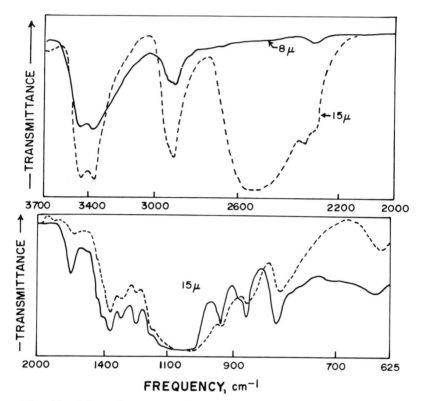

Fig. 15. Infrared spectrum of vegetable ivory mannan: (——) normal;
(- - - -) deuterated [22].

REFERENCES

[1] C.Y. Liang, in <u>Newer Methods of Polymer Characterization</u> (B. Ke,
     ed.), Wiley-Interscience, New York, 1964.
[2] H. Spedding, in <u>Advances in Carbohydrate Chemistry</u> (M.L. Wolfrom
     and R.S. Tipson, eds.), Academic, New York, 1964.
[3] J.W. Ellis and J. Batti, J. Am. Chem. Soc., <u>62</u>, 2859 (1940).
[4] P. Holliday, Nature, <u>163</u>, 602 (1949).
[5] M. Tsuboi, J. Polymer Sci. <u>25</u>, 159 (1957).
[6] R.J.E. Cumberbirch and H. Spedding, J. Appl. Chem., <u>12</u>, 83 (1962).
[7] C.Y. Liang and R.H. Marchessault, J. Polymer Sci., <u>37</u>, 385 (1959).
[8] J. Mann and H.J. Marrinan, J. Polymer Sci. <u>32</u>, 357 (1958).
[9] C.Y. Liang, K.H. Bassett, E.A. McGinnes, and R.H. Marchessault,
     Tappi, <u>43</u>, 1017 (1960).
[10] S.E. Darmon and K.M. Rudall, Discussions Faraday Soc., No. 9,
     251 (1950).

[11] C.Y. Liang and R.H. Marchessault, J. Polymer Sci., 43, 85 (1960).

[12] S. Okajima and K. Inoue, Polymer Letters 1, 513 (1963); J. Polymer Sci., A2, 461 (1964).

[13] W. Punn and S. Yamada, J. Polymer Sci., 56, S14 (1962).

[14] H.J. Marrinan and J. Mann, J. Appl. Chem., 4, 204 (1954).

[15] K.J. Heritage, J. Mann, and L. Roldan-Gonzalez, J. Polymer Sci., A1, 671 (1963).

[16] See, for example, K. Nakamoto, Infrared Spectra of Inorganic and Coordination Compounds, Wiley, New York, 1963.

[17] C.Y. Liang and R.H. Marchessault, J. Polymer Sci., 39, 269 (1959).

[18] S.A. Barker, R.H. Moore, M. Stacey, and D.H. Whippen, Nature, 23, 186, 307 (1960).

[19] F.H. Forziati and J.W. Rowen, J. Res. Natl. Bur. St. 46, 28 (1951).

[20] R.H. Marchessault and C.Y. Liang, J. Polymer Sci., 43, 71 (1960).

[21] K.H. Bassett, C.Y. Liang, and R.H. Marchessault, J. Polymer Sci., A1, 1687 (1963).

[22] W.J. Shell, Master thesis, State University College of Forestry at Syracuse University, 1966.

[23] R.H. Marchessault, Pure Appl. Chem., 6, 107 (1963).

[24] C.Y. Liang and R.H. Marchessault, J. Polymer Sci., 35, 129 (1959); C.Y. Liang, J. Polymer Sci., 62, S5 (1962).

[25] D.W, Jones, J. Polymer Sci., 32, 371 (1958); J. Polymer Sci., 42, 173 (1960).

[26] J. Mann and H.J. Marrinan, Trans. Faraday Soc., 52, 481, 487, 492 (1956).

[27] ϕ. Ellefsen and N. Norman, J. Polymer Sci., 58, 769 (1962).

[28] J. Mann, Pure Appl. Chem., 5, 91 (1962).

[29] H.J. Marrinan and J. Mann, J. Polymer Sci., 21, 301 (1956).

[30] H.J. Wellard, J. Polymer Sci., 13, 471 (1954).

[31] L. Segal, L. Loeb, and J.J. Creely, J. Polymer Sci., 13, 193 (1954).

[32] J. Mann and H.J. Marrinan, J. Polymer Sci., 32, 357 (1960).

[33] R.H. Marchessault and A. Sarko, X-Ray Structure of Polysaccharides (to be published).

[34] R.H. Marchessault and C.Y. Liang, J. Polymer Sci., 59, 357 (1962).

Chapter 3

LIGHT MICROSCOPY IN THE STUDY OF CELLULOSE

Mary L. Rollins and Ines V. deGruy

Southern Regional Research Laboratory
New Orleans, Louisiana

I. INTRODUCTION

In most instances in which the microscope is used to study cellulose, the polymer is in the form of textile fabrics, or yarns and fibers used in the manufacture of fabrics; or of paper sheets or pulp used in the manufacture of paper. Perhaps the widest use of microscopy is in the identification of fibers as to species or origin. The most common cellulose fibers of native origin are kapok, cotton, flax, ramie, jute, hemp, kenaf, sisal, and abaca. Wood pulp fibers used in paper and rayon manufacture are the fibrovascular bundles of coniferous woods such as spruce, fir, and pine, and the deciduous woods, maple, birch, beech, cottonwood, gum, magnolia, and chestnut. Man-made fibers of regenerated cellulose are rayon and acetate. The rayons are manufactured by various processes to produce fibers with properties required for such different purposes as wearing apparel, automobile tires, and carpets. They consist of cuprammonium rayon and various types of viscose rayon including high modulus and polynosic rayons and conjugate filaments. Cellulose acetate of textile usage is the secondary acetate; the triacetate is marketed as "Arnel" or "Tricel" and Fortisan is a hydrolyzed acetate produced under tension for a high crystallinity, high strength rayon.

In addition to fiber identification, the microscope is used in quality control and in research to follow the modification of cellulose by chemical substitution or crosslinking for special effects and specific properties and by coating or impregnation with noncellulosic materials. The detection of the additive, its location in or on the fiber, and its identification often challenge the ingenuity of the microscopist. Comparisons of morphological and structural changes which accompany such experimental reactions as decrystallization, crosslinking, etherification, esterification, and graft polymerization in emulsion or vapor systems yield information useful in altering the properties of cellulose materials for particular purposes.

II. INSTRUMENTATION

A. Microscopes

It is not the purpose of this chapter to delineate all the kinds of microscopical instruments available, or to discuss the theory or construction of microscopes. There are ample treatments of this in standard encyclopedias, and in such texts and reference books as those by Clark [1] , Martin [2, 3] , Needham [4, 5] , Allen [6, 7] , Shillaber [8] , Cosslett [9] , and others [10-14] . A paper by Meyers [15] in the 1969 Scientific Research provides a table of available research microscopes and their manufacturers. Further details may be solicited from the manufacturers. The present objective is to describe briefly those instruments and practices commonly used in the analysis of cellulose materials with emphasis on their appropriate manipulation specifically for cellulose or its more common derivatives.

Any generally satisfactory microscope providing magnification up to 400X will be appropriate for fiber examination. For routine identification and ordinary investigations, transmitted light and bright field are sufficient. Other methods of illumination such as darkfield, polarized light, ultraviolet and fluorescence, and phase contrast and interference microscopes are more in the nature of research tools for special purposes. They are briefly discussed.

1. The Stereoscopic Microscope

The stereoscopic microscope (Fig. 1) is a binocular instrument which provides slightly different images for each eye, thus giving a stereoscopic effect which permits seeing objects in three dimensions, and allows the object to be seen in its natural orientation rather than as an inverted image as in the compound microscope. Binocular stereoscopic instruments are most convenient for examination of whole pieces of fabric, or single yarns, for preparing and mounting specimens for the compound microscope, and for preliminary examination of the specimen. They are usually used with reflected light, but may be used with transmitted light or a combination of transmitted and reflected light depending on the type of specimen. The highest power used with the stereoscopic microscope is approximately 200X. These instruments usually provide magnifications of 2X, 10X, or 15X, 30X, 45X, 80X, and 100X. Modern instruments have a "stereozoom" feature which permits a range of magnification up to twice the initial magnification of the lenses to be dialed from one lens setting, thus providing enlargement to 200X. The microscopes with a built-in holder for the illuminator are most useful.

2. The Compound Microscope

The transmitted light microscope used for higher magnifications is a compound microscope (Fig. 2). The typical compound microscope consists of a stand, a stage to hold the specimen, a mechanical stage with controls for easy movements of the specimen slide in both x and y directions, a movable body tube containing two lens systems, and is equipped with a rack-and-pinion gear for controlled raising and lowering of the body tube for focusing. The lens system nearest the specimen is called the objective; the one nearest the eye is called the ocular or eyepiece. Beneath the specimen stage is a condenser lens which directs light to the specimen. The objective receives the light changed by passing through the specimen and produces the initial magnification of the image; the eyepiece, through which the specimen is viewed, magnifies the image created by the objective.

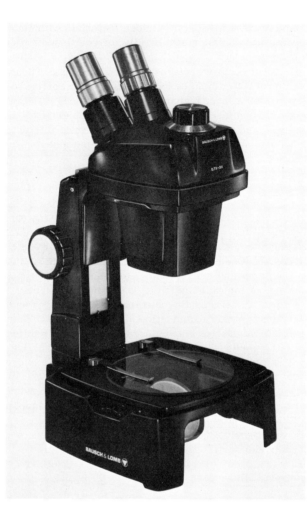

Fig. 1. Stereoscopic widefield microscope. Reproduced by courtesy of Bausch and Lomb, Inc.

The total magnification of the viewed image is the product of the magnification of the objective and the eyepiece. Thus an eyepiece magnifying 10X used with a 40X objective gives a total magnification of 400X. There are many variations in types of objectives, eyepieces, and condensers. Some modern instruments are equipped with a "dynazoom" system which provides a continuously variable magnification range of 1 through 2X, the

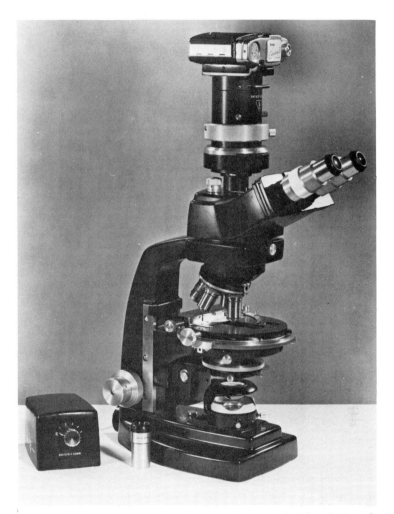

Fig. 2. Research compound microscope equipped with substage lamp and eyepiece camera. Published by courtesy of Bausch and Lomb, Inc.

eyepiece-objective combination, and is controlled by a knob at the top pf the body calibrated in 0.1X divisions. Instruments may also be obtained with fixed power magnification as in previous conventional designs. The light paths through the conventional microscope and through modern "zoom" microscopes are shown in Fig. 3. Choice of the appropriate compound microscope was recently described by Delly [16].

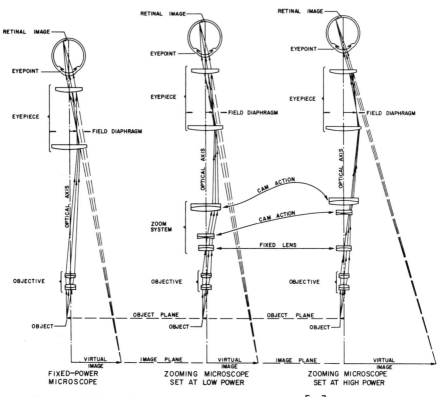

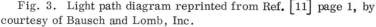

Fig. 3. Light path diagram reprinted from Ref. [11] page 1, by courtesy of Bausch and Lomb, Inc.

a. Objectives. Objectives are classified with respect to optical correction as (1) achromatic, (2) apochromatic, and (3) fluorite or "semi-apochromatic." The lens systems are composed of lenses having spherical surfaces. Since a spherical surface does not form a perfect image, all lens systems have defects or aberrations to a greater or lesser degree. The manufacturers have been able to compensate for both chromatic and spherical aberrations by combining in proper sequence in the same system two or more lenses differently ground, or of different glass formulations, and, in some cases, fluorite lenses with glass lenses.

Achromatic objectives are made with glass only, and are chromatically corrected to bring two wavelengths of light into focus at the same plane; they are also corrected spherically for one wavelength. They are the least expensive series of objectives, and are extensively used.

Apochromatic objectives are made with a combination of fluorite and glass lenses and are superior to achromats in their correction for spherical

and chromatic aberration. In apochromats chromatic aberration is corrected for three colors of the spectrum and spherical aberration for two,
which means that practically all of the images produced by the different
colors of the spectrum lie in the same plane and are equally sharp. These
objectives are used for high quality research work.

Fluorite objectives represent a compromise between the achromat and
the apochromat in their degree of correction.

b. Eyepieces. There are three types of eyepieces usually available
with any microscope: Huygenian, compensating, and hyperplane, each
characterized by distinct properties fitting it for the kind of objective with
which it was intended to be used [17] .

The Huygenian eyepiece is the simplest type, designed principally for
use with achromatic objectives. It consists of two planoconvex lenses
mounted one at each end of the eyepiece tube, with a field diaphragm
between them. These eyepieces effect a certain amount of correction for
chromatic difference of magnification in the achromatic objectives.

Compensating eyepieces are corrected primarily for use with apochromatic objectives. They are properly computed to eliminate color fringes
which would be conspicuous if apochromatic objectives were used with
other eyepieces. The combination of apochromatic objective and compensating eyepiece gives a field free of color to the very margin. They can
be used with the higher powers of apochromatic and fluorite objectives.

Hyperplane eyepieces have a flatter image plane than Huygenian ones.
They have a color compensation about midway between Huygenian and
compensating eyepieces which makes their use with high power achromatic
and fluorite objectives extremely advantageous.

In up-to-date laboratories the compound microscope most frequently
used has a binocular tube with two eyepieces which can be independently
focused to accomodate personal eye differences, and adjusted for inter-
pupillary distance; usually the eyepieces are inclined rather than upright
for comfort of the observer.

c. Magnification and Resolution. Resolution, the ability of the microscope to render fine detail visible, is far more important than magnification.
The term resolution usually refers to lateral resolution, the distance
between discrete particles necessary to form separate and distinct images.
Small objects below the power of the lens to resolve will be seen indistinctly,
if at all, and they will have a hazy, blurred outline. A particle that is
resolved is clearly and sharply defined and the image should be vivid.
Particles below the limit of resolution may be seen as indistinct images;
if the power of the eyepiece is increased these images will be larger but
more indistinct than before, but they will certainly be visible. If the object
has been magnified to the point that its image becomes fuzzy or indistinct
due to the limited resolving power of the lens system, further magnification

can only make the image larger, and less distinct without showing any more detail. Such useless magnification is called "empty" magnification.

The smallest particle that can be resolved with the visual microscope in green light is roughly 0.2μ. This is only a little over 0.9μ smaller than that resolved by the low power (10X) objective. In other words, microscopial resolution involves a particle size variation of only about 1μ from the smallest to largest resolvable particle. Nevertheless, resolution plays an important part in microscopial investigations.

The resolving power of a microscope depends generally on the objective. An objective capable of utilizing a large angular cone of light coming from the specimen will have a better resolving power than an objective limited to a smaller cone of light, and will give considerably more detail in the image. The measure of the maximum cone of light an objective can take and refract to the observer's eye is the numerical aperture (N.A.) of that objective. By definition, numerical aperture is

$$N \sin U,$$

where N is the refractive index of the object space and U is the half-angle of the light cone. The N.A. of each objective is engraved on its collar. The higher the N.A., the more complex the lens system within the objective and the greater its cost. The 0.95 N.A. Apochromat represents a lens design giving the highest value of U which is practicable. N.A.'s higher than 0.95 are achieved by using immersion fluids to change the refractive index in the space between the object and the objective, that is, to replace air with a medium of higher refractive index, nearer to that of the glass of the lens. This is accomplished by placing a drop of immersion fluid on top of the slide and lowering the objective into it. The immersion fluid is usually oil (N 1.52), but water (N 1.33) and monobromonaphthalene (N 1.66) have also been used to some extent. By means of immersion fluids objectives with N.A. as high as 1.6 have been produced but the practicable limit even for immersion objectives is considered to be N.A. 1.40. A useful empirical rule in microscopy is not to let the total magnification exceed about 1000 times N.A. nor fall below 250 times N.A. [18] of the objective used.

d. Condensers. The condenser is a light collecting lens placed below the specimen stage to direct a beam of light through the object. The Abbe condenser has only two lenses and an N.A. of 1.30. It is not corrected for spherical or chromatic aberration, but, because of its simplicity and light-gathering power, it is extensively used in general microscopy.

The achromatic condenser is a 1.40 N.A. condenser and is corrected for both chromatic and spherical aberrations. Because of its high degree of correction it is recommended for research microscopy and color photomicrography.

The variable focus condenser is a two-lens system with 1.30 N.A. The upper element is fixed and the lower one focusable. By this means it is possible to move the lenses apart in order to fill the field of view of low power objectives with light without removing the top element of the condenser. With other condensers one must often unscrew and remove the top lens from the condenser when using lower power objectives.

Darkfield Condensers. A special type of illumination in which the object is bright against a black background is known as "darkfield" and is used principally on transparent unstained material where the contrast between object and background is so low that brightfield illumination fails to make the object visible. Darkfield condensers depend on the use of a high N.A. hollow cone of light and must be oil-contacted to the lower face of the specimen slide. It is also necessary to use an objective with an N.A. less than that of the illuminated N.A. of the condenser so that no direct light gets into the image. If objectives with N.A. higher than 1.0 are used, a "funnel stop" must be inserted into the back of the objective to reduce the aperture of the rear lens surface in the objective. There are two types of darkfield condensers offered with research microscopes: cardioid and paraboloid. These produce a hollow cone of light with the focal point in the plane of the specimen. If the specimen is completely transparent and homogeneous, the light goes directly through and does not enter the objective since the N.A. of the illuminating cone exceeds that of the objective. If the specimen has fine transparent detail which differs in refractive index from the mounting medium, it will scatter light due to refraction and reflection and will appear bright since some of the scattered light will enter the objective. In cellulose work darkfield condensers are seldom used, but for delusterant particles in man-made fibers or films, and for observations of finely divided hydrocellulose they may have their place.

Rheinberg illumination [8] is a modification of darkfield which provides special effects useful with objects of low contrast. It consists of insertion under the condenser of bicolored disks, the center having one color, the margin another. They are available in various combinations: blue with red center, red with blue center, etc. The center portion illuminates the field of view and the border illuminates the object, thus providing contrast to enhance any faint details within the transparent object.

e. Depth of Focus. When one focuses a microscope on an object, there is a finite range above and below which the object goes out of focus and some other object appears in sharp focus. This range is called depth of focus and is of prime importance in photomicrography since the photographic plate is capable of recording in only one plane. In visual use, the range is somewhat greater because the eye is capable of a certain amount of accommodation assumed to be approximately 250 mm. For the most common objectives the depths of focus given by Benford [11] are:

Objective	Eyepiece	Depth of focus	
		Photomicro-graphic	Visual
10 X, 0.25 N.A.	10 X	0.0080 mm	0.0335 mm
40 X, 0.65 N.A.	10 X	0.0010 mm	0.0026 mm
100 X, 1.25 N.A.	10 X	0.0003 mm	0.0005 mm

3. The Polarizing Microscope

The polarizing microscope is essentially an instrument for measuring and analyzing the effects of a transparent specimen on a beam of light which passes through it, and thus determining crystalline and orientation characteristics of the specimen. Ordinary light vibrates in all planes about the axis of propagation; in plane polarized light the vibration directions perpendicular to that of propagation have been confined to one particular plane. In microscopes, polarization of light is achieved by the use of Nicol prisms constructed from naturally occurring calcite prisms or by the use of sheets of disks of Polaroid, a plastic in which oriented crystals of marked absorption properties have been embedded.

If any two polarizing devices are arranged in optical sequence and so oriented that their polarization planes are parallel, light from one will pass through the other. If one of them is rotated so that the planes are perpendicular, no light will pass. Both polarizer and analyzer, as the two devices are called, are arranged to rotate about the axis of the microscope. The polarizer is mounted in the substage, usually beneath the condenser so that all the light reaching the specimen is polarized, and the analyzer either just above the objective, or sometimes above the eyepiece, and so mounted that it can be rotated.

When the polarizer and analyzer are set with planes of vibration parallel, the light polarized by the polarizer may pass and give a bright field of view; if the analyzer is rotated so that the planes are crossed at 90° the light cannot pass through the analyzer and a dark field results. When a non-polarizing object is placed on the microscope stage with the polarizers in the crossed (90°) position, nothing happens and everything in the field of view is dark. But if a polarizing object such as a crystal or a highly oriented fiber is placed on the stage, the parallel polarized light from the polarizer upon striking the specimen is repolarized in two directions corresponding to the directions of vibration of the specimen. These waves, on striking the analyzer, are again repolarized. Wavelengths which were sufficiently retarded in passing through the specimen to interfere with each other are canceled out, and the remaining light will appear colored. Thus, the specimen, if it is highly oriented, may appear brightly colored on a dark background. Retardation colors may often serve as clues to the identity of the fibers in question. When looking through a polarizing microscope one is said to be facing north regardless of his actual orientation. If a sample on the stage of the microscope is viewed with the polarizer in the N-S direction and the analyzer in the E-W direction it is said to be viewed through crossed polars or crossed nicols. If the light vibration direction of the polarizer and analyzer are parallel, the sample is viewed with parallel nicols or parallel polars. If the analyzer is removed and the polarizer remains, the samples is said to be viewed by plane polarized light.

The eyepieces used with the polarizing microscope contain a set of cross hairs in the top element of the eyepiece in a movable sleeve so that the cross hairs can be sharply focused for the individual eye. The top of the body tube will usually have two slots and the eyepiece will have a peg to fit these slots. When the peg is fitted into the slot the cross hairs of the eyepiece will be accurately set in either the NS-EW position or the 45° position.

In conventional polarizing microscopes, an accessory slot is usually found at the lower end of the body tube just above the objective mount and is oriented at 45° from the vibration direction of the polarizer and analyzer. The slot permits the insertion of one of a number of compensators (quarter-wave plate, first-order red plate, quartz wedge, Berek compensator, Ehringhaus compensator, etc.) into the path of the light for observations and measurement of the optical properties of the sample.

Polarized light can be of great assistance in general microscopial work. Under certain circumstances it is helpful to reduce glare. Its principal use is in the determination of orientation by measurements of refractive index and birefringence. The usefulness of polarized light in microscopy lies in the fact that many different materials do polarize light to some extent and that evaluation of this characteristic permits correlation of optical properties with physical behavior.

To realize the full capabilities of the polarizing microscope the microscopist must understand the instrument and be able properly to interpret what is seen through it. Excellent works on the polarizing microscope are by Hartshorne and Stuart [19], Chamot and Mason [20], Johannsen [21], and Hallimond [22].

4. The Phase Microscope

The normal microscopical object is seen because it has regions varying in density. A completely transparent object is very difficult to see in any detail as all parts are equally dense. Darkfield illumination shows up border effects in completely transparent objects due to edge scattering and diffraction; and polarized light is useful if the transparent specimens happen to have directional or crystalline properties. Phase contrast microscopy

employs a method of illumination in which a portion of the light passing
through the object is artificially induced to interfere with the rest in such a
manner that it produces a visible image of an otherwise transparent specimen.
It offers the possibility of observations of fine details of structure without
the necessity of staining. The phase microscope is described in great detail
by Bennett et al. [23].

When a light wave passes through an absorbing object it is reduced in
amplitude and intensity. Since the perfectly transparent object does not
absorb light the intensity remains unaltered, and the object is invisible in
an ordinary microscope. However, the light that has passed through the
transparent object was slowed down by it and arrives at the eye a minute
fraction of a second later than it would otherwise have done. Such slight
differences in optical path are not detected by the eye, and the fundamental
problem in the microscopy of transparent objects is to convert them into
visible intensity changes. In phase contrast microscopy, the diffracted and
undiffracted light are separated by means of controlled interference and
artificially recombined so that the image formed by the interference of the
diffracted and undiffracted portions simulates the image of a specimen
having variations in density rather than merely differences in refractive
index. An annular diaphragm is placed at the front focus of the substage
condenser so that the object is illuminated by a hollow cone of light. A
phase plate is put at the image diaphragm formed by the objective. This
phase plate usually consists of a transparent annular layer evaporated on a
transparent plate, the added layer corresponding in size and shape to the
image of the diaphgram and having a thickness equal to 1/4 of a selected
wavelength of light. A metallic layer is generally superimposed to diminish
the transmittance of the annulus. When diffraction takes place at the object
the central maximum has a phase shift of 1/2 wavelength with respect to the
diffracted light. This phase shift results in an intensity difference in the
image which increases the contrast.

Retarding the direct light by one-quarter of a wavelength brings the direct
and diffracted waves into phase, giving rise to a negative phase-contrast
image in which those portions of the object giving greater retardation are
rendered bright against a darker background. Advancing the direct wave
by one-quarter wavelength, or, what is equivalent, retarding them by three-
quarters of a wavelength, puts the direct and diffracted waves out of phase
by half a wavelength. This again gives rise to destructive interference,
known as positive phase-contrast, in which those portions of the object
giving greatest retardation are now rendered dark against a lighter background. It is absolutely essential that annular diaphragms of condenser and
objective are appropriately matched.

One outstanding feature of the image in phase microscopy is the presence
of a halo around the object. The halo is useful in bringing out the contrast,
but, because of the halo, the phase contrast microscope is unsuitable for

making absolute measurements. It can, however, be used to demonstrate differences in refractive index between adjoining areas in a given specimen.

In cellulose research the phase microscope is occasionally useful in comparison of treated fibers. The nature of thin coatings on paper samples, and textile fibers can be determined, and possibly polymer grafts could be detected within textile fibers. Much can be done to improve visibility by choice of immersion media of correct refractive index relationship to the material.

5. The Interference Microscope

In the polarizing microscope, interference occurs between light rays which follow the same path in the object but vibrate in two mutually perpendicular planes in which the anisotropic nature of the object separates the light. In the phase contrast microscope interference occurs between direct and diffracted rays originating from the same object detail; a phase difference is artificially added between those two beams by means of a phase plate placed in the microscope. Thus, local differences in refractive index of the object are artificially converted into differences in contrast in the image for better observations of transparent detail within the specimen.

In the interference microscope, interference occurs between components passing through the object and those passing through the mounting medium. In contrast to the polarizing microscope and the phase microscope, the interfering beams do not originate from the same object point, but from points of object and surrounding medium. The refractive index of the particular point on the specimen is compared with the refractive index of the mounting medium. The transmission interference microscope is thus essentially a device for measuring the difference in optical path between a ray of light passing through a microscopial specimen and one passing through a reference area containing a medium of known refractive index.

There are over 100 varieties of interference microscopes described in "Contributions to Interference Microscopy" by Krug et al. [24]. Models which have been used in biological research and for textile fibers are the Dyson interference microscope [25], the AO- Baker instrument [26, 27], and the Zeiss interference photomicroscope [28, 29].

It is usual to divide the light from the lamp into two parts by means of a beam splitter, such as a semireflecting mirror. One beam is then made to traverse the object, the other travels along a similar path but does not pass through the object. When the two beams are recombined, under suitable conditions, there is interference, and the differences in optical path introduced by various parts of the object can be seen as variations in intensity of color. The phase change depends on the product of refractive index and specimen thickness; one of these quantities can be calculated if the other is measured.

Faust [25] described the use of the multiple-beam interference microscope for the study of optically heterogenous specimens. The methods
employed involved immersion of the specimen in liquids of different refractive index to obtain two separate interference patterns from which it was
possible to calculate a mean refractive index, and the thickness of the
specimen at any desired point. He pointed out that instead of using two
different immersion liquids one can either alter the wavelength of the
incident monochromatic light to obtain two different interference values, or
alter the temperature of the system with concomitant change in the refractive
index of the immersion fluid. In these ways it is easy to change the refractive index differences by small increments, a procedure equivalent to using
a large number of immersion liquids. This has the advantage that the
specimen is not disturbed during the experiment, and the same part of the
specimen is always presented to the observer in exactly the same manner.
Moreover, different areas of the specimen which may have different refractive indices may readily be detected during the observations. In a subsequent
paper, Faust [27] gave detailed explanations of the application of these
different techniques using the AO-Baker type of interference microscope in
which, instead of multiple beams, there are but two coherent wavefronts
produced by the use of birefringence components in both the condenser and
the objective. He gives detailed descriptions of the temperature variation
method and the wavelength variation method of determining refractive
indices at different points on the specimen; and conversely, determination
of the specimen thickness in samples of known refractive index. He also
listed comparisons between two-beam and multiple-beam interference
techniques, pointing out that, under good conditions, the multiple-beam
interference fringes are often extremely sharp and the multiple-beam
system more sensitive than the two-beam system. However, since the
light passes through the specimen and is reflected back many times, the
phase relationships within the object are not so clearly portrayed as with
the two-beam system. With thicker specimens this phase distortion could
be a serious disadvantage; with thin sections it would be of little consequence.
By gradually changing the index of the immersion liquid relative to that of
the specimen, one can determine the manner in which mean refractive index
varies from region to region of the specimen. These changes in relative
refractive index can be effected either by varying the wavelength of the
incident light or by changing the temperature of the system. The attainable
accuracy varies with the thickness of the specimen; under favorable circumstances, the index at any point on a specimen 10μ thick should not be in error
by more than 0.0003 [27].

6.   Ultraviolet Microscope

Since optical systems with a high N.A. could not be constructed, experiments to try to increase the resolution of microscopes were undertaken with

the use of shorter wavelengths in the ultraviolet region of the spectrum.
Glass lenses are not transparent to these short wavelengths, and in
constructing ultraviolet microscopes it has been necessary to introduce
quartz lenses or reflecting optics. Using ultraviolet radiation of 2000 to
3000 Å, a gain in resolution by a factor of 2 can be obtained, but the main
advantage of a uv microscope is not increased resolution but the fact that
unstained biological specimens show a strong natural absorption which
produces contrast in the image [4, 6] .

In the field of cellulose the ultraviolet microscope has not been used
either as a microscopic or a spectroscopic tool. The absorption curve for
cellulose is practically constant throughout both the visible and near-
ultraviolet regions of the spectrum to the far ultraviolet, turning sharply
upward at 2700 Å until, at about 2200 Å, absorption of the radiation by
cellulose is roughly 5 times what it was at 2700 Å [30] . It might be
possible to get useful differential absorption in the wavelengths below
2500 Å, but the difficulty is to find sources of radiation which produce
those shorter wavelengths in sufficient intensity, and microscopic equipment which can utilize this type of illumination.

7. Fluorescence Microscopy

A fluorescent specimen is one which when illuminated by light of one
color emits light of another color. The color may often be a clue to the
identity of a specimen. Illumination is usually in the near-ultraviolet and
the blue-violet region, and since fluorescence is usually rather weak it is
necessary to use a very intense source and to provide special filtering
techniques to accentuate the fluorescent image. The source is usually a
high pressure mercury arc. Most of the visible spectrum is filtered out
by "exciter" filters which are located in the illuminating beam and pass
only those wavelengths needed to excite fluorescence in the specimen.
Since the deep blue and violet visible energy is not completely removed by
the exciter filter, a second filter, the "barrier" filter, is used in the image-
forming system, after fluorescence has taken place, to remove the blue
and violet but to pass on the longer fluorescent wavelengths emitted by the
specimen.

Fluorescence can be divided into (1) primary or intrinsic fluorescence,
in which the specimen is fluorescent in its natural state, and (2) secondary
fluorescence, in which the specimen may be nonfluorescent, or weakly
fluorescent, but is stained by a fluorescent dye or by incorporation of
fluorescent substances [4, 6, 31] . Primary fluorescence is of greater
fundamental importance as indicating a definite chemical property of the
specimen. Fluorescent dyes increase contrast in probably much the same
way as staining methods for visible microscopy.

B. Illuminators

Successful microscopy and photomicrography depend upon efficient,
controllable, and convenient sources of illumination.

1. Reflected Light

Illumination for the stereoscopic widefield microscope is usually by direct
illumination of the surface of the object. There are several types of lamp
for this, but the Nicholas spotlight is one of the best. It may be placed at
any angle to the object and can be used for both transmitted and reflected
light by positioning the lamp in the holder built into the microscope. Often,
in examination of the surfaces of a fabric, it is advantageous to use two
lamps, one at 90° to the other, both obliquely directed at the fabric surface.
Shillaber [8] gives a number of lighting arrangements for achieving optimum
illumination of surfaces.

2. Transmitted Light

The compound microscope calls for illumination by intense transmitted
light. In the more modern microscope, there is a built-in substage lamp
below the condenser. Usually, this type of lamp has a coiled-filament
100 W, 6-V bulb. It is placed 10-15 in. in front of the microscope, and
the light beam is directed at the substage mirror of the instrument. It is
advantageous to have a base built for the microscope and lamp so that once
satisfactory alignment has been acheived both lamp and microscope may be
fastened firmly in position to prevent later jostling out of alignment.
  The most popular type of illuminator for the compound microscope
is the Orthoilluminator. It has a special base for positioning of the
microscope and clamps to hold it in the proper position, so that once alignment has been achieved it may be maintained throughout all operations.
The Orthoilluminators also provide two switches: one for low intensity
illumination for ordinary visual observations, the other at full intensity
light of approximately 3200° K color temperature for darkfield, polarizing,
phase, interference, and oil-immersion bright field observations, and for
color photomicrography.

  a. Alignment. "Critical illumination" is that form of light arrangement
in which the light source is imaged directly on the specimen. For many
years it was used in high power microscopy, microprojection, and photomicrography where an intense and controlled beam of light was necessary,
but because of unevenness in brightness of the light source itself, this
form of illumination has been replaced in modern practice by Koehler
illumination which presents certain advantages such as field diaphragm
control and elimination of nonuniform brightness in the microscope image.
For photomicrography Koehler illumination is imperative.

Accepted procedure for alignment of microscope and light source is as follows:

(1)   Focus the microscope on the specimen and adjust the field of view diaphragm of the lamp so that it just clears the circular edge of the field as seen in the microscope eyepiece. This adjustment must be made for every change of objective.

(2)   Close the iris diaphragm of the microscope substage condenser to a point where the illumination in the microscope just begins to dim. Remove the eyepiece, and, looking into the back of the microscope tube, adjust the iris diaphragm aperture of the microscope so that the back lens of the objective appears to be filled with light. For each change of objective, checking this aperture opening will ensure optimum resolving power of the optical lens system.

(3)   The intensity of illumination should be adjusted by use of neutral density filters, or by a transformer or rheostat for reducing the lamp voltage. The iris of the microscope substage diaphragm should not be disturbed, except when it is necessary to increase depth of focus, or optical contrast. The sole function of the substage iris is to control the effective N.A. of the condenser. In practice, the N.A. of the condenser is cut to 2/3 or 3/4 that of the objective to get a compromise between good resolution and good contrast. Glare arises from light which traverses the optical system and reaches the eye without playing an essential part in image formation. It is fundamental to satisfactory microscopy that as far as possible the illuminated area of the object should not exceed the area of the object under observation. This condition is regulated by the field stop (iris diaphragm) on the lamp, not the microscope.

b. Filters. The use of color filters between the light and the specimen is a great aid to bring out details and contrasts during visual observations and is essential for most types of black-and-white photomicrography. The Wratten set of gelatin filters offered by Eastman Kodak Company in nine colors [32] or the glass filters containing metals or metal salts such as the Corning glass series are highly desirable additions to every microscopical laboratory. To reduce glare without changing the resolution characteristics of the microscopical system, neutral density filters are essential. Heat absorbing filters are useful in instances where the mounting medium may be volatile or the specimen unusually affected by heat. Color filters are used to:

(1)   Improve resolution by reducing wavelength of light employed. This dictates a blue filter of the shortest wavelength.

(2)   Achieve monochromatic light. Objectives are corrected for the yellow-green portion of the spectrum, indicating that a yellow-green filter can be advantageous.

(3)   Improve detail. With dark specimens of too great contrast, detail in the specimen can be improved by using a filter of the same color as the object.

(4)   Heighten contrast. For an image which is colored, contrast can be increased by using a color complementary to that of the specimen:

Color of object	Filter
Red	Green
Yellow	Blue
Green	Red
Blue	Yellow
Brown	Blue
Purple	Green

C. Cameras

The most satisfactory method for recording microscopial images is by photomicrography. Many modern microscopes can be had with camera attachments which can be fastened to the eyepiece of the microscope. More stable arrangements are those which have a separate camera stand set over the microscope. The camera requires no optical parts or lenses. Any box or bellows provided with a light-tight connection to the microscope eyepiece, a plate holder on the opposite side, and a shutter mechanism could be used, but the devices commonly offered by all microscope vendors make the process much more convenient and results more certain. In the case of microscopes with eyepiece cameras, a separate eyepiece for viewing is desirable, but a camera equipped with a ground glass is sufficient. For very fine structures, it will be necessary to remove the ground glass and view the image with a focusing lens through a clear glass plate. The clear glass plate should have a mark on the side nearest the eyepiece. When this mark is brought into focus with the focusing lens, the image is then focused to the same plane.

Elaborate camera microscopes, which are combined systems of light source, microscope, and camera with all accessories, are most convenient to use since they have built-in automatic exposure meters and are always ready for taking photomicrographs without the necessity of setting up the arrangement. They are, however, extremely expensive. Details of photomicrographic procedures are given in excellent texts by Shillaber [8], Allen [7], Needham [4], and Barnard and Welch [33].

III. TECHNIQUES

A. Micrometry

1. Length Measurements

Measurement of microscopic objects is relatively simple but depends on the availability of the following accessories: a measuring eyepiece, an eyepiece micrometer, and a stage micrometer. The eyepiece micrometer is a round glass disk on which a linear scale is engraved. It fits inside the eyepiece and rides on a collar between the top and bottom lenses of the eyepiece. To insert the micrometer the upper element of the ocular is unscrewed and the micrometer dropped from the opening onto the shelf. The engraved surface must be uppermost. In handling the micrometer touch it only at its edges. It may be cleaned with lens paper whenever necessary, but great care should be exercised not to scratch its surface. If the scale is not in good focus, the top lens of the eyepiece may be screwed in either direction slightly until the lines are sharp. The image of the specimen is formed by the objective at the plane of the eyepiece diaphragm on which the micrometer scale rests, so that the divisions of the scale appear to be superimposed on the image. The eyepiece micrometer is calibrated by comparing its arbitrary divisions with the known length units of a fine glass scale which is placed on the microscope stage. This stage micrometer is the same size as an ordinary slide and carries a series of engraved lines separated at 10μ and 100μ intervals, respectively. Calibration of the eyepiece micrometer is as follows:

(1) Place the stage micrometer on the microscope and focus the microscope on the scale. Both scales will then appear sharply defined. By turning the eyepiece, they are laid parallel to each other.

(2) Determine how many divisions of the eyepiece micrometer correspond to a certain distance on the stage micrometer and calculate the length which corresponds to one division of the eyepiece micrometer. It is advisable to use the whole eyepiece scale, or the longest portion of it which can be made to coincide exactly with any two lines on the stage micrometer. An estimate must usually be made of the number of microns between the last line of the ocular micrometer and the last line of the stage micrometer just before it. However, if the whole 50-unit scale is involved in the calibration, a slight error in the estimate is not likely to affect the value of each unit any significant amount. The microscope must be recalibrated with each objective used. Calibration of each objective and ocular combination should be determined and recorded so that the tedious and time-consuming calibration need not be repeated. However, it must be emphasized that any change in lens combinations must be recalibrated or the measurements will be in error.

Once the ocular micrometer is calibrated it is used in making measurements by being superimposed on the image of the object to be measured and the number of units occupied by the length or width or other desired dimension counted. The total is then multiplied by the known value of each unit of the ocular. Because of the difficulty of making absolutely precise measurement, especially when a fraction of a unit in the eyepiece micrometer must be estimated, it is probably unwise to report any dimension nearer than the nearest tenth micron.

2. Measurement of Thickness

Occasionally it is desirable to know the thickness of a specimen. With modern microscopes the fine adjustment screw is graduated to show vertical travel and thus constitutes a measuring device. The scale on the knurled head of the fine adjustment knob is graduated into approximately 50 divisions; usually each division represents 2μ in vertical travel of the objective. This figure is usually marked on the fine adjustment. The procedure is to find some field where the exact focus on both top and bottom of the object can be obtained. Measurements of this type are usually made with the high-power objective with an aperture of at least 0.85 N.A., so that the optical plane of focus will be sharply defined. After the area to be measured is chosen the number of scale divisions on the fine adjustment which are passed through in going from top focus to bottom focus are determined. The number of divisions is multiplied by the number of microns in each division. Thus, if ten divisions are traversed in going from the bottom focus to the top and the value of each division is 2μ, the total movement has been 20μ. However, this is the actual distance between the two positions when the medium between them is air. To determine a more accurate distance, it is necessary to multiply the 20μ by the refractive index of the medium between the top and bottom focus. If it be water, the actual thickness will be 20 x 1.33 or 26.6μ. The operation should be repeated several times and an average taken, as there are more opportunities for error in this type of measurement than in longitudinal measurements using the eyepiece micrometer. A more accurate measure of thickness can be achieved by interference microscopy.

B. Microtomy

Morphological differences in textile fibers can best be compared in cross sections of bundles of fibers. The cross-sectional view is essential for quantitative estimates of fiber dimensions and for determination of degree of penetration or location of reaction in chemically treated samples. Characteristics of laminated sheet film can also be best studied in cross section.

1. Hand Microtome

Cross sections may be made with the Hardy [34] hand microtome (Fig. 4) or the Calco [35] or Mico [36] modification of it. Details of the procedure are given in ASTM Designation D-1444 [37]. The fiber sample is combed so that fibers lie parallel, with care not to remove any more fibers than absolutely necessary in arriving at a well-paralleled thin flat bundle. The combed fiber bundle is inserted into the slot of the metal plate, and the device is reassembled until the bundle is held snugly, but not tightly, in the slot (Fig. 5). A small amount of collodion or other quick-drying lacquer is applied to the fiber bundle with a camel hair brush, and worked into the bundle. When it has dried, the protruding ends of the bundle are cut off with sharp scissors as closely as possible to the metal plate on both sides. More lacquer is brushed over both cut ends and allowed to harden. The fiber ends are cut flush with the surface of both sides of the metal plate of the device with a sharp razor blade, and the plunger attached in place.

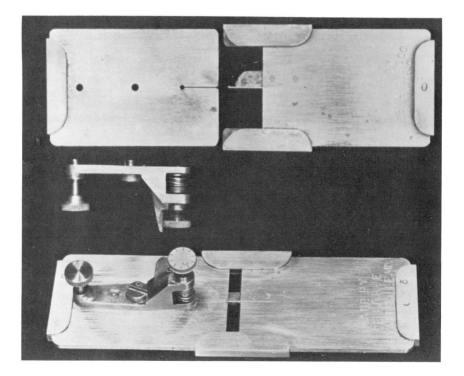

Fig. 4. Hardy hand microtome. Reproduced by courtesy of A. M. De la Rue.

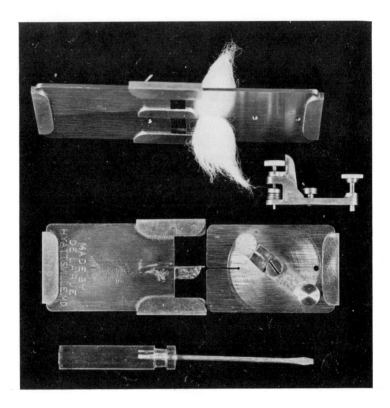

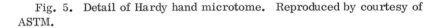

Fig. 5. Detail of Hardy hand microtome. Reproduced by courtesy of ASTM.

Under constant observation through a widefield stereoscopic microscope at a magnification of 45X the lacquer-embedded tuft of fibers is propelled through the slot by turning the thumb screw of the microtome until the complete block of fibers comes up 5 to 8μ. A coat of lacquer is brushed over the protruding ends and allowed to dry until firm and hard (5 to 10 min). Both lacquer and fibers are sliced off with a single stroke of the razor blade. For good results it is recommended that the blade and the device be moved simultaneously toward each other with the blade at an angle of about 45° to the slot holding the fibers. The section is mounted in a drop of mineral oil and a cover glass placed over it. The section is examined by transmitted light at about 250X. If it is not satisfactory in appearance, successive sections are cut, the face of the fiber block being coated with lacquer <u>after</u> <u>each</u> <u>slice</u>. It is absolutely essential that the lacquer be allowed to dry <u>hard</u> before a slice is attempted.

2. Clinical Microtome

A clinical rotary microtome (Fig. 6) equipped with a steel knife may be used for fiber sections and is preferable for sections of plied yarns and of fabrics. Moreover, in sectioning fibers, several different specimens may be cut with one slice if separate bundles are embedded in the same block. There are many approaches to the embedment of fabrics and yarns for sectioning [38-41]. A convenient system is as follows:

Fabrics or yarns may be mounted for embedding by stapling to cardboard frames cut to fit the inside of a number 12 gelatin capsule (veterinary). Bundles of fibers tied securely at each end may be stretched across a frame of Nichrome wire bent to fit inside capsules size number 00. The frames are placed inside the capsules which are then filled with methyl methacrylate monomer. Botty et al. [42] found that material susceptible to methacrylate or its solvents could be protected during embedding if precoated with an emulsion of rubber latex.

Fig. 6. Rotary microtome "Model 820." Reproduced by courtesy of American Optical Corporation.

The open capsules are placed under vacuum several times within an hour to remove air bubbles and to encourage diffusion of the embedding solution into the fiber bundles or the fabric. The monomer is poured off and replaced by partially prepolymerized embedding solution composed of 4 parts methyl methacrylate monomer and one part butyl methacrylate monomer plus 2% of the catalyst solution Luperco CDB (50%, 2, 4, dichlorobenzoyl peroxide with dibutyl phthalate). The capsule is covered and placed in an oven at 55° C until polymerized (6 to 8 h). The gelatin capsule is removed by soaking in hot water and the specimen face is trimmed. The actual cutting face should be a pyramidal protrusion about 2 mm across on the face of the block; the block should be of appropriate size to clamp into the jaws of the specimen chuck on the microtome. The specimen is aligned with the knife of the microtome so that there is a small angle between knife and specimen. Sections are cut approximately 7μ thick and mounted on a slide in mineral oil and covered with a coverslip for microscopical observations.

C. Refractive Index Determination

Refractive indices of fibers can be measured by the immersion method of Becke, using the polarizing microscope. A sodium vapor lamp, light from a monochromator, or white light with appropriate filters, is used to provide monochromatic light. The substage condenser and lamp are adjusted to furnish a narrow pencil of axial illumination. The fiber is observed at as low a magnification as convenient, usually 100X or less, using plane polarized light.

Short lengths of fiber (1 to 2 mm) are cut from the sample, placed on a microscope slide, and covered with a cover glass. A drop of a liquid of known refractive index is run under the coverslip, and the fibers brought into sharp focus. If the index of the liquid and of the fiber are the same, the fiber outline will disappear; thus, the refractive index of the liquid is the refractive index of the fiber. If the refractive index of the sample and the liquid do not match, a bright line, called the Becke line, is seen parallel to the edge of the fiber just inside or just outside the fiber outline. When the microscope is focused upward the Becke line is observed to move toward the fiber or the mounting liquid, whichever has the higher index. Conversely, when the microscope is focused downward the line moves toward the medium of lower refractive index. In this manner it is determined if the fiber refractive index is greater or less than that of the liquid. More preparations are then made on fresh slides with other liquids until a match is obtained or the value is shown to be between two liquids which bracket the fiber in refractive index. In the latter case, the value is reported as midway between the two bracketing liquids. This procedure must be conducted once with the fibers oriented on the microscope stage with the fiber axis parallel to the vibration direction of the polarized light to obtain n, parallel ($n_{\parallel}$) and once with the fiber oriented 90° from this position to obtain n, perpendicular $(n_\perp)$. The birefringence of the fiber is calculated as $(n_\perp - n_\parallel)$ and is indicative of degree of orientation. Obviously, in order to determine refractive indices by the immersion method, it is necessary to have available a graduated series of colorless liquids of known refractive indices. Standard refractive index liquids can be obtained in sets which vary in index by small increments. Sets composed of liquid mixtures of predetermined index are marketed by R. P. Cargille, 118 Liberty Street, New York, New York, and by Eastman Kodak Company, Rochester, New York. The liquids are graded from low to high refractive index in increments of 2 places in the fourth decimal place. For some purposes, however, they are unsuitable because the fiber in question may be soluble in the particular liquid provided and be swollen by it. If it is necessary to prepare such a series it is advisable to use mixtures of 2 liquids whose components are of equal volatility. The refractive indices of the mixtures are measured with a refractometer maintained at 20° C and the liquids kept in a set of labeled bottles fitted with ground glass stoppers. Table 1 gives a list of refractive indices of available liquids as recorded by Luniak [43]. The index of any liquid can be raised or lowered by addition of another liquid with which it is miscible. The resulting index can be foretold with a fair degree of precision by the equation

$$n(v_1 + v_2) = n_1 v_1 + n_2 v_2,$$

in which n is the desired index, n_1 and n_2 indices of the two liquids to be mixed, and v_1 and v_2 volumes of the 2 liquids. Kaiser and Parrish [44] discuss methods for final adjustment.

The refractive index of most cellulose fibers varies between n = 1.56 and n = 1.47 so that water (n = 1.33), although widely used in general fiber microscopy, is far from being an ideal mountant because of its low refractive index. A list of suggested combinations of liquids for use in studying cellulose fibers is given in Table 2 [4, 46-49]. Saylor [50] pointed out that mixtures of volatile with nonvolatile constituents will be unreliable because of differential evaporation under use. He suggests for materials whose refractive indices lie between 1.515 and 1.659 methyl phthalate and 1-bromonaphthalene, because their partial pressures in mixtures at room temperature are so nearly proportional to composition that the refractive index of a mixture does not change detectably during use; their boiling points are about 281° C. Needham [4] suggests combinations of liquids covering several ranges of refractive index.

To make quantitative observations it is necessary to use plane polarized light; monochromatic light of the sodium line 5895 Å is standard illumination. However, for many qualitative purposes unpolarized white light is suitable.

TABLE 1

Refractive Indices of Mounting Liquids for Microscopy
of Cellulose Materials [43, 45]

Liquid	Refractive index (n_D^{20})
Air	1.00
Water	1.33
Acetone	1.35
Ethyl alcohol	1.36
n-Heptane	1.39
n-Decane	1.41
Chloroform	1.44
Butyl stearate	1.445
Kerosene (bp 300°F)	1.45
Glycerine	1.46, 1.47
Liquid paraffin	1.46, 1.47, 1.48
Carbon tetrachloride	1.46, 1.47
Turpentine	1.47, 1.48
Castor oil	1.48, 1.49
Xylene	1.49, 1.50
Toluene	1.50
Benzene	1.50
Cedarwood oil	1.513
Clove oil	1.53, 1.54
Methyl salicylate	1.539
o-Dichlorobenzene	1.549
Tricresyl phosphate	1.556, 1.558
Bromobenzene	1.560
Aniline	1.58

TABLE 1 (continued)

Liquid	Refractive index (n_{20}^D)
Alpha-chloronaphthalene	1.632
1-Bromonaphthalene	1.658, 1.659
Methyl iodide	1.74

TABLE 2

Suggested Combinations of Liquids for Refractive Index Media

Liquids	Reference
Alpha-chloronaphthalene (1.6317) and kerosene (bp 300° F) (1.4500)	Fox and Finch [49]
Alpha-monobromonaphthalene (1.659) and liquid Paraffin (1.482)	Meredith and Hearle [48]
Butyl stearate (1.4446) and tricresylphosphate (1.5586)	Hermans [46, 47]
Diethyl oxalate (1.410) and normal butyl phthalate (1.491)	Needham [4]
Castor oil (1.480) and ortho-tricresylphosphate (1.555)	Needham [4]
Heavy mineral oil (1.483) and Halowax (1.555)	Needham [4]
Heavy mineral oil (1.483) and 1-bromonaphthalene (1.656)	Needham [4]
1-bromonaphthalene (1.656) and methylene iodide (1.74)	Needham [4]

Values for refractive index are reported to the third decimal place, but often, for identification work, such precision is not practical; values within 0.02 or 0.03 of those given in the tables will suffice.

The method is applicable to all fibers and is useful in fiber identification. In estimating degrees of orientation and crystallinity of native and man-made cellulose fibers, and degree of substitution in chemically modified cellulose fibers, an increase in birefringence ($n_{\parallel} - n_{\perp}$) is taken to indicate an increase in orientation.

An isotropic refractive index can be calculated from the equation [44]

$$n = 1/3 \ (n_{\parallel} - 2n_{\perp}).$$

This would represent the refractive index of material without regard to its orientation. In general, in a given series of chemically modified fibers, a change in isotropic refractive index would be taken to indicate a change in degree of crystallinity.

The refractive indices of all fibers vary with the wavelength of incident light and with the moisture content and other substances which may be present; in addition, fibers vary slightly in refractive index from place to place both axially and transversely. Because of these limitations it is impossible to bring the refractive index of the fiber and the medium into exact equality except locally; but these local variations are usually undetectable except by refined methods, and any region of markedly different refraction probably indicates the presence of another substance such as oil, titanium dioxide, air bubbles, or other additive [51].

If there is a large difference between refractive index of fiber and mounting medium (such as a dry mount of air) the specimen appears dark and has low transparency, surface details being more prominent. On the other hand, if the refractive index difference is small the fiber is transparent and the internal structure predominates [51]. Fig. 7 shows the difference in effect when a fiber is mounted in media of lower and equivalent refractive indices.

The Becke line method is probably the most commonly used for determination of refractive index microscopically, but there are several modifications. In the modification of a fiber refractometer reported by Preston and Freeman [52], the microscope is set up the same as for the Becke method but uses a monochromator for the light source. The sample filaments are immersed in a liquid with a known dispersion curve and the filaments observed as the wavelength of the light is varied. When the fiber outline and the Becke line disappear the refractive index can be taken from the dispersion curve for the specific liquid. The present series of Cargille liquids [8] are available with the dispersion at 5893 Å, 4851 Å, and 6563 Å on each bottle.

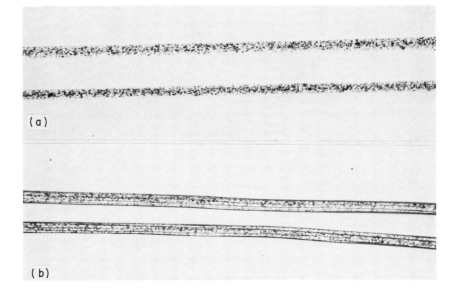

Fig. 7. Longitudinal view of cellulose acetate fibers. Mag., 200X.
(a) Mounted in mineral oil, refractive index 1.47; (b) mounted in heptane,
refractive index 1.39. Reproduced by courtesy of ASTM.

IV. APPLICATIONS TO CELLULOSE

A. Fabrics and Yarns

The microscope has been used to good advantage in the study of fabric
constructions and yarn characteristics. Goldberg's book "Fabric Defects"
[53] and the more recent paper by Westbrook on "Controlling Loom Related
Defects" [54] point up the need for methods of evaluating discrepancies in
the performance of the weaving loom, and "The Nature of the Break-Spun
Yarn" by Lord and Senturk [55], the changes in spinning procedures currently
underway in textile technology. For the most part fabrics are examined
with the stereoscopic microscope at rather low magnifications. However,
Shillaber [8] describes lighting arrangements for photomicrography of fabrics
and yarns taken at low power with the compound microscope. The essentials
of such work are the reduction of glare and production of uniform illumination, neither of which is easy to achieve in a white fabric with hairy surface.
Surface illumination from two sources placed at 180° to each other is recommended, and it is suggested that filters of complementary colors be used on

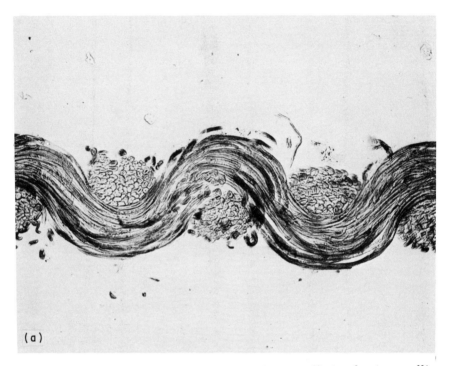

(a)

Fig. 8. Sections of high-pickage fabric showing effects of water swelling on fabric construction of tight weave. (Filling yarns in cross section.) (a) In dry state; (b) in wet state. Mag., 136X. (U.S. Dept. of Agriculture photographs)

the two lamps. Polarized light is also suggested as a means of reducing glare from individual fibers in the fabric surface. Fig. 8 shows sections of a high-pickage fabric before and after wetting.

The character of a yarn is often reflected in a cross section; and Fig. 9 shows a yarn spun from a cotton-rayon blend.

B. Fiber Identification

As early as 1836 [56] a microscopial study was included in a report submitted to the Privy Council of King William IV on the status of the textile industry of Great Britain. Comments on the typically convoluted shape of a cotton fiber, and observations as to fiber diameter, maturity of cell wall development, and variability of fiber properties were made. The description was illustrated by excellent line drawings of coarse Indian and fine Eygptian cotton fibers under the microscope.

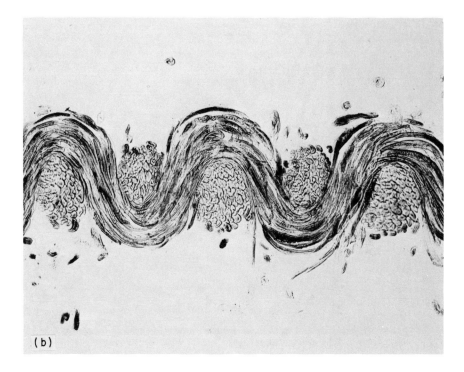

(b)

Fig. 8, continued.

By 1907, Hanausek and Winton were able to picture in their "Microscopy of Technical Products" [57] line drawings of salient microscopial features for the identification of all of the then common textile fibers; and 25 years later the microscope was so much used in textile work that between 1928 and 1934 Lawrie [58], Skinkle [59], Heerman and Herzog [60], Preston [61], and Schwarz [62] all produced textbooks giving techniques for the practice of microscopy in textile testing and research. In the intervening years a sizable body of literature on microscopial techniques has developed in technical journals, and there is an excellent collection of texts and atlases on fiber microscopy per se published since World War II. The 6th edition of Mauersberger's revision of "Matthews Textile Fibers" [63] and the "Textile Atlas" of von Bergen and Kraus [64] were both issued in 1946. Luniak's excellent 1949 thesis on "The Identification of Textile Fibers" produced in English in 1953 [43], Heyn's 1954 text on "Fiber Microscopy" [65], Stoves' 1957 "Fiber Microscopy" [18], and the British Textile Institute's "Identification of Textile Materials" of which the 6th edition came off the press in 1970 [45] are classical reference handbooks indispensable to any microscopist working with cellulosic materials.

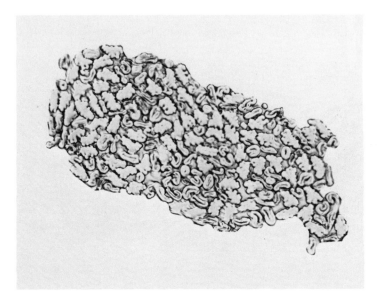

Fig. 9. Cross section of cotton-rayon blended yarn. Mag., 250X.
(U. S. Dept. of Agriculture photograph)

Other useful sources are Cook's "Handbook of Textile Fibers" [66] and
"The Fiber Encyclopedia" [67] ; Harris' "Handbook of Textile Fibers" [68],
"The Structure of Textile Fibers" edited by Urquhart and Howitt [69] , and
"Fiber Structure" edited by Hearle and Peters [70] . The annual Technical
Manual of the American Association of Textile Chemists and Colorists [71]
and the American Society for Testing and Materials Standards on Textile
Materials published annually by ASTM Committee D-13 on Textile Materials
[72] are revised frequently to keep abreast of new developments. There
are also annual charts on cottons [73] and on the man-made fibers [74-76]
published in various journals.

Fiber identification of textiles is made on the basis of morphological
features observed in longitudinal and cross-sectional views, compared
with standard samples of known origin or with photomicrographs published
in the literature [43, 45, 63, 64, 71, 72]. Additional aids are staining
reactions, swelling patterns, and microsolubility characteristics observed
at magnifications of 100X to 400X, and measurements of refractive index
and birefringence. Deviations from the typical in morphology, staining,
swelling, solubility, or birefringence indicate alteration of fiber properties.
To determine if the changes are associated with microbial deterioration,
mechanical damage, or such purposeful chemical modifications as coating,
impregnation, substitution, or crosslinking constitutes the art of fiber
microscopy. In the research investigations of special chemical processes,

comparison of treated samples with the control from the process under consideration permits determination of location, homogeneity, and distribution of the reaction.

Many types of paper, particularly bonds, ledgers, index, and book papers, are bought on the basis of fiber composition. Wood fibers used in the manufacture of paper, paperboard, and in dissolving pulps for rayon and cellulose acetate production, have morphological features which permit distinction of fibers of coniferous woods from those of deciduous wood, and, in most cases, even the identification of species of trees. Detailed descriptions of the various classes of paper fibers are found in a photomicrographic atlas by Carpenter and Leney [77] and a text by Isenberg [78]. Additional information on fiber identification of paper fibers is given in Tappi Standards [79] and ASTM Standards for Paper [80] and in the "Fundamentals of Papermaking Fibers" published by the British Paper and Board Makers Association [81]. Special stains for paper fibers were described by Plitt [82,], Graff [83], Lofton and Merritt [84], and Merritt [85].

1. Morphology

a. Textile Fibers (1) Longitudinal Views. The simplest approach to fiber identification is examination of fibers in longitudinal view at a magnification of approximately 250X. A few fibers are laid as nearly parallel as possible in a drop of a mounting medium on a glass slide and covered with a coverslip. If needed, more mounting medium liquid is run under the coverslip at the side until the entire area is evenly filled. Excess liquid is removed by blotting at the edge of the coverslip with a small piece of filter paper with care that the top of the coverslip remain clean. These operations are best carried out under a widefield stereoscopic. Fibers are often mounted in water, but its refractive index of 1.33 is not optimum for cellulose fibers whose indices range from 1.47 to 1.59. Mineral oil (refractive index 1.47) or glycerine (r.i. 1.45) are more suitable. Preston [61] recommends a ratio of 1.06 between the refractive index of fiber and mounting medium for adequate contrast. The liquid should have no swelling or dissolving effects on the fiber under consideration. A list of refractive index liquids is given in Table 2, but it is usually more satisfactory to use one of the graduated series commercially available [8].

(a) Seed hairs. (i) Kapok. Kapok fibers grow on the inner wall of the fruit pods of the tree. The fiber is smooth, cylindrical, hollow, and thin-walled. It tapers to a point at one end, and the other forms a slightly bulbou base with annular or reticulate markings. The cellulose walls are highly lignified. The average length varies from 20-30 mm and the average diameter is 20μ. The lumen is very broad and filled with air.

(ii) Cotton. Cotton has a characteristic shape; it is a collapsed convoluted tube which at low magnification resembles a twisted ribbon. Very immature fibers are thin-walled and may show few convolutions; mature fibers will be thick-walled and variously convoluted. The lumen may be very irregular in shape and often not visible for the entire length of the fiber.

Mercerized cotton fibers have a smooth, almost cylindrical, appearance, but show some evidence of having been twisted. The lumen width is exceedingly irregular. In fully mercerized fibers, it is not uncommon for the lumen to have disappeared due to the inward swelling of the cellulose wall during mercerization.

(b) Bast fibers. (i) Flax. Flax (linen) fibers are straighter than cotton and taper to an exceedingly slender threadlike tip. Flax fibers usually show horizontal bars at fairly regular intervals along the length of the fiber, and have been described as having the appearance of a cane fishing pole. The cell walls are so thick that the lumen is a narrow line, often not visible at all in longitudinal view. Each fiber is a sharply pointed cell with a mean length of 27 mm and diameter of 23μ.

(ii) Ramie. Ramie has a broad flat fiber with rather blunt, rounded ends and a wide lumen. Ramie has cross markings more prominent than those in flax; they are more numerous and deeper, often appearing as angular bars not extending entirely across the fiber. The ultimate fibers vary in length from 100 mm to 200 mm and in diameter from 25 to 75μ, averaging 30μ.

(iii) Hemp. Hemp has cross markings similar to those of flax, but the fibers are more irregular in size, and the fiber ends are characteristically blunt and thick-walled in contrast to the long tapered point of the flax fiber end. The lumen of hemp is broader than that of flax. The fiber averages 20 mm in length, and from 16 to 50μ in width. Unlike flax, the cell walls are partly lignified. The ends of the fibers frequently fork, and whether forked or not, are blunt and thick-walled.

(iv) Jute. Untreated jute fibers show few if any cross markings, but cell wall thickness and lumen diameter are exceedingly variable along the length of a single fiber. Jute fiber tips are pointed but not as pronouncedly attenuated as those of flax. The length is 1.6 mm up, and the mean length 2.5 mm, the diameter 12 to 18μ.

(v) Kenaf. Individual fiber cells of kenaf are cylindrical with many surface irregularities. The lumen is irregular having such marked contractions that at some points it becomes discontinuous. Fiber ends are blunt and irregular. Length varies from 2 to 6 mm with a mean length of 3 mm. The fibers are from 14 to 33μ wide, with an average width of 21μ.

(c) Leaf fibers. (i) Abaca ("manila hemp"). Abaca fibers are similar to those of jute in having irregular lumens but they taper to a fine hairline tip. The fiber cells have smooth walls without cross markings. They are 2.5 to 12 mm long, and 16 to 32μ in diameter. The fibers are often accompanied by rows of silicified cells which, when ashed, show prominently in rows of scalloped blocks, each with a hole or depression in the center.

(ii) Sisal. Sisal fibers are stiff with a broad variable lumen. The ends are blunt and rounded, rarely forked. Parenchyma cells, which often accompany the fibers, contain crystals of calcium oxalate. The fibers range from 1.8 to 3.2 mm in length and from 18 to 24 μ in width.

(d) Regenerated cellulose. (i) Cuprammonium rayon. Filaments of cuprammonium rayon are small and uniformly cylindrical. They have no particular distinguishing features.

(ii) Viscose rayon. Viscose rayon is produced in various forms and deniers for various purposes, and includes high modulus rayons, polynosic fibers, and conjugate filaments [86-94]. The filaments of conventional continuous filament textile viscose are fairly uniform in diameter and usually have continuous striations running parallel to the fiber edges. Staple fibers are similar except that they are short and may sometimes have crushed or distorted ends where cut into staple lengths. Pigment delusterants are visible in "dull" or delustered rayons as particles speckled throughout the filament. Occasionally they are sufficiently numerous to partly obscure the striations.

(iii) High modulus viscose rayon. These fibers are more nearly round than the ordinary textile viscose. They often show only one striation or none.

(iv) Polynosic rayon. These fibers are usually round and highly delustered. They resemble wool fibers in size and appearance.

(v) Conjugate rayon ("crimped rayon"). These filaments are made of two different rayons extruded from the same spinneret. Their appearance is variable, one side having prominent striations, the other none.

(vi) Cellulose acetate. Cellulose acetate fibers are smoother than textile viscose rayon fibers, with fewer longitudinal striations on the filament. In acetate, delustering is frequently accomplished by inclusion of air bubbles or oil droplets rather than solid particles. Commercial acetate fibers are usually the secondary acetate, but cellulose triacetate fibers are marketed as Arnel and Tricel. The fibers are larger than those of the secondary acetate and more irregularly striated.

(vii) Saponified cellulose acetate fiber (fortisan). Fortisan fiber is very fine in diameter, with few longitudinal striations, but is not a round fiber.

Average diameters of representative native fibers are given in Table 3 and of typical man-made fibers in Table 4. To determine relative amounts of fiber in a sample of mixed fibers, fiber diameter is measured for at least 100 fibers of each type represented in the sample (or average diameters

TABLE 3

Fiber Fineness of Native Cellulose Fibers [63]

Fiber	Width in Microns
Cotton (Upland)	16 to 21
Cotton (Egyptian)	9 to 18
Flax	15 to 17
Jute	15 to 20
Hemp	18 to 23
Kapok	21 to 30
Ramie	25 to 30

from Table 4 or 5 may be used). The mean of the squares of the diameters of each fiber type is multiplied by the specific gravity (Table 5) and by the number of fibers of that type counted in the mixture. Percentage fiber composition of the sample is calculated from these data [72].

(2) Cross-Sectional Views. Each of the natural cellulose fibers has a distinctive cross-sectional shape. Cross sections are made with a razor blade by means of the hand microtome [34, 35] or with a conventional clinical microtome. For either procedure it is necessary to immobilize the bundle of paralleled fibers by embedment in a polymeric solution which will solidify to a block hard enough to hold the fibers tightly while they are being cut, and which is of proper consistency to slice smoothly [38–41]. In general practice, it is believed that hardness of the embedment should be equivalent to that of the substance to be sliced. Sections from 5 to 10μ in thickness are suitable for light microscopy.

(a) Seed hairs. (i) Kapok. Cross sections of kapok are round, partly flattened, with very thin walls and extremely large empty lumens (Fig. 10a).
(ii) Native cotton. Fiber sections of cotton are kidney-shaped to round with a linear lumen. Immature fibers are often folded and contorted into "U" or "C" shapes; sometimes even flattened (Fig. 10b).
(iii) Mercerized cotton. Fiber sections of mercerized cotton are almost circular with practically no lumen (Fig. 10c). In samples from yarns which have been mercerized under extreme stretching, some fiber sections will have almost straight edges with angular corners due to interfiber pressures exerted during mercerization.

TABLE 4

Filament Denier of Man-Made Cellulose Fibers [71]

Filament denier	Average fineness in microns	
	Viscose rayon	Acetate
1	9.6	10.3
2	13.6	14.5
3	16.7	17.8
4	19.3	20.6
5	21.6	23.0
6	23.6	25.2
7	25.5	27.3
8	27.3	29.1
9	28.9	30.9
10	30.5	32.6
12	33.4	35.7
14	36.1	38.5
16	38.6	41.2
18	40.9	43.7
20	43.1	46.1

TABLE 5

Specific Gravity of Some Celluosic Fibers [71]

Fiber	Specific gravity
Acetate (secondary)	1.31
Acetate (triacetate), "Arnel"	1.33
Cotton	1.55
Flax (linen)	1.50
Fortisan	1.52
Hemp	1.48
Jute	1.48
Ramie	1.51
Rayon (cuprammonium)	1.52
Rayon (viscose)	1.52 to 1.54

(b) Bast fibers. (i) Flax. Flax fiber sections are sharply polygonal with a small dot or line for the lumen (Fig. 10d). In some samples there also will be sections with rounded oblong forms with larger lumens.

(ii) Ramie. Fiber sections of ramie are more flattened or oval than those of flax and considerably larger (Fig. 10e). The thick wall has radial fissures; the lumen is long and narrow or of the same shape as the section, and often the lumen wall is deeply fissured.

(iii) Hemp. Fiber sections of hemp are variable in shape from triangular to polygonal and the lumen is frequently wedge-shaped, but sometimes barely visible at all (Fig. 10f).

(iv) Jute. Jute fiber sections are sharply polygonal in shape, like flax, but the lumen is a wide open ellipse of varying dimensions (Fig. 10g). A few fibers may have rounded corners, but the variable size of the lumen is characteristic.

(v) Kenaf. Kenaf cross sections are polygonal, with a thick cell wall which appears to be surrounded by a thick median layer of lignin. The lumen is very irregular in shape and size (Fig. 10h).

(c) Leaf fibers. (i) Sisal. Sections of sisal fiber bundles are crescent to horse-shoe shaped, and often split. Sections of individual fibers are polygonal with more uniform lumens than in abaca. Cell walls are moderately thick (Fig. 10i).

(ii) Abaca ("manila hemp"). Sections of abaca fiber bundles are round to elliptical and slightly indented. Sections of individual fibers are polygonal with slightly rounded corners; cell walls are medium to thick (Fig. 10j).

(d) Regenerated cellulose. (i) Cuprammonium rayon. Cross sections of cuprammonium rayon are almost round; filament diameter is small and regular (Fig. 10k).

(ii) Viscose rayon. Sections of textile viscose are irregularly circular with finely serrated edges. Delusterant particles are often prominent (Fig. 10l).

(iii) Polynosic rayon. Sections of polynosic rayon are round and uniform, but coarser than cuprammonium rayon (Fig. 10m), and usually heavily pigmented.

(e) Cellulose acetate. (i) Acetate (secondary). Sections of cellulose acetate fiber are roundly lobed to cloverleaf shaped, and smoothly crenate (Fig. 10n).

(ii) Triacetate (Arnel and Tricel). Sections of cellulose triacetate fibers are more serrated than those of secondary acetate, and usually larger (Fig. 10o).

(iii) Saponified cellulose acetate (Fortisan). Sections of Fortisan are finely serrated and much smaller in diameter than fibers of either cellulose acetate or viscose rayon (Fig. 10p).

b. Wood and Paper Fibers. Detailed descriptions of the various classes of paper fibers are found in a text by Isenberg [78], in Tappi Standards [79], and in ASTM Standards [80]. Typical fibers are shown in a photo-micrographic atlas by Carpenter, and Leney [77]. Accepted methods for preparation of slides for determination of fiber content of paper are given in Tappi Standard T 401m-53 [79]:

Lines 1 in. from each end of a clean glass slide are drawn with a glass-marking pencil to keep the fiber suspension inside the 1-in. square at each end of the slide and provide duplicate test areas. The slide is placed on a warming plate, and a portion of fiber suspension is withdrawn from the test tube containing the defibered specimen after vigorous shaking to disperse the fibers. A 0.5-ml portion of the fiber suspension is deposited on the square at each end of the slide. The water is allowed to evaporate until there is just sufficient to float the fibers which are now distributed evenly by gently tapping the suspension with a needle. The slide is left on the warming plate until completely dry. An appropriate stain is now applied,

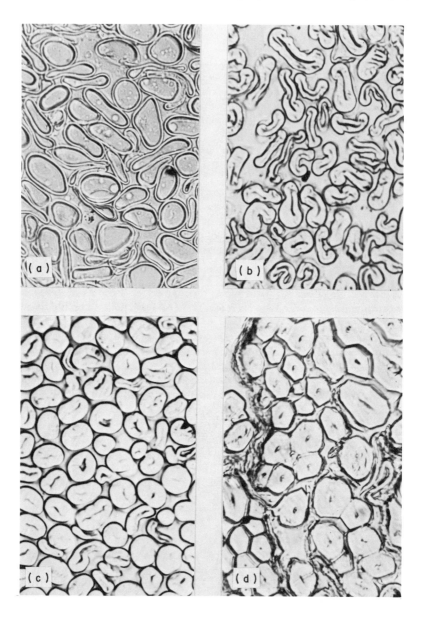

Fig. 10. Cross sections of representative cellulose fibers. Mag., 500X.
(All U. S. Dept. of Agriculture photographs). (a) kapok, (b) cotton,
(c) mercerized cotton, (d) flax.

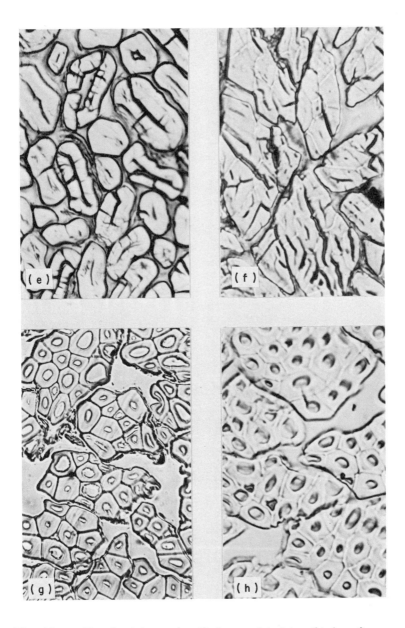

Fig. 10, continued. (e) ramie, (f) hemp, (g) jute, (h) kenaf.

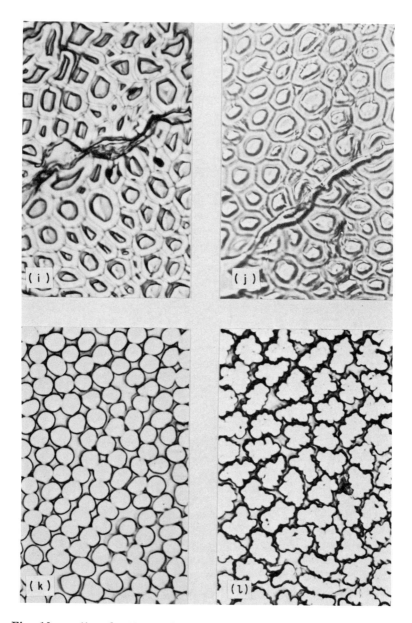

Fig. 10, continued. (i) sisal, (j) abaca, (k) cuprammonium rayon, (l) viscose rayon.

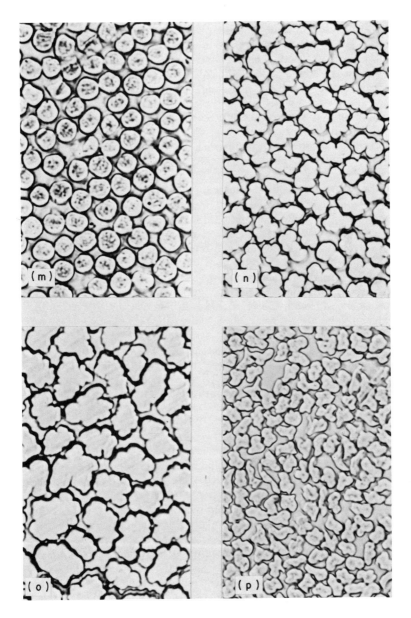

Fig. 10, continued. (m) polynosic rayon, (n) cellulose acetate
(secondary), (o) cellulose triacetate ("Arnel"), (p) Fortisan.

a coverslip placed over the stained fibers, and the slide examined at a magnification of at least 100X to determine kind and relative number of each fiber type present. The procedure is to begin at a point 2 or 3 mm from one corner of the coverglass and move the stage gradually so that the observed field moves across the slide, and the fibers of each kind are counted as they pass under the center of the crosshairs or the end of a pointer in the eyepiece. It is best to use a tally counter of the multiple type so that all the different types of fibers can be counted at one time. If a fiber passes the center of the crosshairs more than once, it is counted each time, unless the fiber follows the center for some distance, in which case it is counted only as one. With fiber bundles every fiber in a line is counted as it passes under the center. When every fiber in a line has been counted the slide is moved vertically 5 mm to a new line on which the fibers are counted in the same manner. This procedure is followed until about 300 - 500 fibers have been counted. The total of each kind of fiber is multiplied by its weight factor (Table 6) to obtain equivalent weights, and the percentage by weight of the total fiber composition is calculated for each type of fiber present in the specimen.

The prominent feature of deciduous woods is the presence among the narrow fibers of broad, lacy tracheid cells; so characteristic are their pit patterns as to be almost infallible for identification of wood species (Fib. 11a). In confierous woods, number, size, and arrangement of pits, and lengths of individual fibers are distinguishing marks [77] (Fib. 11b). Distinctions between uncooked and cooked wood pulp, between sulfate and sulfite, between bleached and unbleached sulfite, and between soda-cooked and sulfite-cooked fibers depend upon application of appropriate stains.

2. Refractive Index

The Becke line method of determining refractive index has proven useful in providing a relatively quick method of fiber identification and for following the course of cellulose modifications. Table 7 gives representative refractive indices of cellulose from a number of sources. In fibers the direction of maximum and minimum refractive indices normally corresponds with the long axis of the fiber and the direction perpendicular to that axis. Birefringence is expressed as the numerical difference between maximum and minimum refractive indices. In general, an increase in birefringence indicates an increase in orientation, and an increase in the isotropic refractive index indicates an increase in the degree of crystallinity.

The two main factors which control refractive index are the chemical nature of the molecules of the fiber substance and the arrangement of those molecules. For example, the acetyl side chains of cellulose acetate fibers reduce the intrinsic anisotropy of the molecule compared to that of the parent substance cellulose; thus cellulose acetate fibers show only small

TABLE 6

Weight Factors for Fiber Analysis of Paper [80]

Fibers	Weight factor
Rag	1.00
Cotton linters	1.25
Bleached flax and ramie	0.50

Coniferous fibers
  Unbleached and bleached sulfite and
    kraft (except western hemlock
    douglas fir, and southern kraft)
	0.90

Western hemlock	1.20
Douglas fir	1.50
Southern kraft	1.55
Alpha (northern)	0.70

Deciduous fibers
  Soda, sulfate, or sulfite (except
    gum and alpha)
	0.60

Gum	1.00
Alpha	0.55
Groundwood	1.30
Unbleached bagasse fibers as prepared for boards	0.90
Bleached and unbleached bagasse fibers as prepared for papers	0.80
Esparto	0.50
Manila and jute	0.55
Sisal	0.60
Straw for board	0.65
Bleached straw	0.35

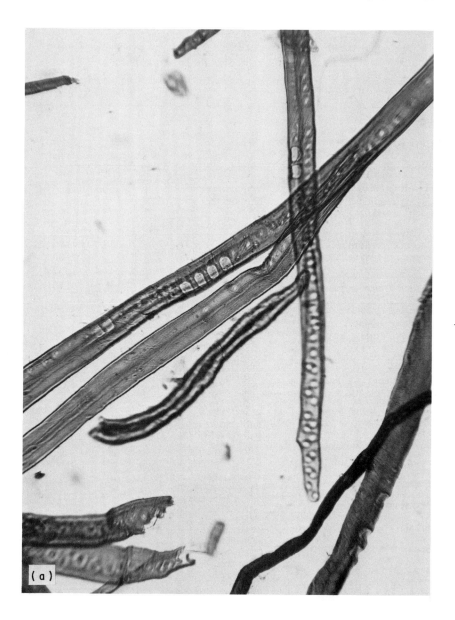

Fig. 11. Characteristic view of wood pulp fibers used in paper:
(a) coniferous; (b) deciduous. Mag., 180X. (U.S. Dept. of
Argriculture photographs)

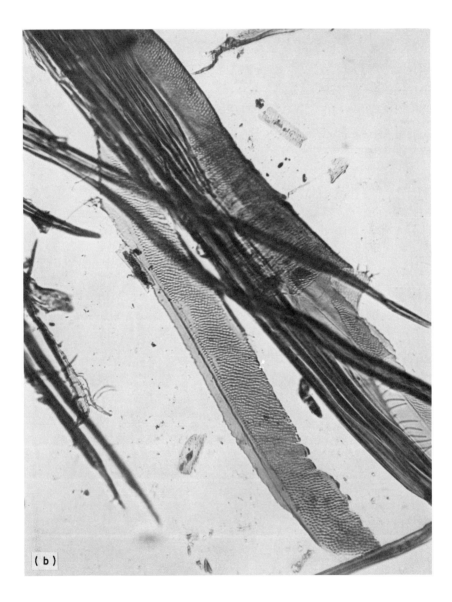

(b)

Fig. 11, continued.

TABLE 7

Refractive Indices of Some Selected Cellulose Fibers

	$n_{\parallel}$	$n_{\perp}$	$n_{\parallel} - n_{\perp}$	Reference
Acetate (secondary)	1.476	1.473	0.003	45
	1.478	1.473	0.005	100
Acetate (triacetate), "Tricel"	1.469	1.469	0.000	45
Acetate (triacetate), "Arnel"	1.472	1.471	0.001	101
	1.472	1.472	0.000	
Cotton, partially acetylated				
(45% acetyl)	1.470	1.472	0.002	96
(17% acetyl)	1.545	1.512	0.033	96
(8% acetyl)	1.565	1.522	0.043	96
Cotton, native	1.577	1.529	0.048	45
	1.578	1.532	0.046	51, 100
	1.577	1.532	0.045	97
Cotton, mercerized slack	1.554	1.524	0.030	100
	1.556	1.524	0.032	97
Cotton, mercerized tensioned	1.586	1.522	0.064	51
	1.566	1.522	0.044	97
Cotton (cellulose III)	1.532	1.517	0.015	97
Cotton (cellulose IV)	1.554	1.532	0.022	97
Rayon, cuprammonium	1.552	1.525	0.027	101
	1.553	1.519	0.034	45
	1.548	1.527	0.021	100
Rayon, viscose	1.547	1.521	0.026	100
Rayon, viscose,				
normal tenacity	1.542	1.520	0.022	45
high tenacity	1.544	1.505	0.039	45
high wet modulus	1.551	1.513	0.038	45
Rayon, viscose (no skin)	1.549	1.523	0.026	99
(thick skin)	1.541	1.516	0.025	99
Rayon, viscose (no skin)				
50% godet stretch	1.554	1.526	0.028	99
30% godet stretch	1.552	1.526	0.026	99
0% godet stretch	1.548	1.526	0.016	99
Fortisan	1.547	1.523	0.024	100
Fortisan 36	1.551	1.520	0.031	100
Flax	1.596	1.528	0.068	45
	1.594	1.532	0.062	45

TABLE 7 (continued)

	$n_{\parallel}$	$n_{\perp}$	$n_{\parallel} - n_{\perp}$	Reference
Hemp	1.581	1.521	0.055	43
	1.591	1.530	0.065	43
Jute	1.577	1.536	0.041	43
Ramie	1.596	1.528	0.068	43

birefringence even if highly oriented. Tripp et al. [96] found that in the partial acetylation of cotton the average refractive index of the fiber showed a continuous decrease as the degree of subsitution increased until, in the case of fully substituted cotton, or of Arnel, the birefringence approached a negative value. Ziifle and co-workers [97] reported changes in refractive index upon the esterification of cotton with long chain fatty acids and of butadiene diepoxide crosslinking products with cotton. Cannizzaro and co-workers [98] showed that refractive indices of cellulose esters of aromatic compounds were consistently higher than those of aliphatic compounds (Table 8).

The effect of tension on mercerization is shown in the respective indices of slack- and tension-mercerized cotton fibers. In like manner, the effect of stretch in the manufacture of viscose rayon filaments was illustrated by Royer and Maresh [99] from whose work the figures in Table 8 are quoted.

3. Phase Microscopy

Although the phase microscope has been one of the most important instruments available for studying living cells in biological and medical research, it has found little use in cellulose investigations. Reumuth [102] in 1947 showed particles, such as silicates, embedded in fibers, fissures due to laundering, and details of native and regenerated cellulose fibers during swelling in cuprammonium hydroxide; but he found that the phase microscope was not useful with thick objects, or in samples with large differences in refractive index, such as between fiber and delustering agent. Köhler and Loos [103] , and Royer and Maresh [99] also summarized applications of phase microscopy in the textile field through discussions of thickness variations of rayon cross sections, investigation of pigments, emulsions, and dispersion, and of thin coatings on textiles, paper, and leather. In recent years there has been little in the literature to indicate that the phase microscope is very useful in studies of cellulose. Stoves [18] reported that bright contrast has proved useful in studying the structure of the cotton fiber; while in ramie, details of the lumen were found to show up more clearly than with the ordinary microscope. In rayon yarns and filaments mounted in paraffin oil the skin and core were differentiated quite clearly without the necessity of staining. Surface saponification of cellulose acetate is readily visible

TABLE 8

Refractive Indices of Cellulose Esters of Low Degree of Substitution [a]

Cellulose ester	No. of Carbons	D.S.	Ref. $n_\parallel$ index	Ref. $n_\perp$ index	Isotropic refractive index $1/3\,(n_\parallel + 2n_\perp)$	Biref. $n_\parallel - n_\perp$
Aliphatic						
Acetate	2	0.10	1.552	1.532	1.539	0.020
Palmitate	16	0.10	1.552	1.534	1.540	0.018
Ricinoleate	18	0.09	1.560	1.534	1.543	0.026
12-OH stearate	18	0.10	1.544	1.534	1.537	0.010
Stearate	18	0.13	1.544	1.530	1.535	0.014
Linoleate	18	0.15	1.552	1.540	1.544	0.012
Substituted alkyl						
Phenylundecanoate	17	0.13	1.564	1.546	1.552	0.018
Aromatic						
Benzoate	7	0.08	1.562	1.544	1.550	0.018
		0.20	1.562	1.560	1.561	0.002
		0.50	1.562	1.558	1.559	0.004
		1.00	1.562	1.560	1.561	0.002
		1.50	1.562	1.564	1.563	-0.002
Cinnamate	9	0.17	1.560	1.562	1.561	-0.002
		1.13	1.594	1.593	1.593	0.001
		1.56	1.594	1.594	1.594	0.000

Aromatic (cont)						
Naphthoate	11	0.31	1.584	1.578	1.580	0.006
		0.86	1.598	1.586	1.590	0.012
		1.30	1.626	1.627	1.627	-0.001
		0.00	1.572	1.538	1.549	0.034
Control, scoured	0					
Control, CH_3OH ext. and scoured	0	0.00	1.570	1.540	1.550	0.030

[a] Taken from Table I of Cannizzaro et al. [98].

under phase contrast, since the saponified area has a refractive index of
1.54 while that of normal cellulose acetate is 1.47. For identification of
paper fibers, however, it was found that staining was more valuable. Phase
contrast microscopy was useful for estimating freedom from lignin, and for
showing the condition of beating of the fibers, and location of surface finishing agents, or penetration of ink into the paper surface.

4. Interference Microscopy

Both phase and interference microscopes can be used for quantitative
work. Image intensity and contrast depend on the phase change of optical
path difference which is defined as

$$\emptyset = (n - n_m)t ,$$

where n is the refractive index of the object, n_m that of the surrounding
medium, t the object thickness, and $\emptyset$ the phase difference between the two
beams measured by an appropriate compensator in the microscope system.
If either thickness or refractive index is known, the other can be calculated.

Heyn [104] , working with the AO-Baker interference microscope, made
quantitative measurements of phase differences between skin and core regions
of viscose rayon fiber and showed a distinct cuticle at the edge of the acetate
fibers. Cotton fiber cross sections were shown to exhibit a different interference color at the primary wall region from that of the main body of the
fiber. Peck and Carter [105] used the same model of interference microscope
to measure the optical properties of external skin layers of cellulose acetate
experimental fibers, modified acrylic, and polyester fibers, and to observe
particulate matter in solutions of polymeric materials for fiber manufacture.

Hollies [28] and Poling and Hollies [29] , using a Zeiss two-beam interference microscope developed a technique for determining refractive indices
and thus location of resins used in manufacture of permanent press cotton
fabrics. A combination of improved embedding and sectioning techniques
made it possible to determine quantitatively the refractive index of impreg-
nating or coating material in situ. This technique employs reference fibers
in the same section as the fiber system under study. From the path length
difference between mounting medium and reference fiber, the section thickness was determined. Subsequently, from section thickness and path length
difference between mounting medium and fiber system, very small differences
in refractive index across the fiber section could be determined. The fibers
in question were embedded in a polymeric matrix of known refractive index.
Cross sections were cut on an ultramicrotome, and with the Zeiss interference microscope optical path differences were determined between embedding
medium, the synthetic fibers, and the fibers whose properties were being
investigated. The sections were mounted in mineral oil-heptane of the same
refractive index as that of the n-butyl methacrylate embedding medium and

the cover glass sealed in place to avoid evaporation of the heptane. Phase differences introduced by the synthetic reference fibers were determined, and from this the exact thickness of the section was determined. Phase differences were then determined for the treated cotton fiber and for specific areas of that fiber of interest in the investigation. Refractive indices of the different areas of the treated fiber were calculated. By comparison of these refractive indices with tables of known refractive index, the type of impregnant could be approximated. By color photomicrographs of interference colors developed in different regions of the section of the unknown fiber, distribution of the impregnant could be recorded.

5. Staining Characteristics

For optical microscopy various dyes can be used to develop contrast between one portion of the specimen and another. If the dyeing characteristics of the substance under investigation are known certain useful deductions can be made. Fibers are usually stained on the slide by immersion in 1% aqueous or alcoholic solutions of the dye at room temperature, or warmed at 50°C for 20 min on a warming table. Washed free of excess dye, they are dried and mounted in an appropriate mounting medium for microscopical examination.

a. Preparation of Specimen for Staining. Before staining or microsolubility tests can be effectively applied to textile or paper samples, it will be necessary to remove from the fibers such foreign materials as starch, gelatin, oil, wax, resins, or rubber. Appropriate methods are described in several of the handbooks [18, 43, 45]. Gelatin and oil sizes can normally be removed by boiling the sample in 0.5% soap solution for 1 h, followed by washing. Linseed oil sizes should be treated for 30 min at 50°C with 0.5% soap and 0.2% sodium carbonate, followed by boiling in the same solution for 30 min. In cases where highly oxidized sizes are present on the fibers the above scour should be preceded by soaking for 15 min in trichloroethylene. Rayon fibers can often be freed of finishes and oils by extracting with dichloroethane and washing with alcohol. Wax sizes are removed by boiling in 0.5% soap plus 0.5% sodium hydroxide. Starch size is removed by treatment for 1 h with 3% diastase solution at 60° C.

Many finishing agents can be removed by boiling in an aqueous solution of detergent followed by rinsing, drying, and subsequent boiling for 2 min in carbon tetrachloride, followed by rinsing and drying.

The removal of dyes calls for more precise measures. Boiling in 20% aqueous pyridine strips the majority of dyes. Boiling 5% sodium hydrosulfite plus 1% sodium hydroxide is an effective agent for removal of a wide variety of dyes. Boiling orthochlorophenol strips nearly all vat and azoic dyes. Cellulose fibers should be swollen by boiling for 1 min in 10% aqueous urea prior to stripping. Dichloromethane and benzene, 50% volume for volume, at room temperature will remove many acetate dyes. Acetate fibers can often be stripped with 90% alcohol and 10% acetone applied cold.

b. Types of Stains. Staining procedures are empirical, but based on the following general classification:

(1) Stains for the noncellulosic constituents of the fiber. This includes such naturally associated substances of vegetable origin as pectin, lignin, or waxes; or additives in the form of sizes, impregnants, and coatings.

(2) Dyes based on the chemical nature of the cellulose. This includes such substitution products as esters and ethers of cellulose and the products of graft polymerization.

(3) Dyes based on the physical character of the cellulose. This involves porosity characteristics and diffusion rates.

(1) Noncellulosic Constituents. Pectin and its associated compounds, which yield glucoronic residues on hydrolysis, are stained with basic dyes. Popular ones are methylene blue (C. I. *Basic Blue 9; C. I. 52015) and ruthenium red (ruthenium oxychloride, ammoniated $(Ru_2(OH)_2Cl_4 . 7NH_3 . 3H_2O)$). Ruthenium red is used as a 0.01% aqueous solution and should be freshly prepared.

Lignin is a constituent of many of the natural fibers. Jute is highly lignified as are both abaca and sisal, and, of course, all of the wood fibers. There are several characteristic color tests used to detect lignin, but none of them are specific; probably it is aldehydes rather than particular groupings that are being stained [107] .

(1) Aniline chloride or aniline sulfate in 1% aqueous solution will stain lignocellulose yellow or yellowish green.

(2) A 2% to 10% solution of phloroglucinol in alcohol is mixed with an equal volume of concentrated HCl. The sample is immersed at room temperature, then examined in the solution. Jute is stained red, hemp and sisal dark brown. Cotton, flax, and ramie do not stain but disintegrate into a tan mass. Unpurified wood fibers are stained dark red.

(3) Safranin (C. I. Basic Red 2; C. I. 50240). Both water-soluble and alcohol-soluble dyes are available. Fibers are soaked in a 1% aqueous solution. Lignified fibers are stained deep orange-red.

(4) Malachite green (C. I. Basic Green 4; C. I. 42000). An aqueous solution stains lignified fibers a deep blue-green color.

---

*Colour Index [106] Classification identifies a specific formula of dye. Dyes should always be specified by Colour Index Number.

Waxes, in natural fibers or when added as coatings, may be colored with solvent dyes such as (Sudan red (C.I. Solvent Red 24; C.I. 26105); or Sudan black (C.I. Solvent Black 3; C.I. 26150). Boiling the fibers briefly in 95% ethanol prior to staining enhances the depth of shade.

Proteins, as protoplasmic residues in native fibers or proteinaceous size applied to textiles or paper, may be stained with acid or basic dyes depending on conditions of pH.

Degree of success achieved in purification process of scouring and bleaching of native fibers can be evaluated by examining under the microscope individual fibers from the purified sample that have been stained for wax, pectin, lignin, or nitrogenous matter. Lack of color indicates good purification, and depth of shade progressive stages in the process.

Starch can be colored dark blue or black by exposure to fumes of iodine or an aqueous solution of it. This enables the microscopist to detect the presence of starch and to evaluate the uniformity of its distribution on yarn, fiber, or paper sheet; or the effectiveness of its removal by enzymes.

Butadiene rubber and similar coatings can be stained with Oil Red O (C.I. Solvent Red 27; C.I. 26125). In cross sections of yarns taken from the fabric and dyed it is possible to see whether the material actually has penetrated the fiber walls or has been deposited evenly among the fibers of the yarn.

Melamine resins can be dyed with acid dyes, notably Kiton Pure Blue V (C.I. Acid Blue; C.I. 42045) [45, 108] , and thus presence and location of polymeric impregnant can be demonstrated in or on yarns or fibers taken from the fabric (Fig. 12). Royer and Maresh [99] stated that extent of cure of melamine formaldehyde types of resin can be tested by staining with Calcocid Alizarine Blue SAPG (C.I. Acid Blue 45; C.I. 63010) in acid solution. Extent of cure can be tested by dyeing a sample of the fabric before and after washing in water. Washing will remove any uncured resin, and the difference in dye pickup between samples before and after washing determines effectiveness of the cure. Cross sections show whether resin in the fabric had been sufficiently cured or whether there was a general improper cure throughout the thickness of the fabric [99].

Another method of determining extent of cure is to use a direct dye which normally dyes cellulose fibers quite well. It has been found that with melamine and urea resins, uncured resin does not limit the dye uptake, but properly cured resin-treated rayon or cotton dyes slowly if at all in normal dyes for cellulose; the degree of dye-resist is a clue to the effectiveness of the cure [99].

Ethylene urea formaldehyde resin is not stained by acid dyes but can sometimes be indicated by Tollens' reagent (ammoniacal silver hydroxide) or by osmic acid (osmium tetroxide) because of residual groups which can react with these reagents [108] .

Phenolformaldehyde resins may be located in or on the cellulose fiber by staining with neutral red (C.I. Basic Red 5; C.I. 50040) [108, 109] .

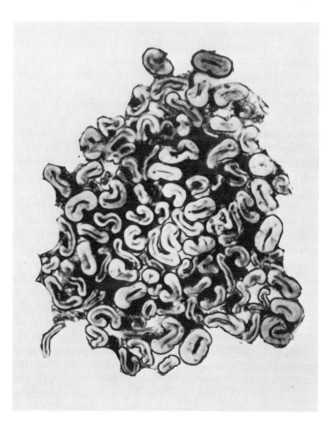

Fig. 12. Cross section of yarn from melamine resin treated cotton fabric. Stained with 2% Kiton Blue (C.I. Acid Blue 1) at pH 3. Unstained areas contain no resin; darkly stained areas contain more resin than lightly stained areas. Mag., 500X. (U.S. Dept. of Agriculture photograph)

(2) Chemical Nature of Cellulose. In substituted cellulose appropriate staining techniques with disperse dyes permit observations as to uniformity and distribution of substitution in the fiber cell walls. Tripp et al. [96] found that at early stages of acetylation of cotton and up to 14% acetyl content, nonuniformities in substitution could be detected by dyeing the yarn in a mixture of direct, Pontamine Fast Heliptrope B (C.I. Direct Violet 51; C.I. 27905) and disperse, Celliton Fast Yellow RRA (C.I. Disperse Yellow 1; C.I. 10345) dyes and then examing the dyed fibers

under the microscope. Another popular dye for this purpose is a mixture of Celliton red (C.I. Disperse red 17; C.I. 11210) and Solantine blue (C.I. Direct Blue 78; C.I. 34200); acetylated areas take the red stain. Royer and Maresh [99] pointed out that saponification of acetate fibers could be followed by staining with cellulose (direct) and acetate (disperse) dyes, and cross-sectioning. The original acetate will be stained one color and those parts which have been converted to cellulose by saponification will be stained a contrasting color.

Aminoethyl cellulose responds to acid dyes. The cyanoethylation process can be followed with disperse dyes. Carboxymethylcellulose derivatives can best be dyed with basic dyes [108].

(3) Sorption Rate. Differentiation among the more nearly pure cellulose fibers may be made on the basis of differential absorption rate; for example, mercerized cotton will stain several shades darker than untreated or scoured cotton with substantive dyes such as Congo red (C.I. Direct Red 28; C.I. 22120). The structure of cuprammonium rayon is more open than that of viscose rayon so that direct dyes diffuse more rapidly into the fiber; this difference is the basis of numerous staining tests which have been employed to distinguish between these two cellulose man-made fibers.

A special case of the sorption-difference method of staining is a technique used for demonstrating "skin" and "core" regions in cross sections of rayon fibers [110-122]. These procedures are all based on difference in effective pore size of the skin and core areas which, in turn, result from the method of formation of the filament [86-94]. When viscose solution is extruded into the coagulating bath, a skin of cellulose forms on the outside of the filament. As coagulation continues, the core of the filament hardens and shrinks, causing a wrinkling of the outer skin of the filament. The result of this can be seen in the serrated cross section of the normal viscose filament. The skin itself can be distinguished by examining a dyed fiber, as the core dyes more readily than the skin. The skin and core are both cellulose, but they differ in the nature and orientation of the crystallites. In the skin, the crystallites are smaller than in the core and are aligned at an angle to the long axis of the fiber as well as in a direction parallel to it.

In the production of high-tenacity rayons, the coagulation and stretching of the fiber are controlled in such a way as to influence the internal structure of the filament. This is accomplished by an increase in the proportion of the skin and a decrease in the proportion of core, to the point at which core disappears completely. The fiber is coagulated in a more uniform way, and the cross section becomes less serrated as the core-shrinkage effect is diminished, until a "whole skin" fiber is developed with an almost circular cross section.

Direct or basic dyes are used in aqueous solution, and the technique usually involves drying on a hot plate, followed by destaining or leaching out of the dye with dilute ethanol or pyridine; the destaining is stopped with dioxane, or with 95% alcohol, or some other shrinking agent. The core absorbs the dye at a more rapid rate than the skin. If, after being dyed and fully dried, the cross section is immersed in a deswelling agent, the dye can be leached out of the core, but the dye molecules are trapped in the skin. The effective pore size of the skin is critical for the particular dye used; the dye can diffuse into the fiber through the water-swollen skin structure, but cannot diffuse out through the deswollen structure. Dyes which have been used for this purpose are: Oxamine Blue 4 R (Direct Blue 3; C.I. 23705); Solophenyl Fast Blue Green BL (C.I. Direct Green 27; C.I. "Polyazo" dye); Congo Rubine (C.I. Direct Red 17; C.I. 22150); Victoria blue (C.I. Basic Blue 26; C.I. 44045); and Azine Brilliant Blue 6B (C.I. Direct Blue 1; C.I. 24410).

Simon's stain [123] is a combination of two dyes which has been used in paper technology to investigate degree of fibrillation during beater operations. It is a combination of two direct dyes, Pontamine Sky Blue 6 BX (C.I. Direct Blue 1; C.I. 24410) and Pontamine Fast Orange (C.I. Direct Orange 15; C.I. 40002) used in equal proportions in 1% aqueous solution. Whole fibers are stained blue, but fibrillated portions of fibers and extremely thin fibers are stained orange. DeGruy et al. [124] found a modification of Simon's stain useful in studies of abrasion in cotton. In mechanically damaged cotton fibers unaffected areas were blue, but mashed and frayed fiber parts stained orange. When previously stained fibers are swelled in cupriethylenediamine hydroxide, the primary wall shows bright orange whereas the rest of the fiber is blue, including the lumen border and its contents.

Another dye combination which reflects differences in effective pore size is the differential dye test of Goldthwait et al. [125] for determination of maturity in swatches of cotton. The dyes used are Chlorantine Fast Green BLL (C.I. 34045) and Diphenyl Fast 5 BL Supra I Red (C.I. 28160). The fiber swatches are boiled in the dye, set with sodium chloride, and subsequently rinsed in boiling water. Immature, thin-walled fibers are dyed green; mature, thick-walled fibers, red (ASTM Designation D-1464) [72] . Cross sections observed in the microscope confirm the differential dyeing.

(4) "Universal" Stains. There are a number of dye mixtures which have been developed for the identification of textile fibers with a single stain application. The more modern ones are Shirlastain A and Shirlastain E, developed by Shirley Institute, Manchester, England; Calco Identification Stain #2 developed by American Cyanamid Company, Bound Brook, N.J.; and DuPont's Identification Stain 4. They are developments similar to the older German Neocarmine W, and Calco Identification Stain #1. When these stains are applied according to instructions provided by the manufacturer, fibers will be stained distinctive colors which can be identified from the chart supplied or from comparisons with the staining reactions of fibers from a standard fiber collection [43, 45] .

c. Paper Fibers. Paper furnished from the most common methods of pulping can be identified by the application of various conventional staining procedures. Stains used in the identification of paper fibers are mostly based on determination of different levels of lignin content. They contain various proportions of the iodine-potassium iodide complex combined with alkali salts. Ligneous materials always absorb iodine to give a yellow color, while the colors developed in carbohydrate materials vary from orange to red to violet to blue to almost black depending on the degree of hydration of the sample under swelling by the alkali. Herzberg's stain [78, 79, 80, 82], Graff's "C" stain [78, 83], Selleger's [79, 80], and Wilson's [79, 80] stains are more or less universal stains for fiber identification by iodine coloration after swelling of the cellulose with metal salts such as zinc chloride, or potassium, aluminum, cadmium, or calcium hydroxides.

(1) Herzberg's stain, used in both paper and textile technology, depends on saturated zinc chloride as swelling and hydration agent. The formula for Herzberg's stain is:

Solution A: Prepare zinc chloride solution (sp gr 1.80 at 28°C) by dissolving 50 g dry $ZnCl_2$ (fused sticks in sealed bottles, or crystals) in approximately 25 ml water.

Solution B: Dissolve 0.25 g of iodine and 5.25 g of KI in 12.5 ml water. Mix 25 ml of solution A with the entire solution B. Pour into a narrow cylinder and let stand until clear (12 to 24 h). Decant the supernatant liquid into an amber-colored, glass-stoppered bottle and add a leaf of iodine to the solution. Avoid undue exposure to light and air. Fibers on the slide are covered with the stain and after 2 or 3 min are blotted dry, covered with fresh stain, and a coverslip and examined. Results are as follows:

Fibers staining red: Entirely nonlignified natural fibers such as cotton, linen, bleached abaca, and "alpha cellulose" from wood pulp.

Fibers staining blue: Partially delignified fibers such as chemically purified paper pulps low in lignocellulose; viscose and cuprammonium rayons, mercerized cotton, and oxidized cellulose.

Fibers staining yellow: Those high in lignocellulose, such as ground wood, jute, unbleached sisal, and unbleached abaca. Cellulose acetate fibers become yellow but dissolve.

(2) Graff's "C" stain gives a wider range of color differentiations among the chemical wood pulps:

Solution A: Prepare an aluminum chloride solution (sp gr 1.15 at 28°C) by dissolving about 40 g $AlCl_3 \cdot 6H_2O$ in 100 ml water.

Solution B: Prepare a calcium chloride solution (sp gr 1.80 at 28°C) by dissolving about 100 g $CaCl_2$ in 150 ml water.

Solution C: Prepare a zinc chloride solution (sp gr 1. 80 at 28°C) by dissolving 50 g dry $ZnCl_2$ (fused sticks in sealed bottles or crystals) in approximately 25 ml water. (Do not use $ZnCl_2$ from a previously opened bottle.)

Solution D: Prepare an iodide-iodine solution by dissolving 0. 90 g dry KI and 0. 65 g dry iodine in 50 ml water. Dissolve the KI and iodine by first thoroughly intermixing and crushing together, then adding the required amount of water dropwise with constant stirring. Mix well together 20 ml solution A, 10 ml solution B, and 10 ml solution C; add 12. 5 ml solution D and again mix well. Pour into a tall, narrow vessel and place in the dark. After 12 h, when the precipitate has settled, pipet off the clear portion of the solution into a dark bottle and add a leaf of iodine. Keep in the dark when not in use. Fresh stain should be made every 2 or 3 months.

Selleger's stain is a mixture of $Ca(NO_3)_2$ with KI, and Wilson's stain a mixture of iodine, cadmium iodide, formaldehyde, calcium nitrate, and cadmium chloride. Their special formulas are found in ASTM Standard D-1030 [80].

(3)   Alexander's stain [80, 82] distinguishes between coniferous and deciduous wood fibers in a soda-cooked paper pulp.

Solution A: 0. 2 g Congo red (C. I. Direct Red 28) dissolved in 300 ml water.

Solution B: 100 g calcium nitrate in 50 ml water.

Solution C: Normal Herzberg stain.

Fibers are treated on the slide for 1 min in 2 drops of solution A. Excess dye is removed with filter paper and the sample allowed to air dry. Three drops of solution B are added for 1 min, and then one drop of solution C is added. The whole is quickly and thoroughly mixed and the coverglass replaced. After 2-3 min, the sample is examined. Fibers of coniferous wood are stained pink; those of deciduous wood, blue.

(4)   The Kantrowitz-Simons stain [80, 82] permits indication of degree of cooking and bleaching.

Solution A: Dissolve 2. 7 g of $FeCl_3 . 5H_2O$ in 100 ml water.

Solution B: Dissolve 3. 29 g $K_3 Fe (CN)_6$ in 100 ml water.

Solution C: Dissolve 0. 5 g of benzopurpurin 48 (C. I. Direct Red 2; C. I. 23500) in 100 ml of 50% ethyl alcohol. Warm the solution until the dye is completely dissolved. This solution may be used indefinitely; some of the dye will precipitate on cooling but can be redissolved on warming. Mix equal parts of solutions A and B just before using; apply for 1 min at room temperature; thoroughly wash the stain from the fibers, and then stain them for 2 min with solution C. After staining, again thoroughly wash the fibers before observation. A well-cooked, well-bleached pulp will be red, while poorly cooked, unbleached pulp will be blue. All stages between will be found with different degrees of cooking and bleaching.

(5)   The Lofton-Merritt stain [82, 84, 85] was devised to differentiate between unbleached sulfate and unbleached sulfite pulps.

Solution A: Dissolve 2 g of malachite green (C. I. Basic Green 4; C. I. 42000) in 100 ml water.

Solution B: Dissolve 1 g of basic fuchsin (C. I. Basic Violet 14; C. I. 42510) in 100 ml water.

Solution C: 0. 1% HCl.

Add a mixture of solutions A and B in equal amounts to the fibers on the slide. After 2 min remove excess dye by blotting with filter paper. Add a few drops of solution C and, after 30 sec, remove excess with coverglass. Unbleached sulfite fibers will be reddish to lavender; unbleached sulfate fibers will be green to blue. If the pulp is entirely free of lignin the fibers will be colorless.

All of the above classic tests for paper pulp are subject to alteration by special treatments to which the paper may have been subjected. They therefore should be used with full knowledge that they apply to pulps in the unaltered state. If special finishes, coatings, or bleaches have been applied, these staining methods may be invalidated.

6. Microsolubility

Microsolubility tests are accomplished by watching under the microscope the effects on the fibers of solvents or swelling agents. By observing the pattern of fiber disintegration upon swelling it is possible to draw conclusions with respect to morphology of natural fibers, and their state of preservation, or of modification.

Dry fibers are placed on a glass microscope slide and covered with a coverglass. Solvent is applied at the edge of the cover glass, and its effect observed under the microscope. Tables showing solubilities of textile fibers are given in ASTM Designation D-276 [72] and in Ref. [45 and 123].

The natural cellulose fibers have characteristic and distinctive swelling patterns as they go into solution. When cotton fibers are immersed in cellulose dispersing agents, such as cupriethylenediamine hydroxide (0. 5 M Cu^{2+}, 1. 0 M ethylenediamine) (ASTM Method D-539-53) [72], the fiber rapidly untwists, and the primary wall, ruptured by the swelling pressure of the cellulose of the secondary wall beneath it, breaks, often into a spiral pattern, and peels back to form constricting bands. The "ballooning" thus produced is characteristic. The central canal (lumen) contains the protoplasmic residue, composed, for the most part, of coagulated proteins, (Fig. 13). Flax and hemp also swell and dissolve in these reagents, and the pattern

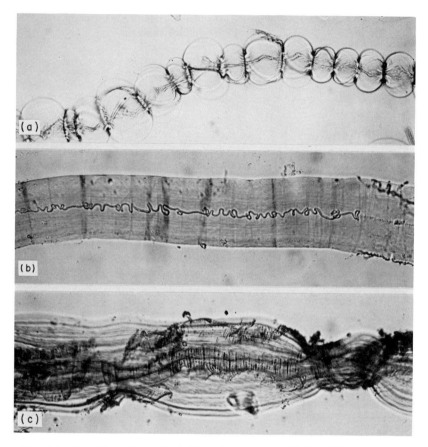

Fig. 13. Swelling characteristics of common cellulose fibers in cupriethylenediamine hydroxide: (a) untreated cotton; (b) flax; (c) hemp. Mag., 212.5X. (U. S. Dept. of Agriculture photographs)

made by the dried protoplasm in their central canals is an identifying feature (Fig. 13 b, c). In flax it is invariably vermiform and continuous; in hemp, a flat, fluted ribbon. In ramie it is a wide band of uneven diameter.

The pattern of fiber morphology produced in swelling contributes information useful in interpreting the results of chemical treatment. Fully crosslinked cotton does not swell in 0.5 M cupriethylenediamine hydroxide after 30 min exposure, but fibers less fully reacted show progressively more and more swelling. Untreated cotton will first form balloons and then go into solution.

Regenerated cellulose fibers such as viscose rayon and cuprammonium rayon have solubility properties similar to those of purified cotton. Among the cellulose derivatives, cellulose triacetate is soluble only in dichloro-methane, m-cresol, or 90% phenol; it softens in 100% acetone, hot.

Secondary acetate is soluble at room temperature in 80% acetone, in m-cresol, and in 90% phenol, but not in dichloromethane .

Sixty percent H_2SO_4 dissolves both viscose and cuprammonium rayon, but not acetate fibers. Hot calcium thiocyanate, concentrated, dissolves viscose and cuprammonium rayon, but not cotton.

Seventy percent H_2SO_4 dissolves cotton and linen [18] .

V. SPECIAL TESTS

A. Fiber Damage in Cotton

In many cases evidence obtained microscopically may be useful in the study of fiber damage and its causes. Types of damage which may be sus-pected are abrasion, compression, tensile break, heat damage, microbial damage due to enzymatic attack of microorganisms, or chemical damage such as acid tendering of cotton fibers, and saponification of acetate due to action of alkalies. Some common tests for damage follow.

1. Benzopurpurin 10 B (Color Index Direct Red 7; C.I. 24100)

The fibers are boiled in 0.1% aqueous solution for a few minutes and rinsed with water. The dyestuff stains the damaged fiber parts more in-tensely than the rest of the fiber.

2. Boiling Sodium Hydroxide

Damaged fibers are partly soluble in boiling 5% NaOH.

3. Congo Red (Clegg Test) [126]

Fibers on a microscope slide are soaked for 5 min in 9% NaOH, washed, and blotted dry; then soaked for 6 min in a saturated aqueous solution of Congo red (CI Direct Red 28; C.I. 22120), washed, and mounted in 18% NaOH for microscopical observation. Damaged portions of the fibers will be swollen and stained much darker red than the undamaged portions.

4. "Dumbbell" Test for Cotton Fiber Damage [45, 65].

Short lengths of cotton fibers about 0.5 mm are cut from a bundle of fibers. Two single-edged razor blades taped together make a convenient tool. The short lengths are mounted on a slide in 15% NaOH and examined microscopically. In undamaged fibers, the secondary wall cellulose will swell and be extruded from the cut ends of the fiber forming dumbbell shapes or mushroom heads at each end of the fiber section. If the primary wall of the fiber has been damaged it will be weakened and unable to withstand the internal pressure generated by the swelling secondary wall. In damaged fibers the whole fiber swells more or less uniformly and there are no mushroom heads or dumbbell formations. By counting the number of each type of fiber present, a quantitative expression of the amount of damage in the cotton sample can be estimated (Fig. 14).

5. Detection of Fungal Growth within the Cellulose Fiber

Fibers are warmed for 1 min in a lactophenol solution and then for several minutes in an aqueous solution of Cotton Blue (C. I. Basic Blue 10; C. I. 51190 or Direct Blue 8; C. I. 24140). Lactophenol solution:

Phenol, 1 part by weight
Lactic acid, 1 part by weight
Glycerol, 2 parts by weight
Water, 1 part by weight

The fibers are then rinsed in water and mounted in 5% NaOH for microscopical observation. Fungal hyphae and spores are stained blue to lavendar, cotton fibers remain pale blue or colorless.

6. Heat Damage or Acid Tendering in Cotton

Cotton fibers which are suspected of damage by scorching may be mounted in 18% NaOH and examined after 10 min. The fibers will be swollen, but also nicked at short intervals by horizontal cracks across the axis of the fiber (Fig. 15).

A similar but more pronounced response is observed in cotton fibers which have been tendered by exposure to mineral acids. The fibers are frequently segmented into short fragments upon swelling in the caustic. This phenomenon is sometimes referred to as "chemical sectioning."

7. Detection of Oxycellulose

(a) Soak the fibers for 1 h in 10% ferrous sulfate rinse. Add 1% potassium ferrocyanide. Oxycellulose will turn blue.

(b) Place the fibers for 1 h in a 1% solution of crystallized stannous chloride containing 1-2 drops of glacial acetic acid. Rinse with distilled water; add a very dilute solution of gold chloride. Damaged parts of the fibers will turn reddish brown.

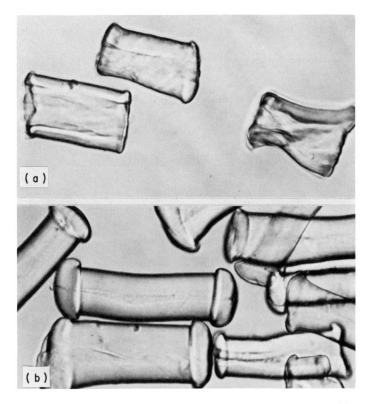

Fig. 14. "Dumbbell" swelling test for fiber damage in cotton fibers: (a) damaged; (b) undamaged. Mag., 500X. (U.S. Dept. of Agriculture photograph)

8. Detection of Hydrocellulose

No specific test is known for hydrocellulose. Its presence is deduced from loss of strength, and negative tests for oxycellulose.

9. Detection of Saponification

(a) Cellulose acetate fibers which are suspected of having been saponified may be examined by cross section after being stained with a mixture of direct and disperse dyes. Suitable dye mixtures are:

1% Pontamine Fast Heliotrope B (C.I. Direct Violet 51; C.I. 27905) and 1% Celliton Fast Yellow RRA (C.I. Disperse Yellow 1; C.I. 10345); or 1% Solantine Blue (C.I. Direct Blue 78; C.I. 34200) and 1% Celliton Red (C.I. Disperse Red 17; C.I. 11210).

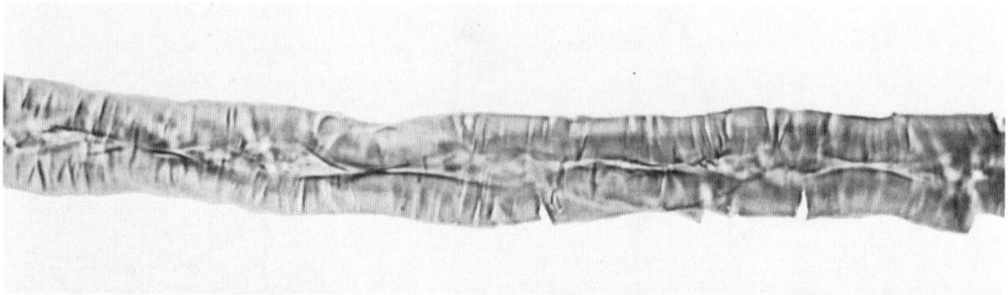

Fig. 15. Longitudinal view of heat damaged cotton fiber swollen in solium hydroxide. Mag., 375X. (U. S. Dept. of Agriculture photograph)

(b) Acetone will dissolve out the cellulose acetate, and leave the saponified cellulose as an outer shell. This is best achieved with 90-100% acetone, warmed.

B. Fiber Maturity in Cotton

Cotton fibers do not all grow to the same degree of maturity even on the same seed (Fig. 16). In every commercial sample of cotton there are approximately one-fifth of the fibers which may be classed as immature. Conventional methods for determining maturity in a gross way on relatively large samples from the cotton bale were originally derived from data developed by microscopical methods.

1. Caustic Soda Method (ASTM Designation D-1442-68T) [72]

A group of fibers is combed and laid on a dry slide, flooded with a solution of 18% sodium hydroxide, and observed with the microscope. Mature fibers will have swollen to lie almost straight and will appear to be almost circular in cross section. Immature fibers will have swollen and twisted into a corkscrew shape, and extremely immature ("dead") fibers will not be swelled at all. The fibers are counted as mature or immature and the percentage of each reported. In cases of doubt, fibers whose lumen width is more than twice the thickness of either cell wall are counted as immature whether or not they may have twisted in swelling.

2. Cross-Sectional Method (ASTM Designation D-1444) [72]

In a photomicrograph of cross sections of a small tuft of cotton fibers it is possible to make measurements of both long and short diameters of a flattened or irregular cross section and to calculate the circularity ratio of individual fibers for a determination of maturity fiber to fiber. Usually it is adequate to count the number of poorly developed, thin-walled fibers in any given area of the cross section without the necessity of making

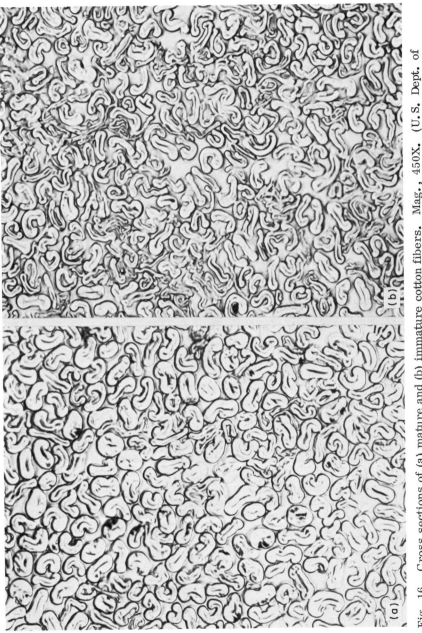

Fig. 16. Cross sections of (a) mature and (b) immature cotton fibers. Mag., 450X. (U. S. Dept. of Agriculture photographs)

measurements. The degree of maturity is expressed as percentage of mature fibers in the sample. Details of the method are given in ASTM D-1444[72].

3. Polarized Light Method [127-129] (ASTM-D-1442-68) [69, 72]

This method utilizes the interference colors displayed by cotton fibers when viewed in polarized light. Essential equipment includes a polarizing microscope with a magnification of about 100X. It should be equipped with a rotating stage to which a mechanical stage is attached and a red of the first-order selenite retardation plate to step up the interference colors for more positive identification. The microscope is adjusted so that the polars are crossed, the cross hairs of the ocular are fixed at 45° with the plane of vibration of the polarizer and the analyzer, and the selenite plate is placed in position. With the fiber in a parallel position with relation to the arrow of the selenite plate, each fiber is classed in turn according to its color. Those cotton fibers which appear violet or indigo are very immature; blue fibers are immature; green or yellow-green fibers are mature; and yellow fibers are extremely mature. When the stage is rotated 90° the violet fibers become orange, blue fibers change to yellow, yellow-green fibers do not change, and yellow fibers change to yellow-green. These color changes which occur on rotation of the fibers with respect to the incident polarized light serve as a check on doubtful fibers. When rotated through 360°, immature fibers show practically complete parallel extinction in contrast to mature fibers which show no extinction.

C. Mercerization in Cotton

The oldest and best known modification of cotton is that of mercerization, which is employed commercially to improve luster and strength of cotton yarns and to enhance the dyeability of cotton materials. It consists of immersing cotton yarn or fabrics in 20% sodium hydroxide solution. The effect is instantaneous and the cellulose of the fiber is swollen rapidly. After mercerization, the fiber is rinsed, then washed in dilute acid, washed again, and dried. The fully mercerized fiber resembles a solid cylinder of cellulose (Fig. 17). Commercial mercerization is done often at such high speeds that this effect is incomplete. Although x-ray analysis for cellulose II is necessary to fully confirm degree of mercerization, a number of relatively simple microscopical tests can give tentative estimates of considerable usefulness.

1. Refractive Index

Refractive indices of individual fibers may be measured by the Becke line immersion technique. Native cotton has a refractive index of 1.578 parallel to the fiber axis; cotton mercerized under tension, 1.566; cotton mercerized in the slack state, 1.556.

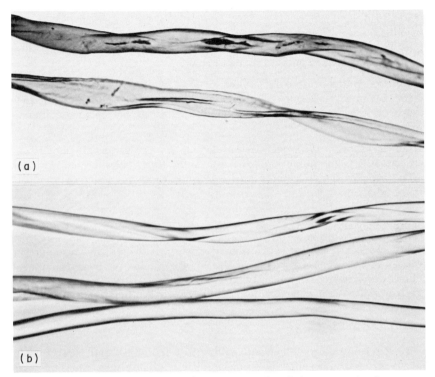

(a)

(b)

Fig. 17. Longitudinal view of (a) untreated and (b) mercerized cotton fibers. Mag., 400X. (U. S. Dept. of Agriculture photographs)

2. Deconvolution Count

Fibers are cut into short lengths, mounted in mineral oil, and observed for their twisted or twistless character. In untreated cotton approximately 15% of the fibers will show no conolutions, but all the others will resemble twisted ribbons. In fully mercerized cotton the deconvolution count is approximately 70-80%.

3. Swelling

In longitudinal view mercerized cotton fibers exhibit a smooth cylindrical appearance with a small or missing lumen in many fibers. They show uniform swelling in cuprammonium hydroxide or cupriethylenediamine hydroxide.

4. Staining

(a) Hubner's Reagent. 20 g iodine are dissolved in 100 ml of a saturated solution of potassium iodide. The fibers are immersed on the slide for a few seconds in this reagent, then washed several times in water. Unmercerized cotton is only slightly stained; mercerized cotton becomes black or dark blue.

(b) Brilliant Blue 6 BA (Color Index Direct Blue 1; C.I. 24410). Fibers are stained for 3 min in a 1% aqueous solution of the dye, then washed and examined in water on the slide. Both untreated and mercerized cotton stain blue, but after repeated washings the untreated cotton fibers become white while the mercerized cotton remains blue.

(c) Congo Red (C.I. Direct Red 28; C.I. 22120). The fibers are brought to the boil in a 1% aqueous solution of the dye, then washed thoroughly in water. Untreated cotton fibers are pink to light red; mercerized cotton fibers are dark red.

(d) Herzberg's Stain (see Section B5, C-1). The sample of fiber is wet out, squeezed, and immersed for 3 min in the Herzberg solution. In a drop of this stain on the slide, untreated cotton fibers are red, mercerized cotton fibers purple to dark blue.

5. Cross-Sectional Shape

The mercerized fiber in cross section has lost the kidney shape of native cotton and has become almost round with a tiny lumen, or no lumen at all. Fibers which have been mercerized under excessive tension may have some straight sides and angular corners somewhat resembling flax fibers in shape.

6. Fluorescence

Mercerized fibers stained with eosin 3Y (Color Index Acid Red 87; C.I. 45380) give a bright yellow fluorescence in ultraviolet light, while unmercerized cotton does not fluoresce under these conditions.

D. Distinction between Cuprammonium Rayon and Viscose Rayon

1. Staining with Brilliant Blue 6 B (C.I. Basic Blue 1; C.I. 24410)

Fibers are stained for 3 min in a 0.2% aqueous solution, washed, and examined in water or glycerin. Cuprammonium rayon is stained blue; viscose rayon is almost colorless.

2. Staining in Methylene Blue (C.I. Basic Blue 9; C.I. 52015)

Fibers are stained in a 0.1% aqueous solution, then washed in warm water until no more dye can be washed out. Viscose rayon stains dark blue, cuprammonium rayon a very light blue.

3. Cross-Sectional Shape

Viscose rayon fibers are usually irregularly crenated at the edges. Cuprammonium fibers are round and uniform.

4. Longitudinal View

Viscose rayon fibers are strongly striated, and often speckled with delusterant particles. Cuprammonium rayon fibers have no striations or other markings. They are seldom delustered.

E. Distinction between Flax and Hemp Fibers

1. Refractive Indices

The maximum refractive index for flax is 1.594; that for hemp 1.585.

2. Staining

Fibers are warmed in a 1% aqueous solution of Cyanine (C.I. Acid Blue 118; C.I. 26410). Hemp is stained a bright blue to green color which persists on washing. Flax is only faintly stained and the color washes out.

3. Swelling Pattern

Dry fibers on a slide are flooded with 0.5 M cupriethylenediamine hydroxide. Flax fibers swell to show a thin vermiform lumen residue which is continuous; hemp fibers swell to show a wide, fluted ribbon of residue in the lumen, which may or may not be continuous along the fiber length.

4. Twist Test

Fibers are soaked in water, until thoroughly wet. A single fiber is withdrawn from the mass with forceps, and held over a hot plate to dry, with fiber tip pointing toward the operator. Flax fibers will twist as they dry in a clockwise direction. Hemp fibers on drying twist counterclockwise.

F. Distinction between Sisal and Abaca Fibers

1. Swett Test [43, 65]

The sample is washed in ether to remove oil, then soaked in 5% sodium hypochlorite for 10 min. After washing in water, and rinsing in ethyl alcohol, the sample is exposed to fumes of concentrated ammonia until fully colored. Abaca fibers stain brown while sisal and other vegetable fibers are magenta red. The stain is not permanent and fades after a short time.

2. Billinghame's Test [45]

The sample is washed in methylene chloride to remove oil and allowed to dry. The sample is then boiled in 5% nitric acid for 5 to 10 min. Excess acid is washed out with water and the sample is immersed in cold 0.25 N sodium hypochlorite for 10 min, then dried. Abaca fibers stain orange; sisal and other leaf fibers are stained a pale yellow.

REFERENCES

[1] G. L. Clark, ed., The Encyclopedia of Microscopy, Reinhold, New York, 1961.

[2] L. C. Martin, The Theory of the Microscope, American Elsevier, New York, 1966.

[3] L. C. Martin and B. K. Johnson, Practical Microscopy, Chemical Publishing Company, Brooklyn, N.Y., 1951.

[4] G. H. Needham, The Practical Use of the Microscope, Thomas Publisher, Springfield, Ill., 1958.

[5] G. H. Needham, The Microscope, A Practical Guide, Thomas Publisher, Springfield, Ill., 1969.

[6] R. M. Allen, The Microscope, Van Nostrand, Princeton, New Jersey, 1940.

[7] R. M. Allen, Photomicrography, Van Nostrand, Princeton, New Jersey, 1943.

[8] C. P. Shillaber, Photomicrography in Theory and Practice, Wiley, New York, 1945

[9] V. E. Cosslett, Modern Microscopy, Cornell Univ. Press, Ithaca, New York, 1966.

[10] F. J. Munoz and A. Charipper, The Microscope and Its Use, Chemical Publishing Company, Brooklyn, N.Y., 1943.

[11] J. R. Benford, The Theory of the Microscope, 4th ed., Bausch and Lomb, Rochester, New York, 1965.

[12] F. K. Möllring, Microscopy from the Very Beginning, Carl Zeiss, Oberkochen/Wuerttemberg, West Germany, 1968.

[13] N. Meyers, Rediscovering the Microscope (Color Filmstrip, Tape Recording and Manual), Univ. of Michigan Audio-Visual Education Center, Ann Arbor, Mich., 1966.

[14] A. Barer and V. E. Cosslett, Advances in Optical and Electron Microscopy, Vol. I, Academic, London and New York, 1966.

[15] N. Meyers, Sci. Res., 3 (24), 37 (1968).

[16] J. G. Delly, Am. Lab., p. 8 (April 1969).

[17] L. V. Foster, J. Biol. Phot. Assoc., 2 (3), 140 (1934).

[18] J. L. Stoves, Fibre Microscopy, National Trade Press, London, 1957.
[19] N. H. Hartshorne and A. Stuart, Crystals and the Polarizing Microscope, 2nd ed., Arnold, London, 1950.
[20] E. M. Chamot and C. W. Mason, Handbook of Chemical Microscopy, Vol. I, Wiley, New York, 1958.
[21] A. Johannsen, Manual of Petrographic Methods, McGraw-Hill, New York, 1948.
[22] A. F. Hallimond, Manual of the Polarizing Microscope, Cook, Troughton and Simms, York, 1953.
[23] A. H. Bennett, H. Jupnik, H. Osterberg, and O. W. Richards, Phase Microscopy, Wiley, New York, 1951.
[24] W. Krug, J. Rienitz, and G. Schulz, Contributions to Interference Microscopy (translated by J. Homer Dickson), Helger and Watts, London, 1964.
[25] R. C. Faust, Proc. Phys. Soc. (London), B65, 48 (1952).
[26] O. W. Richards, Reference Manual AO-Baker Interference Microscope, 2nd ed., American Optical Co., Instrument Div., Buffalo, New York, 1963.
[27] R. C. Faust, Quart. J. Microscope. Sci., 97, Pt. 4, 569 (1956).
[28] N. R. S. Hollies, Textile Res. J., 37, 277 (1967).
[29] F. D. Poling and N. R. S. Hollies, Microscope, 16 (3), 201 (1968).
[30] J. D. Dean, C. M. Fleming, and R. T. O'Connor, Textile Res. J., 22, 609 (1952).
[31] H. Zukriegel, Textile-Praxis, 23 (10), 702 (1968).
[32] Eastman Kodak Co., Photography through the Microscope, Rochester, 1952.
[33] J. E. Barnard and F. V. Welch, Practical Photomicrography, 3rd ed., Longmore, London, 1936.
[34] J. I. Hardy, Textile Res., 3, 189 (1933).
[35] G. L. Royer, C. Maresh, and A. M. Harding, Calco Tech. Bull. No. 770, American Cyanamid Co., Bound Brook, N. J., 1945.
[36] E. R. Schwartz, Textile Res., 6, 270 (1936).
[37] American Society for Testing and Materials, Textile Standards Part 25, American Society for Testing and Materials, Philadelphia, Pa., 1971.
[38] H. L. Steedman, Section Cutting in Microscopy, Blackwell Scientific Publications, Oxford, 1960.
[39] Anon., Embedding Specimens in Methacrylate Resins, Special Products Bull. 46, Rohm and Haas Special Products Department, Philadelphia, Pa., December, 1962.
[40] Anon., Thin Sectioning and Associated Techniques for Electron Microscopy, 2nd ed., Ivan Sorvall, Norwalk, Conn., 1965.
[41] I. H. Isenberg and J. D. Hankey, Tappi, 49, 373 (1966).
[42] M. C. Botty, C. D. Felton, and E. Anderson, Textile Res. J., 30, 959 (1960).

[43] B. Luniak, The Identification of Textile Fibers, Isaak Putnam Sons, London, 1953.

[44] E. P. Kaiser and W. Parrish, Ind. Eng. Chem. Anal. Ed., 31, 560 (1939).

[45] Anon., Identification of Textile Materials, 6th ed., Textile Institute, Manchester, 1970.

[46] P. H. Hermans, Contributions to the Physics of Cellulose Fibers, Elsevier, Amsterdam and New York, 1946.

[47] P. H. Hermans, J. Textile Inst., 38, 63 (1947).

[48] R. Meredith and J. W. S. Hearle, Physical Methods of Investigating Textiles, Textile Book Publishers, New York, 1959.

[49] K. R. Fox and R. B. Finch, Textile Res. J., 11, 62 (1946).

[50] C. P. Saylor, J. Res. Natl. Bur. Std., 15, 277 (Res. Paper 829), (1935).

[51] J. M. Preston, J. Textile Inst., 38, T78 (1947).

[52] J. M. Preston and K. J. Freeman, J. Textile Inst., 34, T19, (1943).

[53] J. B. Goldberg, Fabric Defects, McGraw-Hill, New York, 1950.

[54] W. Westbrook, Textile Ind. (Atlanta), 133 (6), 102 (1969).

[55] P. R. Lord and N. Senturk, Textile Ind. (Atlanta), 133 (2), 89 (1969).

[56] A. Ure, The Cotton Manufacturers of Great Britain Systematically Investigated, Vol. I, Charles Knight, London, 1836, p. 56.

[57] T. F. Hanausek and A. L. Winton, Microscopy of Technical Products, Wiley, New York, 1907.

[58] L. G. Lawrie, Textile Microscopy, Ernst Benn, London, 1928.

[59] J. H. Skinkle, Elementary Textile Microscopy, Howe Publishing, New York, 1930.

[60] P. Heerman and A. Herzog, Mikroskopische un mechanische-technische Textiluntersuchungen, 3rd ed., Springer, Berlin, 1931.

[61] J. M. Preston, Modern Textile Microscopy, Emmott and Co., Manchester, 1933.

[62] E. R. Schwarz, Textiles and the Microscope, McGraw-Hill, New York, 1934.

[63] J. M. Matthews, Textile Fibers, Their Physical, Microscopic and Chemical Properties, (H. R. Mauersberger, ed.), 6th ed., Wiley, New York, 1954.

[64] W. von Bergen and W. Kraus, Textile Fiber Atlas: Revised Edition, Textile Book Publishers, New York, 1949.

[65] A. N. J. Heyn, Fiber Microscopy, Wiley-Interscience, New York, 1954.

[66] J. G. Cook, Handbook of Textile Fibres. Vol. I. Natural Fibres. Vol. II. Man-Made Fibres, Merrow, Watford, England, 1968.

[67] J. G. Cook, The Fibre Encyclopedia, Merrow, Watford, England, 1968.

[68] M. Harris, Handbook of Textile Fibers, 1st. ed., Harris Research Laboratories, Washington, 1953.

[69] A. R. Urquhart and G. O. Howitt, eds., The Structure of Textile Fibers, The Textile Institute, Manchester, 1953.

[70] J. W. S. Hearle and R. H. Peters, eds., Fibre Structure, Butterworth's and the Textile Institute, Manchester and London, 1963.

[71] American Association of Textile Chemists and Colorists, Technical Manual, Howes Publishing Co., New York, 1969.

[72] American Society for Testing and Materials, ASTM Standards on Textile Materials, Part 25, American Society for Testing and Materials, Philadelphia, Pa., 1968.

[73] Anon., "The Textile World 1969 Cotton Chart," Textile World, 119 (11), 68 (1969).

[74] Anon., "The Textile World 1968 Man-made Fiber Chart," Textile World, 118 (8), 63 (1968).

[75] Anon., "Man-made Fiber Identification Chart," Mod. Textiles Mag., 50 (3), 37 (1969).

[76] Anon., "Properties of Man-made Fibers," Textile Ind. (Atlanta), 133 (8), 69 (1969).

[77] C. H. Carpenter and L. Leney, Papermaking Fibers, State University of New York College of Forestry, Syracuse, N.Y., 1952.

[78] I. A. Isenberg, Pulp and Paper Microscopy, 3rd ed., The Institute of Paper Chemistry, Appleton, Wis., 1959.

[79] Technical Association of Pulp and Paper Industry, Fiber Analysis of Paper and Paperboard, Tappi Standard T 401m- 53 (1968).

[80] American Society for Testing and Materials Committee D-6 on Paper and Paper Products, ASTM Designation D-1030-65, American Society for Testing and Materials Paper Standards Part 15, Philadelphia, Pa., April 1969.

[81] Anon., Fundamentals of Papermaking Fibres, The British Paper and Board Makers Association, Kenley, England, 1967.

[82] T. M. Plitt, Microscopic Methods Used in Identifying Commercial Fibers, U. S. Department of Commerce, National Bur. Std. C423 (1939).

[83] J. H. Graff, A Color Atlas for Fiber Identification, The Institute of Paper Chemistry, Wisc., 1940.

[84] R. E. Lofton and M. F. Merritt, Methods for Differentiating and Estimating Unbleached Sulphite and Unbleached Sulphate, Natl. Bur. Std. Tech. Paper 189 (1921).

[85] M. F. Merritt, Pulp and Paper Fiber Composition Standards, Natl. Bur. Std. Tech. Paper 250 (1924).

[86] R. L. Mitchell and G. C. Daul, "Rayon" in Kirk-Othmer Encyclopedia of Chemical Technology, 2nd ed., 17, 168 (1968).

[87] S. M. Atlas and H. F. Mark, Cell. Chem. Tech., 1 (4), 431 (1967).

[88] J. D. Griffiths, Textile Inst. Ind., 3, 54 (1965).

[89] G. C. Daul, Am. Dyestuff Reptr., 54 (22), 48 (1965).

168   M. L. ROLLINS and I. V. deGRUY

[90] N. S. Wooding, "Rayon" and "Acetate Fibers, " in Fibre Structure,
     (J. W. S. Hearle and R. H. Peters, eds.), Butterworth and the
     Textile Institute, Manchester and London, 1963.
[91] W. A. Sisson, Textile Res. J., 30, 153 (1960).
[92] L. H. Welch and W. S. Sollenberger, Am. Dyestuff Reptr. 49, p695
     (1960).
[93] W. E. Horton and J. W. S. Hearle, Physical Properties of Textile
     Fibers, Butterworth and Co., London, 1962.
[94] R. W. Moncrieff, Man-Made Fibers, 3rd ed., National Trade Press,
     London, 1957.
[95] Deleted in proof.
[96] V. W. Tripp, R. Giuffria, and I. V. deGryy, Textile Res. J., 27,
     14 (1957).
[97] H. Ziifle, R. Benerito, R. J. Berni, and A. M. Cannizzaro,
     Textile Res. J., 38, 1101 (1968).
[98] A. M. Cannizzaro, W. R. Goynes, M. L. Rollins, and R. J. Berni,
     Textile Res. J., 38, 842 (1968).
[99] G. L. Royer and C. Maresh, Textile Res. J., 17, 477 (1947).
[100] D. C. Felty, Textile Ind. (Atlanta), 131, 51 (1967).
[101] Modern Plastics Encyclopedia, 1969, Modern Plastics, New York, 1969.
[102] H. Reumuth, Textile Res. J., 17, 69 (1947).
[103] A. Köhler and W. Loos, Textile Res. J., 17, 82 (1947).
[104] A. N. J. Heyn, Textile Res. J., 27, 449 (1957).
[105] V. G. Peck and W. L. Carter, Textile Res. J., 34, 4 (1964).
[106] Anon., Colour Index, 4 Vols, The Society of Dyers and Colourists,
     Dan House, Picadilly, Bradford, Yorkshire, England, 1956-1958, and
     Color Index Classification, the American Association of Textile
     Chemists and Colorists, Lowell Technological Institute, Lowell,
     Massachusetts, 1958.
[107] L. E. Wise and E. C. Jahn, eds., Wood Chemistry, 2nd ed.,
     Rheinhold, New York, 1952.
[108] M. L. Rollins and V. W. Tripp, in Methods in Carbohydrate Chemistry
     Vol. III, Cellulose (R. L. Whistler, ed.), Academic, New York, 1963.
[109] C. M. Gordon, J. Soc. Chem. Ind. (London), 63 (9), 272 (1944).
[110] H. Takasawa, K. Sato, and N. Kuroki, Seni-i Gakkaishi, 24, 481
     (1968).
[111] G. Jayme and K. Balser, Melliand Textilber., 45, 1208 (1964).
[112] K. Kato, Textile Res. J., 27, 803 (1957).
[113] H. Hara, H. Sado, and I. Hashimoto, Textile Res. J., 26, 44 (1956).
[114] R. W. Berry, Textile Res. J., 24, 397 (1954).
[115] G. D. Joshi and J. M. Preston, Textile Res. J., 24, 971 (1954).
[116] E. Elöd, Melliand Textilber., 34, 1 (1953).
[117] P. H. Hermans, Textile Res. J., 20, 553 (1950).

[118] J. M. Preston and K. I. Narasimhan, J. Textile Inst., 40, T327 (1949).
[119] P. H. Hermans, Textile Res. J., 18, 9 (1948).
[120] E. Elöd and H. G. Frohlich, Melliand Textilber., 27, 103 (1946).
[121] F. F. Morehead and W. A. Sisson, Textile Res. J., 15, 443 (1945).
[122] K. Ohara, Sci. Papers, Inst. Phys. Chem. Res. (Tokyo), 25, 152
       (1934).
[123] F. L. Simons, Tappi, 33, 312 (1950).
[124] I. V. deGruy, J. H. Carra, V. W. Tripp, and M. L. Rollins, Textile
       Res. J., 32, 873 (1962).
[125] C. F. Goldthwait, H. O. Smith, and M. P. Barnett, Textile World,
       97 (7), 105 (1947).
[126] G. G. Clegg, J. Textile Inst., 40, T449 (1949).
[127] M. A. Grimes, Textile World, 95, 161 (1945).
[128] C. L. Pattee, Textile World, 84, 2012 (1934).
[129] W. E. Stevens, Textile World, 85, 1475 (1935).

Chapter 4

ELECTRON MICROSCOPY OF CELLULOSE AND
CELLULOSE DERIVATIVES

Mary L. Rollins, Anna M. Cannizzaro and Wilton R. Goynes

Southern Regional Research Laboratory
New Orleans, Louisiana

I. INTRODUCTION

The development and application of the electron microscope was one of the most dramatic advances of the highly charged era of technology immediately following World War II [1 - 3]. The first transmission electron microscope was built at the Technical University of Berlin in 1931 by Knoll and Ruska who used two magnetic lenses to achieve a magnification of 12,000X. In 1938 Von Borries and Ruska constructed an electron microscope of advanced design which was capable of resolving 100 A. In the same year Mahl constructed the first successful electrostatic instrument which resolved 70 Å and limited production was begun in 1940; meanwhile Siemens and Halske began commercial production of the Von Borries-Ruska instrument in 1939. The first electron microscope operated in North America was built by Hillier and Prebus in Canada in 1940. At the same time Marton in Brussels constructed an electron microscope and Metropolitan Vickers in England undertook to manufacture an instrument designed by Martin at Cambridge University. The advent of World War II curtailed further developments in Europe. In 1940 Hillier built for the Radio Corporation of America the first U.S. electron microscope, the RCA "Type A." In 1941 the RCA "Type B" designed by Hillier, Vance, and Zworykin became the first instrument commercially produced in America. In 1944 von Ardenne in Germany built an instrument capable of from 12 to 15 Å resolution, but it was never marketed. In 1946, Hillier at RCA attained a resolution of 10 Å which was considered the border of practical resolution as determined by spherical aberration. By 1970 nearly all electron microscopes on the market guaranteed resolution of 10 Å and several instruments even guaranteed 5 Å or less.

The wide acceptance of the transmission electron microscope as a useful tool in science is demonstrated in Table 1 which lists the manufacturers and indicates the range of resolutions offered by respective companies in their different models.

In the first three decades after its commercial introduction, the electron microscope achieved resolution equivalent to the theoretical, an accomplishment that required 300 years for light microscopy. The advent of the electron microscope opened up an entirely new frontier in morphology affecting most dramatically medicine and biology but having also significant impact in the fields of metallurgy and industrial materials analysis. From the beginning cellulose has been an object of investigation in electron microscopy. In modern materials science, structural characterization is basic to a thorough understanding of the nature of a material; it is considered as important as knowledge about physical and mechanical properties or chemical composition. Electron microscopy allows the study of structure in the "ultrastructural" domain that lies below the limits of resolution of the light microscope. In this submicroscopic region below 0.2μ, lie explanations for some of the behavior patterns of cellulosic materials. Comparisons of

TABLE 1

Transmission Electron Microscopes Available in 1970[a]

Marker or vendor	Guaranteed resolution[b] (Å)
1. Akashi, Ltd., Tokyo, Japan (marketed by Materials Analysis Corporation, Palo Alto, Calif.)	–
2. Associated Electronics Industries, Ltd. Harlow, England	10, 6, 5
3. Hitachi, Ltd., Tokyo, Japan (marketed by Perkin-Elmer Corporation, Palo Alto, Calif.)	10, 8, 7, 5
4. Japan Electron Optics Laboratory Company, Ltd., Tokyo, Japan (marketed by Jeolco, Ltd., U.S.A., Medford, Mass.)	100, 10, 7, 6, 4, 3
5. Philips Electronics, Eindhoven, Netherlands (marketed by Philips Electronics, Inc., Mt. Vernon, N.Y.)	30, 10, 5
6. Forgflo, Sunbury, Pa.	10, 8
7. Siemens, A.G., Karlsruhe, Germany (marketed by Siemens America, Inc., New York)	35, 8, 5
8. Carl Zeiss, Oberkochen, Germany (marketed by Carl Zeiss, Inc., New York)	9
9. Tesla, Prague, Czechoslovakia	25, 4.5
10. V.E.B., Jena, Austria	30

[a]Informal communication [3a].
[b]Resolutions of individual models.

morphological details of native, chemically modified, and regenerated cellulose at this level have produced vast volumes of information.

It is not the purpose of this chapter to describe all types of electron microscopes, nor to delineate the details of all the techniques that have been devised by the ingenuity of users within the period from 1940 to 1970. However, attempts will be made to describe the instruments adequately for an understanding of their performance and to provide information on those techniques specifically useful in the investigation of cellulosic materials.

II. INSTRUMENTATION

A. Microscopes

Electron microscopes used in the study of cellulosic materials are the transmission electron microscope (Fig. 1) and the scanning electron microscope (Fig. 2 and 3). The former produces an image by passing an electron beam through a thin specimen and presenting the shadow image for observation, contrast being dependent upon the relative absorbing and scattering powers of different portions of the specimen. The latter microscope relies upon secondary electrons emitting from the surface of a specimen bombarded with a narrow electron beam continuously scanned across its face; the emitted electrons are amplified in a scintillator and the image projected on a fluorescent screen synchronously with the original beam scan of the specimen. This instrument offers tremendous depth of focus but limited resolution of about 100 Å; its great utility lies in three-dimensional rendition of topographical detail of extremely rough surfaces, a feature impossible with either light or transmission electron microscopes. On the other hand, some transmission electron microscopes offer resolutions below 5 Å which allows direct magnification to 500,000X. It permits observations of very fine detail in thin specimens with a sharpness which allows photographic enlargement to magnifications well over a million times natural size.

1. The Transmission Electron Microscope

Although other types of electron microscopes are available, the term commonly refers to the transmission type of instrument in which electrons produced by a hot tungsten filament are passed through extremely thin specimens. As they pass through, the electrons are diffracted by atoms in the sample; they are then focused and projected on a fluorescent screen for viewing or on a photographic plate for recording. Just as a conventional light microscope forms a magnified image by focusing a beam of light, an electron microscope forms an image by focusing a beam of electrons, but in the electron microscope the focusing is accomplished with electromagnetic lenses instead of glass lenses. Because of the

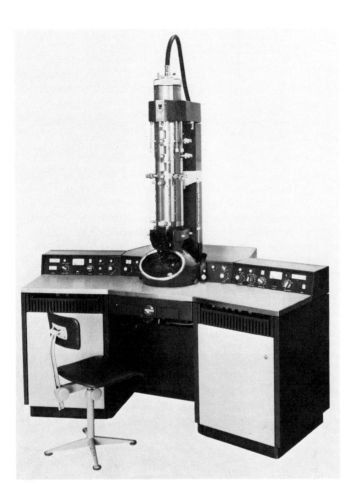

Fig. 1. Philips transmission electron microscope Model EM-300.
Reproduced by courtesy of Philips Electronic Instruments.

wavelength of visible light, the resolution of the best lenses for optical
microscopy is approximately 0.2μ (2000 Å), restricting the magnification
of light microscopes to about 1500X. The shorter wavelength of the electron
beam permits much higher resolution, and in modern transmission electron
microscopes particles 10 Å apart can be separated readily. Indeed some
of the commercially offered microscopes guarantee resolutions of 5 Å to
3 Å permitting direct magnification of 500,000X and greater.

In general, objects must be less than 2000 Å thick, must withstand a
vacuum, and should not undergo morphological changes when irradiated
by an electron beam. All atoms absorb and scatter electrons to some
extent; the amount absorbed or scattered increases with the atomic weight
of the atom. This scattering is the main source of contrast in the electron

image; the relative darkness of different parts of the specimen depends on ability to scatter electrons out of the imaging pencil or beam.

a. Difference between Light and Electron Microscopes. Although light and electron microscopes are more or less analagous in construction and performance there are certain important differences which must be kept in mind (Fig. 4).

The mode of image formation is different in the two microscope systems. In the light microscope the image is formed by differential absorption of light while in the electron microscope it is formed by electron scattering by the individual atoms which results in variations of intensity in the image.

In the light microscope different magnifications are achieved by the use of a series of ocular or objective lenses of different power, whereas in the electron microscope, changes in magnification are brought about by varying the focal length of the projector lens. Only electrostatic microscopes have interchangeable lenses.

Magnetic electron lenses do not have fixed focal lengths like optical lenses. Their focal lengths, which are dependent upon the strength of the field, can be varied by changes in the current that passes through the coil.

Fig. 2. Stereoscan scanning electron microscope. Reproduced by courtesy of Kent-Cambridge Scientific Instruments.

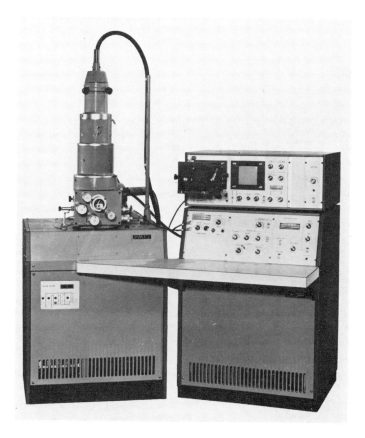

Fig. 3. JSM-U3 scanning electron microscope. Reproduced by courtesy of Japan Electron Optics Laboratory Co. Ltd.

In the light microscope differences of the depth of the field can be directly seen by manipulation of the fine focus control. These cannot be appreciated in electron microscopy by visual observation.

Many textbooks and manuals on the design and operation of transmission electron microscopes exist; among these Heidenreich's [4] and Wischnitzer's [3] are outstanding. Early texts by Zworykin et al. [5] Wyckoff [6, 7], Hall [8], Siegel [9], Fischer [10], Drummond [11], and Cosslett [12, 13] made very significant contributions to the development of electron microscopy and encouraged its use by many who had no formal knowledge of electronics. Other useful texts on general electron microscopy are by Haine and Coss- lett [14] and Barer and Cosslett [15].

M. L. ROLLINS, A. M. CANNIZZARO, and W. R. GOYNES

LIGHT MICROSCOPE　　　　**ELECTRON MICROSCOPE**

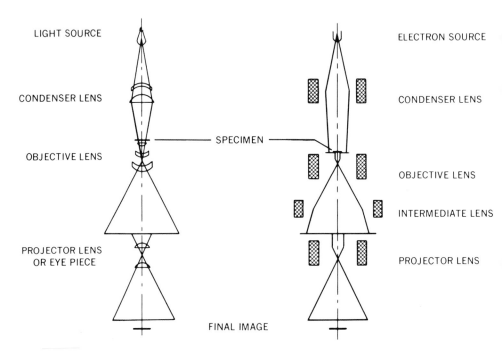

LIGHT SOURCE

CONDENSER LENS

OBJECTIVE LENS

PROJECTOR LENS
OR EYE PIECE

ELECTRON SOURCE

CONDENSER LENS

SPECIMEN

OBJECTIVE LENS

INTERMEDIATE LENS

PROJECTOR LENS

FINAL IMAGE

Fig. 4. Comparison of lens systems of light and electron microscopes. Reproduced by courtesy of Eastman Kokak Company.

b. Functional Aspects. From the point of view of function an electron microscope can be divided into 3 essential components: (1) an illuminating system, (2) an imaging system, and (3) a recording system. The illuminating and imaging systems are analagous to the components of the light microscope.

(1) The illuminating system contains two units: the electron gun which is the source of electrons, and the condenser which regulates the intensity of the beam and directs it onto the specimen. The function of the illuminating system is to produce an electron beam and to direct it onto the object to be viewed. The electron gun and the condenser, respectively, fulfill these purposes. The beam must be sufficiently intense for recording purposes and should converge on the object at an angle that allows optimal use of the imaging system. The total beam current and the illuminated area at the specimen should be as small as possible to minimize heat damage and contamination of the specimen by evaporated materials from the electron source. The electron gun, which has the components of and

acts like an electrostatic lens, consists of three parts: the filament, the shield, and the anode. The filament is a V-shaped piece of tungsten wire, about 0. 1 mm in diameter (4 mil), which can be heated to incandescence. Tungsten has the highest melting point of all metals, supplying a copious stream of electrons even below its melting point. However, some of the metal is evaporated in vacuo and so the useful life of a filament averages only 10 to 20 h. When a current is passed through the filament its apex becomes the hottest part, the effective electron source. By the use of apertures and lens systems the electrons are "channeled" so that they can be utilized in illuminating the specimen.

The shield is an apertured cylinder that lies immediately in front of the filament. It serves to produce a small source whose intensity is equal to that normally obtained without a shield, thereby reducing the load on the high voltage source. The anode, or positive electrode, is an apertured disk, which, like the rest of the microscope, is grounded (being at zero potential) but is positive with respect to the cathode. It is coaxially aligned with the hot filament. The electrons emitted from the tip of the hot tungsten filament are accelerated in the space between the filament and anode by the potential difference across this gap. After leaving the electron gun the electrons pass down the column at a constant velocity. The filament, shield (Wehnelt cylinder), and anode together constitute a cathode lens and form an image of the filament tip, " the crossover." This cathode lens assembly is called the "electron gun" and corresponds to the lamp in a light microscope assembly.

The condenser consists of a magnetic lens (in some instruments two magnetic lenses) that condenses or focuses an image of the crossover at some distance beyond crossover, depending on the strength of the current, through windings within the metal collar or "lens. " When high or low currents pass through the condenser coil the intensity of illumination at the object increases or decreases and the illuminated area changes correspondingly. Thus the condenser lenses perform a function similar to that performed by the substage condenser in the light microscope. The diameter of the aperture in the condenser lens is usually 0. 1 to 0. 3 mm. The size of this aperture controls the convergence angle of aperture of illumination at the specimen. This angle is greatest when crossover is focused at the object plane. In the newer models of transmission microscopes two condenser lenses are used to achieve a wider range of spot size, angular aperture, and intensity of illumination. Thus, the basic function of the condenser lens is to focus the electron beam emerging from the gun onto the specimen; this provides optimal illuminating conditions for visualizing and recording the image. The condenser lens of the electron microscope is a relatively weak lens with a focal length of the order of a few centimeters, but the length can be adjusted to produce a considerable range of intensities.

(2) The imaging system consists of the objective lens and two additional

magnifying lenses called projector lenses. The objective lens forms a magnified image of the object while utilizing only those electrons that have passed through the object without being scattered beyond the field. A small circular objective from 10 to 50 μ in diameter enhances contrast in the image by removing from the beam all electrons scattered through angles greater than a certain value. The magnification produced by a single lens (usually 50X to 200X) is approximately equal to the distance from lens to image divided by the focal length of the lens. In order to achieve direct magnification of about 200,000X two additional magnifying lenses called projector lenses are usually employed beyond the objective lens. By adjusting the amount of current in the coil of the objective lens the final image can be brought into a sharp focus on a fluorescent screen as judged by visual observations with either the unaided eye or an optical binocular magnifier that gives an additional magnification of 5X or 10X.

(3) The recording system encompasses a camera for photographing the image viewed on the fluorescent screen. In most microscopes the photographic plate or film is inserted above or below the viewing screen and exposed directly to the electron beam. Exposure time is usually from 1 to 5 sec. Micrographs are rarely recorded at the highest magnification that the instrument is capable of producing since it is more convenient to use smaller film with fine grain and subsequently enlarge the photograph.

c. Operational Aspects. From the point of view of operation, the major components of a transmission electron microscope are: (1) the imaging system; (2) the vacuum system; and (3) the electrical system.

(1) The imaging system, described above, consists of a series of electromagnetic lenses of different strengths assembled in sequence in a column with their magnetic fields (but not necessarily their physical dimensions) aligned. In some of the earlier microscopes the lenses were electrostatic, but nearly all modern microscopes use electromagnetic lenses.

(2) The vacuum system exhausts the column. The microscope must be exhausted to remove residual air, vapors, and gases. It is necessary that the column be evacuated for the following reasons:

(i) To avoid collisions between electrons of the beam and gas molecules which would result in a spread of electron velocities, thus creating chromatic aberration, decreased resolution, and electron scattering which reduces the contrast of the image;

(ii) To avoid high voltage breakdown and gas discharge between anode and cathode;

(iii) To increase the life of the tungsten filament which decreases when excessive amounts of gas, especially oxygen, are present.

It has been found that the above requirements are met if the gas pressure in the lens column is reduced to 10^{-4} Torr (mm Hg). Under these conditions the mean free path of electrons is about 2.5 m, which is the minimum necessary for modern microscopes whose column length is of the order of 1 m. To attain and maintain the desired level of vacuum a very efficient pumping system must be incorporated into the microscope system. This usually consists of a rotary mechanical forepump which is used to reduce the pressure to 10^{-2} Torr at which level activation of the high-speed high vacuum pump can be initiated. The high vacuum pump is usually an electrically heated oil diffusion pump which has been developed to operate quietly and with a minimum of vibration so that with proper shock mounting it can be incorporated into the physical structure of the electron microscope.

Specimens and photographic materials are inserted into and removed from the vacuum through a system of vacuum gates, valves, and airlocks. Some modern instruments contain specimen airlocks which permit change of specimen with the introduction of such a small volume of air that the function of the pumps is not impaired. The high speed vacuum system restores the desired vacuum well within minutes.

(3) The electrical system of the transmission electron microscope must provide:

(i) Current to heat the tungsten filament and thus generate the image-forming-electrons;

(ii) High voltage to accelerate the emitted electron beam so that its particles have a sufficiently short wavelength;

(iii) Current to each magnetic lens so that the focal lengths can be adjusted;

(iv) Current for the vacuum indicators.

The high voltage supply permits selection of from 2 to 4 values of accelerating beam potential in the range 2 to 200 kV with an output of 10 to 400μ A. Since the focal length of the lenses varies directly with the beam potential, the output of the voltage supply must be stabilized so that it does not vary by more than 1 V for even short periods of time. Similarly, the currents for the lenses, once they have been set for focus and magnification, should not vary by more than 1 part in 300,000 during the focusing and photographing operations. In the newer models of microscopes the circuitry has been extensively transistorized in standard exchangeable modules.

Figure 5 shows schematically the principal electron-optical components inside the microscope tube. Apart from various diaphragms, these components are, from top to bottom: the electron gun, two condenser lenses, the objective lens, the magnification lenses (diffraction lens and intermediate lens) and the projector lens. Altogether there are six lenses, all of them electromagnetic. The specimen is located approximately at the upper focal plane of the objective lens.

Applications of transmission electron microscopy to investigations of cellulose produced a voluminous literature which was extensively reviewed and summarized. The survey of Usmanov and Nikonovich [16] covered much of the literature through 1960. Two brief reviews by Rollins 17, 18 presented electron microscopical findings on the architecture of plant cell walls and provided cursory information on investigations into the synthesis of cellulose. More extensive coverage of the structure of plant cell walls was provided in the general texts by Frey-Wyssling [19], Roelofsen [20], Treiber [21], and Preston [22]. The early electron microscopy of wood was ably reviewed by Liese and Cote'[22a] to include work done in the period 1950 to 1960. Papers by Rånby and Rydolm [23], Preston [24], Mühlethaler [25], and Frey-Wyssling [26] presented summaries of the more significant work on cell wall structure with the electron microscope in the first 20 years of endeavor. In the period 1960 to 1970 Warwicker et al. [27], Betrabet [28], Jayme and Balser [29], and Peterlin and Ingram [30] presented concepts of cellulose ultrastructure based in large part on electron microscopical findings, and comprehensive discussions of present thought on fibrillar morphology in native cellulose are found in recent papers by Preston and Goodman [31], Mühlethaler [32, 33], Fengel [34], and Frey-Wyssling et al. [35, 36].

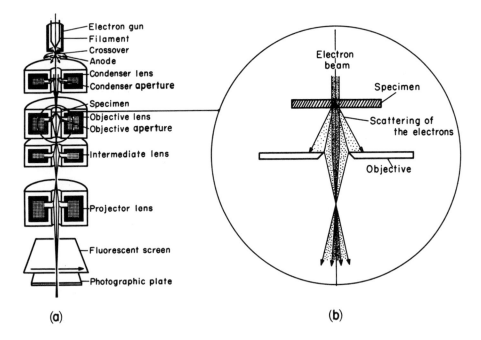

Fig. 5. Schematic diagram of transmission electron microscope. Reproduced by courtesy of J. T. Black.

2. The Scanning Electron Microscope

The first use of a scanning electron microscope to study cellulose was at the Pulp and Paper Research Institute of Canada where Smith [37] in 1959 used a prototype instrument constructed for the Institute by the Engineering Department at the University of Cambridge. It was the third instrument produced by Cambridge but the first to use electromagnetic instead of electrostatic lenses. This type of scanning electron microscope became available commercially at the end of 1965 (Figs. 2 and 3); in five years its use grew explosively, and by mid-1970 there were 8 different companies offering scanning microscopes (Table 2).

TABLE 2

Scanning Electron Microscopes of 1970

1. Cambridge	Kent-Cambridge Scientific Company Morton Grove, Illinois
2. JSM-2	Japan Electron Optics Laboratory Company, Inc. (USA) Medford, Mass.
3. Hitachi Scanscope SSM-Z	Perkin-Elmer Corporation Palo Alto, California
4. MAC Model 5	Materials Analysis Company Palo Alto, California
5. AMRC	Advanced Metals Research Corporation Burlington, Mass. and Philips Electronics Mt. Vernon, New York
6. Ultrascan	Ultrascan Company Cleveland, Ohio (first commercial microscope equipped with an electron-ion pump and to claim vacuum of 10^{-8} Torr)
7. Applied Research	Applied Research Laboratory Sunland, California
8. "Cameca"	C.A.M.E.C.A. Courbevoie, Paris, France

The literature concerning scanning electron microscopy is growing rapidly. Thornton [38] gives the theory of the scanning electron microscope and Oatley et al. [39], Smith and Oatley [40], Kimoto and Russ[41, 42], Everhart [43], and Black [44] discuss theory, practice, and applications.

In the conventional transmission electron microscope the scattering of some of the electrons as the beam passes through the material gives rise to the image. The scanning electron microscope works on a somewhat different principle: an electron beam produced by a conventional gun and focused by magnetic lenses is scanned across the surface of the specimen in a regular fashion by means of a set of scanning coils, somewhat as the beam in a television tube is scanned in raster fashion across the face of the tube (Fig. 6). When the electron beam hits the specimen, low energy secondary electrons are generated at the point of impact and are emitted along with reflected primary electrons from the surface of the specimen. These electrons are attracted by means of a small positive potential to a scintillator crystal which converts every electron impact into a flash of light. Each of these flashes in the scintillator crystal is then amplified by

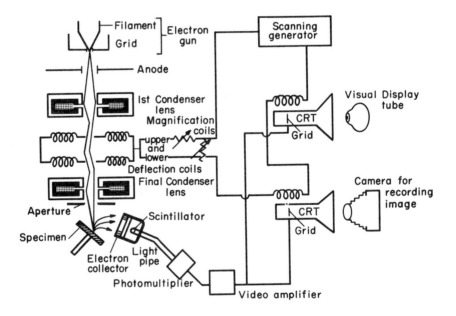

MICROSCOPE COLUMN   CRT DISPLAY

Fig. 6. Schematic diagram of scanning electron microscope and cathode-ray tube display. Reproduced by courtesy of J. T. Black.

means of a photomultiplier tube to produce a signal voltage, which in turn modulates the brightness of a cathode-ray tube whose screen is scanned in synchronism with the scanning of the object. The picture appears on the cathode-ray tube so that the resulting image is an exact representation of the surface of the specimen as imaged by the output of its secondary electrons. Image magnification is equal to the ratio of the scanning amplitudes on the screen to those on the object; magnification can be changed very easily by altering the amplitude of the original scanning beam.

The scanning rate for "visual" observations is about two frames (i. e., the cathode-ray tube is scanned twice) per second, and therefore a long persistence phosphor is used in the viewing screen. Varying the current in the final lens adjusts its focal length and hence is used to focus the microscope. If the rate of scanning is fast enough (e.g., 30 images per second) the observer sees a complete image of the object, owing to the persistence of the image on the fluorescent screen, even though only a single image element is given by the electron probe at any one time. To record a picture with a camera, however, the frame time is increased to about one frame per minute; a separate high resolution cathode-ray tube is used for the image to be photographed, and the phosphor on the screen is shorter lived than the one on the viewing screen.

The scanning electron microscope offers surface images which are supplementary to those obtained by optical microscopy and transmission electron microscopy. Images obtained by scanning electron microscopy resemble those formed by reflected light optical microscopy but show much greater depth of focus (by a factor of at least 300 at equivalent magnifications) and much higher magnifications, being useful in the range 14X to 10,000X by continuously variable steps. It provides easily interpreted, realistic, three-dimensional images of surface details and a resolution approximately 10 times that of the light microscope. Fine details can be detected as a consequence of the curve trajectories emerging from the scanned object. This allows for visualization of points not in direct line of sight of the electron collector, such as areas behind protruding surfaces, or within depressions and reentrant holes.

The resolution depends largely on the size of the electron beam and the penetration of incident electrons. In currently available commercial models image resolutions of approximately 100 Å with magnifications of 13,000X are feasible, but a more realistic resolution for organic material is 200 Å. That the resolution of scanning microscopes will be improved is indicated by results obtained with a laboratory-built instrument that achieved a resolution of 5 Å [45-48]. Dr. A. V. Crewe at the University of Chicago, by decreasing probe diameter to a very fine point and increasing beam intensity, was able to photograph uranium atoms attached to a molecule of DNA [49, 50]. This instrument is not commercially produced, but its successful performance presages the improvement of resolution in commercial scanning microscopes.

a. Electron Probe Analysis. Several of the newer instruments were designed more particularly for electron probe analysis of the specimen by back-scattered x rays than for photographic reproduction of surface features. The same exciting electron beam which provides direct visualization of surface structure by means of electron emission also causes emission of characteristic x radiation from the elements which compose the specimen. It is thus possible to measure specific x rays elicited from each element so that a physical analysis of the specimen may be performed. These scanning x ray microanalyzers permit analysis of specific areas in the specimen which are also photographed with the secondary electron emission mode. Most instruments offer both capabilities, and the potential of analyzing the exact area represented by a specific topographical feature is the most attractive property of this new instrumental development. For unmodified cellulose and for most samples of chemically treated cellulose this is an empty advantage since carbon, hydrogen, and oxygen have low atomic numbers. However, the use of metal-containing additives may enhance the potential application to cellulose.

b. Operational Aspects. In spite of the simplicity of the scanning electron microscope as compared with the transmission instrument, several problems occur in its use. A build-up of electrical charges on any nonconducting sample prevents focus on that portion of the image; moreover, if the surface charge becomes large, the sample jumps toward the anode during the scan. This problem can be eliminated by vapor deposition of gold on the surface of the sample; gold yields a low resistance path to ground so that the charge can leak off rapidly. Since heat from the high-energy electrons can cause changes in the surface being investigated, it is important not to scan a surface for extended periods, especially at high magnifications. This type of damage can happen even to heat stable polymers. However, if the surface is coated with gold, or, if probe scanning techniques are used, this problem does not occur.

The instrument is sensitive to outside fields. Once installation is complete, the environment should not be changed. Any change giving rise to increase in mechanical vibration or alternating magnetic fields should be avoided. Such devices as fluorescent lighting fixtures and transformers should not be placed within about 15 ft of the microscope column.

Publications on applications of the scanning electron microscope specifically to cellulose have just begun to appear. A 1959 publication by Smith, well before scanning microscopes were commercially available, showed excellent pictures of the surface of paper pulp fibers [37]; a more ambitious study was the 1962 Technical Report 294 of the Pulp and Paper Research Institute of Canada on tensile fractures of chemical fiber hand sheets [51]. The early papers of Sikorski et al. [52, 53] showed cotton fibers in comparison with wool and synthetics, and similar cursory observations were shown by Hearle [54], Reumuth [55, 56], and Billica [57].

By 1968 actual use of the scanning electron microscope as an investigative tool in research was reported by the Pulp and Paper Research Institute of Canada [58]; other publications including a study of adherent natural soils on fiber surfaces [59]; the effect of pigment particle size on the surface profile of coated papers [60]; on binder migration in coating structure[61]; a description of woody anatomy of pine and willow specimens [62]; the effect of refining on wet fiber pads and on air-dried fiber building boards [63]; and structure and properties of nodules in flash-dried pine [64] indicated that the instrument was being applied in several different directions. Papers by Goynes [65, 66] reported the effect of abrasion on chemically modified cotton fibers, and a paper by Rollins et al. [67] demonstrated the relationship between vinyl monomer coatings and the cotton fabric substrate in several types of coated cotton fabrics.

3. Reflection Electron Microscope

The reflection electron microscope was used in the period 1953 to 1956 for examination of paper, paperboard, and wood pulp fibers [68-74]. Here, the specimen was mounted at a small angle to the axis of the microscope so that the electron beam would strike it at near grazing incidence. A commercial transmission electron microscope was used for reflection at high angles, the specimen being illuminated by the electron beam at a glancing incidence and examined by imaging electrons scattered by the specimen at a small angle with the specimen plane. The advantage was enhanced depth of field compared to conventional transmitted electron microscopy, but the disadvantages were a foreshortened image and poor resolution (about 600 Å).

4. High Voltage Electron Microscope

Since the 1940's it was known that increased accelerating voltage would improve penetration of the specimen and increase resolution. Considerable attention was given in the 1960's to development of microscopes capable of accelerating and focusing electrons up to 3 MV, and by 1969 three high-voltage instruments were operating in the United States, with several in Europe and Japan. The University of Virginia had an RCA instrument capable of 500 kV; the University of California at Berkeley has a Hitachi instrument capable of 750 kV, and U. S. Steel was operating a high voltage microscope at their Research Center, Monroeville, Penn. In Japan, a 300-kV instrument built in 1950 was the forerunner of commercial instruments of 650- and 1000-kV capability; these instruments have been mostly used to determine atom displacement and crystal dislocation in metals. At Cambridge University a laboratory-made 750-kV instrument was the prototype for an instrument built by Associated Electron Industries designed for 1 MV. In France, Dupuoy at Toulouse was operating an instrument

since 1962 up to 2.2 MV [75-77]. Before 1970 no experiments were reported of high voltage electron microscopy of cellulose, probably because
at the higher kilovolts the low atomic weight specimens give little or no
contrast.

B. Vacuum Evaporator Systems

An essential ancillary item for preparing specimens for any electron
microscope is a vacuum evacuated bell jar in which carbon or metal may
be evaporated onto the specimen. This equipment must provide facilities
for:

(1) Preparation of support films on specimen grids;
(2) Preparation of carbon (or silica) replicas;
(3) Shadowing specimens with any of various heavy metals;
(4) Possible evaporation of two different materials successively without breaking the vacuum.

The system should pump down as rapidly as possible to an operating
vacuum of at least 10^{-5} Torr (mm Hg). Vacuum gauges for accurate
measurements of the vacuum inside the bell jar are essential. Electrical
current of at least 24 V capable of delivering up to 30 A should be provided.
The size of the bell jar should be such as to permit the specimen to be
placed at a distance of up to 15 cm from the filament from which materials
are to be evaporated, and some easy means of measuring the angle of
evaporation should exist. Both large and small evaporating devices specifically designed for preparation of electron microscope specimens are
commercially available. These units contain a mechanical forepump and
a high-capacity diffusion pump, suitable valving systems, and Penning
gauges which indicate the vacuum. The electrical supply varies from 20
to 40 V, but most units can yield a current of 30 to 50 A across the electrodes within a few seconds of operation. The modern instruments provide
up to 4 or 5 pairs of electrodes separately controlled by switches so that
more than one material may be evaporated without braking the vacuum.
Those evaporating devices specifically designed as ancillary to the scanning electron microscope are provided with a specimen stage which can be
rotated during the evaporation for even deposit of the metal on all areas of
the specimen, and which can be tilted in both X and Y directions to an angle
of 40° to accommodate coating of extremely rough surfaces. To assure
accurate control of the specimen film thickness the more up-to-date evaporating devices also have a thickness monitor in which the sensing element
is a quartz crystal located on the same plane as the specimen surface. The
build-up of the evaporant on the specimen is indicated on a meter which
can be pre-set to switch off power to the heater when a predetermined
film thickness has been reached.

Vacuum evaporation equipment in both simple and highly sophisticated
models are available from the following:

(1) Denton Vacuum, Inc., Cherry Hill Industrial Center
     Cherry Hill, N. J.
(2) Kinney Vacuum Division of New York Air Brake Co.,
     Boston, Mass.
(3) Velco Instruments, Inc., New York, N. Y.
(4) Varian Associates, Vacuum Division, Palo Alto, California.
(5) Japan Electron Optical Laboratory, Inc. (USA), Medford, Mass.
(6) Perkin-Elmer Corporation, Palo Alto, California.

C. Microtomes

1. Instruments

Microtomes conventionally used for slicing specimens for light microscopy are required to cut sections down to a minimum thickness of about
1μ although sections of 10μ are often acceptable; consequently, the method of embedding and sectioning were developed to suit these conditions.
Ultramicrotomes are a necessity for electron microscopy since the maximum section thickness which can be tolerated is of the order of 0.5μ
(5000 Å). Thinner sections are required for high resolution work, and
special microtomes have been developed capable of cutting slices as thin
as 250 Å if the specimen is suitable. Because of the toughness of natural
cellulose fibers, sections as thin as this are not obtained, but sections of
the order of 1000 Å are routinely prepared with the ultramicrotomes available. Several different models of microtome are on the market specifically designed for preparation of sections thin enough for examination in the
transmission electron microscope. The Porter-Blum-MT-I (Fig. 7) enjoys
the widest popularity with cellulose investigators [78, 79]. Its mechanical
advance is of great value in practical microtomy. The motor driven
Porter-Blum-MT-II is often desired for biological tissue. A more complex
instrument with variable speed motor drive and thermal expansion specimen advance is the LKB "Ultratome" based on a design by Sjostrand [80,
81]. A 1970 model, the "Pyramitome," incorporates devices for shaping
the specimen pyramid on the embedding block as well as for sectioning to
thicknesses appropriate for either light or electron microscopy. Other
popular makes of microtomes are those offered by Cambridge, Leitz,
Reichert, and Japan Electron Optics Laboratory Co. In all of them cutting
is accomplished by striking the moving specimen against a fixed knife.
Nearly all have some arrangement for permitting the specimen to by-pass
the knife edge on the return stroke; most have thermal advance mechanisms,
and many have variable speed motor drives. For sectioning cellulose,
experience has indicated that a slow knife stroke controlled by hand operation is most satisfactory.

Fig. 7. Porter-Blum Ultra-Microtome MT-1. Reproduced by
courtesy of Ivan Sorvall, Inc.

2. Knives

A great deal of experimentation has been carried out to find a suitable
cutting edge for thin sectioning. The first successful sections were cut by
Baker and Pease [82] with a modified conventional steel microtome knife.
Glass knives made by breaking 1/4-in. thick commercial plate glass into
1-in. strips at 45° to a scored edge are very satisfactory for most biological tissues [78, 83-85]. However, they dull quickly and fresh knives
must be broken for nearly every group of sections, so that the cost in time
and manpower becomes rather large. The majority of microscopists use
diamond knives [78, 83, 86, 87] for sectioning cellulose fibers. These
industrial diamonds mounted in holders to fit the microtome mounting and
sharpened to a suitable bevel for thin slices can be purchased from a few
companies in Europe and from vendors of special scientific equipment in
the United States. A diamond knife and sharpening service are also marketed by Dupont. Diamond knives can be obtained mounted in the convenient
trough-type holder assembly designed for the particular microtome

specified. The knives have cutting edges from 1 mm to 3.5 mm, and the edge may be used a little at a time and moved laterally as a used portion becomes dulled. Diamond knives are durable and if properly cared for may be used for comparatively long periods before being repolished by the manufacturer. Although knives of a fairly acute angle of about 42° to 45° are best suited for most biological tissues, knives with an obtuse angle of about 50° to 55° are more suitable for such hard materials as bone, metals, and fibers. Diamond knives are relatively hydrophobic so that, when the specimen trough behind the knife is filled with water, the edge of the knife should be wet with a sliver of wood or the meniscus of the liquid raised to permit wetting of the knife edge prior to sectioning.

III. TECHNIQUES

A. Transmission Electron Microscopy

1. Microscope Operation

To obtain successful electron micrographs, certain preliminary adjustments are essential.

a. Alignment. The gun and condenser are aligned by manipulation of the translational adjustments of both until symmetrical expansion of the illumination spot on either side of crossover is attained. The gun is centered with respect to the condenser axis, that is, its beam-defining aperture. When a filament is replaced the new filament is centered and the height adjusted relative to the anode. Then gun-condenser alignment is usually not necessary. It is a good practice to have a spare gun in which the filament has been aligned so that replacement of the filament can be effected without loss of time in operation. The best sequence of operation for alignment of the magnetic lenses is determined by the mechanical adjustments available on the particular instrument. Operators' manuals furnished by the microscope company should be consulted.

Testing and Correction of Astigmatism. Astigmatism is defined as an optical defect causing the image of a point to be seen as lines, and causing the lines to be less distinct in one direction than in the opposite direction. To correct astigmatism in the transmission electron microscope a "holey" film grid is used. It is made of carbon-coated collodion or of carbon film etched with an electrical spark. A range of hole sizes in the film permits its use at all magnifications. When symmetry of Fresnel fringes on all sides of the hole is achieved, astigmatism is eliminated.

b. Adjustment of Beam Intensity. In viewing and photographing the specimen image, the beam intensity should be adjusted to avoid excessive electron bombardment of the specimen. This can result in severe specimen damage. Therefore, to minimize the effect on the specimen both the maximum intensity of the beam and the length of exposure time during preliminary visual examination and photography need careful control. The level of beam intensity must be such that an image of sufficient brightness is formed to permit location of the desired field of view, precise focusing, and examination of the electron image. Short exposure of the specimen during photography reduces the possibility of obtaining a blurred picture. Exposures of longer than 3 sec usually result in losses of resolution due to specimen drift and circuit instability.

c. Calibration of Magnification. Since actual magnification depends on such factors as position of the specimen, voltage employed, image distortion, etc., it is reasonable that normal calibrations may have a probable error of ±10%. For routine work, determination of magnification is made with simplicity and reasonable accuracy by the use of standard reference objects such as fine diffraction gratings or polystyrene latex spheres.

d. Determination of Resolution. The methods most commonly used for determination of resolution are:

Measurements of the minimum distance between distinct particles on the photograph.

Measurement of the diameter of the smallest particle which can be sharply focused.

A classical method is measurement of dispersed latex spheres or other particles on a specimen grid, shadowed lightly with gold or platinum. Within the shadow are many small particles of the metal free of distracting shadows or contaminants; these can be measured with confidence. It is imperative that the photographic enlargement of negatives be done with meticulous accuracy.

Another resolution test object is the tetrad of the ferritin protein molecule. The test specimen is made by pipetting a very pure suspension of ferritin onto a Formvar substrate supported by a 200-mesh standard copper screen.

Potassium chloroplatinate crystals are also used as a resolution test specimen. The material is placed on a "holey" Formvar film supported by a 200-mesh round-hole grid. A picture taken where the crystalline material is stretched across a hole forms a good test specimen. The resultant micrograph should give a pattern with clear line spacings of 5.6, 6.9, or 12.6 Å, depending on the orientation of the crystals.

Other crystals suggested for testing the resolution of electron microscopes are made by boiling the dyestuff Indanthrene Olive TWP (ex. con. pdr., General Aniline and Film Corporation) in quinoline. Tiny, lathe-like crystals form having a molecular plane spacing of 2.49 Å in the direction of their widths. Deposited on carbon-coated grids, these crystals can be used for checking resolution [88].

e. Common Operational Difficulties. Many disturbances can interfere with successful photography of the object under study. Under ideal instrument conditions resolutions of from 10 to 5 Å can be obtained. In actual practice, a number of operating difficulties must be prevented to achieve high resolution.

(1) Image Drift. Image drift, which is a shift of the image in the field, can be caused by stray magnetic fields from an ac carrying cable, a transformer, or a voltage stabilizer in the vicinity of the installation. This makes it impossible to align the column; therefore, electrical tests should be made to find and eliminate the interfering current. The most sensitive part of the beam path is just on the image side of the specimen, and this disturbance manifests itself as a shift in the image produced after the beam is focused, and is consistent from specimen to specimen. If the region between condenser and objective is affected, then the presence of the transverse alternating magnetic field manifests itself as a beam wobbler. Image shift from mechanical vibration may be due simply to improper insertion of the holder into the column.

(2) Specimen Drift. Specimen drift may be due to thermal drift from absorption of heat by the specimen grid. If image drift is apparent immediately upon activation of the beam, the thermal drift is eliminated as soon as thermal equilibrium is attained between the specimen grid and its holder. Thermal drift can be minimized by slowly increasing beam intensity from a minimum until thermal equilibrium is reached. Thermal contact must be provided by fitting the specimen grid accurately into its holder, and the holder must be inserted correctly into the instrument.

(3) Contamination. Contamination is due to the slow deposit on the beam-exposed surfaces within the column of organic matter from the residual gases, from greased areas, from embedded polymers, or from vapors from specimen deterioration by beam impact; hydrocarbons from diffusion pump oil are a major cause of contamination. To reduce contamination, cooling devices are now used in most transmission electron microscopes. A copper rod, cooled at the outer end by immersion in liquid N_2, placed in contact with the specimen holder reduces the temperature of the specimen so that the specimen may be irradiated for a relatively long time without contamination problems.

To minimize this disturbance, cleanliness of all the beam contacting surfaces, especially the objective aperture, must be assured. The aperture opening is often about 25μ in diameter, and alignment is performed by various means in different instruments. The usual objective lens is provided with a separate aperture to insure optimal aperture angle and to provide adequate contrast. A physical aperture introduces a possible source of error after a short period of use because a nonconducting film is deposited around the opening. Scattered electrons which strike the lens elements may also form deposits on the surfaces of the lenses. These non-conducting contaminations become charged and act as electrostatic lenses to distort the field. Platinum apertures can be cleaned by sonication in acetone, followed by flaming in a Bunsen burner. Molybdenum apertures, however, are best cleaned after sonication in acetone by heating in the vacuum evaporator while supported in a molybdenum boat. Other cleaning procedures are described by Pease [83] and Horner [88].

(4) Vacuum Leaks. Vacuum leaks can be very troublesome. Maintaining an adequate vacuum level in the column is essential. A serious vacuum leak is indicated by the vacuum gauges, but a minor leak can be very difficult to detect. The exact site of the leakage is usually found at the junction between components of the column and can be located by a process of elimination. In rare cases the difficulty lies within a component, and this requires radical dismantling of the columns before the leak can be sealed.

After preliminary adjustments of the microscope have assured optimum performance of the instrument, the specimen is placed in the column and a field of view is selected at low magnification and at as low beam intensity as possible to obtain a preliminary focus.

The magnification is raised to the desired level of resolution of detail in the image on the viewing screen.

The intensity of illumination is increased to a level just adequate for critical focusing and for reasonable exposure time.

The final image is brought into critical focus by varying the current in the objective lens.

A series of photographs is taken with the objective lens current varied slightly in each direction from the assumed critical focus.

2. Specimen Preparation

Actual performance of an electron microscope system depends in large part on the specimen preparation as well as the microscope. A high performance instrument is worthless if the specimen preparation is poor. Valuable guides in an area of endeavor, which is still more art than science, are "Techniques for Electron Microscopy" by Kay [89], Pease's "Histological Techniques for Electron Microscopy" [83], and Drummond's "The Practice of Electron Microscopy" [11]. Other useful publications are [15, 84, 90].

a. Specimen Supports. Specimens for the transmission electron microscope are mounted on metal screen wire disks ("grids") 3 mm in diameter with holes or slots sufficiently small that the thin film placed over it to support the specimen remains intact over the openings. The grids most commonly used are 200-mesh copper, nickel, or stainless steel. A variety of other patterns are available from instrument supply houses as illustrated in Fig. 8. To prepare grids to receive specimens the grids must be coated with a material of sufficiently low atomic number that electron scattering by the coating or support film does not confuse the pattern of the specimen.

The requirements for the support film itself are that it should be less than 100 Å thick, it should be amorphous and structureless, it should be strong and capable of withstanding the effects of the electron beam, and its preparation should be relatively simple. For a long time collodion films were in common use. These were prepared by spreading a dilute solution of collodion in a suitable solvent onto a water surface. When the solvent evaporated a thin film remained on the water and the grid was raised through the film to become covered with the thin film of collodion. A common technique consists of applying a drop of 2% or 3% solution of collodion in amyl acetate to the surface of a dish of water. The drop spreads spontaneously over the surface and, when the solvent evaporates, a thin film of collodion 5 to 20 Å thick remains floating on the surface. A number of other plastic films can be prepared in similar fashion. If they do not spread on water, they can be spread on a smooth glass surface and stripped after drying; however, this method can be quite time consuming. The films are picked up over the metal disks by immersing the disks under the film or by dropping the disks onto the film and then deftly inverting a glass slide to lift grids and film together from the water. An easier method is to lay a strip of filter paper over the grids on the film, immerse the sandwich of grids, and lift them with the film and grids on top of the filter paper.

Inorganic films are stronger and better able to withstand high temperatures. Such films are usually of carbon or silicon monoxide evaporated in a vacuum bell jar onto collodion-filmed grids; the collodion is then removed by immersing the grids in solvent. Silicon monoxide is vaporized from a conical helix of tungsten wire that is heated electrically [86]. Carbon is evaporated from 2 electrodes tapered or necked down at their point of contact and heated to a high temperature by the passage of alternating current [91, 92]. To estimate the thickness of the carbon film, a small piece of white porcelain or milk glass is placed next to the slide containing the grids, and a drop of vacuum oil put on it. With this indicator about 8 cm from the source, the thickness of the film is about 100 Å when darkening of the porcelain in comparison to the colorless oil droplet is first observed. The carbon films are almost amorphous, are extremely strong, withstand high temperatures, and can be supported across a 200-mesh grid even when their thickness is less than 50 Å. They are also chemically inert and have a relatively high electrical conductivity so that they are not affected by electron bombardment. It is general practice to prepare several dozen grids at a time and store them for future use away from atmospheric humidity. A nearly structureless support film can be produced by evaporating platinum and carbon simultaneously from the same source [93].

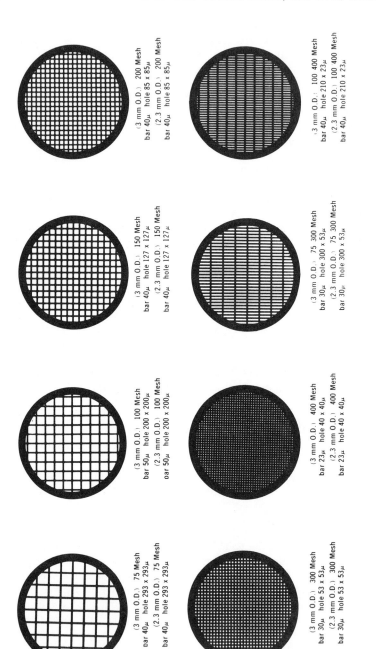

(3 mm O.D.) 200 Mesh
bar 40μ hole 85 x 85μ
(2.3 mm O.D.) 200 Mesh
bar 40μ hole 85 x 85μ

(3 mm O.D.) 100 400 Mesh
bar 40μ hole 210 x 23μ
(2.3 mm O.D.) 100 400 Mesh
bar 40μ hole 210 x 23μ

(3 mm O.D.) 150 Mesh
bar 40μ hole 127 x 127μ
(2.3 mm O.D.) 150 Mesh
bar 40μ hole 127 x 127μ

(3 mm O.D.) 75 300 Mesh
bar 30μ hole 300 x 53μ
(2.3 mm O.D.) 75 300 Mesh
bar 30μ hole 300 x 53μ

(3 mm O.D.) 100 Mesh
bar 50μ hole 200 x 200μ
(2.3 mm O.D.) 100 Mesh
bar 50μ hole 200 x 200μ

(3 mm O.D.) 400 Mesh
bar 23μ hole 40 x 40μ
(2.3 mm O.D.) 400 Mesh
bar 23μ hole 40 x 40μ

(3 mm O.D.) 75 Mesh
bar 40μ hole 293 x 293μ
(2.3 mm O.D.) 75 Mesh
bar 40μ hole 293 x 293μ

(3 mm O.D.) 300 Mesh
bar 30μ hole 53 x 53μ
(2.3 mm O.D.) 300 Mesh
bar 30μ hole 53 x 53μ

(3 mm O.D.) single hole
1.0 mm diameter
(2.3 mm O.D.) single hole
1.0 mm diameter

(3 mm O.D.) single hole
0.8 mm diameter
(2.3 mm O.D.) single hole
0.8 mm diameter

(3 mm O.D.) single slot
0.2 × 1.0 mm

(3 mm O.D.) single hole
0.6 mm diameter
(2.3 mm O.D.) single hole
0.6 mm diameter

(3 mm O.D.) rect. opening
1 × 2 mm
(2.3 mm O.D.) rect. opening
1 × 2 mm

(3 mm O.D.) 100 special Mesh
bar 56/32μ   hole 100 × 100μ
(2.3 mm O.D.) 100 special Mesh
bar 56 32μ   hole 100 × 100μ

Fig. 8. Types of metal grids used for specimen supports in transmission electron microscopy. Reproduced by courtesy of LKB Instruments, Inc.

b. Shadow casting. To have sufficient contrast in biological material
a layer of a heavy metal is usually deposited upon the material at an oblique angle, a technique referred to as "metal shadowing" [94-97] . The
specimen on the grid is placed in a bell jar containing a heavy gauge (15
mil) tungsten wire filament carrying the metal to be deposited. The bell
jar is evacuated to a vacuum pressure of at least 10^{-4} Torr and the filament is heated until the metal evaporates. In a vacuum the metal atoms
travel in straight lines from the filament in all directions. Some of the
atoms fall on the sample surface but most are deposited behind surface
projections or in depressions. In the microscope, electrons are scattered
less where no metal is deposited and a difference in exposure results permitting surface irregularities to be seen more clearly. In a photographic
negative, or a negative print, the shadows are black and the picture appears as if the objects were brilliantly illuminated by light from the side.
If the angle of shadowing and the length of the shadow are known, the
height of the object which produces it can be calculated.

When the metal comes in the form of wire, ribbon, or foil, it can be
wrapped around or clamped over the tungsten wire at a region where the
tungsten wire has been bent to a V-shaped point or coiled to form a basket,
to hold the metal fragments. When the tungsten wire is heated electrically with currents from 50 to 100A it will heat most intensely where it has
been kinked, and the molten evaporating metal will form a liquid drop at
the base of the "V."

Metals used for shadow casting are: gold, chromium, palladium, platinum, tungsten, uranium, and germanium. Gold and chromium have low
boiling points and are thus easy to use but both produce rather coarse
grain in the deposit, confusing fine details of specimen structure. Platinum has a finer grain size but requires much higher temperature of evaporation. Uranium has the very smallest grain size and, because of its a-
tomic weight, can be used in extremely thin layers to produce excellent
contrast. However, it often has an oxide film which causes undesirable
sputtering and so should be etched momentarily in nitric acid just before
use. A mixture of gold-palladium gives best results, and a wire of suitable alloy is available, but it still shows grain of 25 Å and is thus undesirable
for use at the highest magnifications. The carbon platinum mixture devised by Bradley [93, 97] appears to be the only structureless shadowing
material.

It is important that the thickness of the deposit should be the minimum
required to give contrast or the resolution may be seriously reduced. To
calculate the amount of metal to be deposited for a layer of the required
thickness, evaporation is assumed to take place spherically so that

$$W \sin \phi = \pi d^2 \, px$$

where W is the weight of metal, p the density of metal, d the distance

betweentungsten filament and specimen,   the angle of inclination of specimen toward beam, and x the thickness of metal layer on surfaced specimen.
Generally, an angle between 20° and 30° gives adequate shadows. This
is achieved by a 2:1 arrangement with the horizontal specimen to electrode
distance twice that of the vertical distance from base plate to metal. For
specimens of very fine particles or of very flat contours an angle close to
10° gives more satisfactory pictures. The size of the angle of shadow-
casting depends on the nature of the surface of the object. Very long shadows which darken the structure are generally undesirable.

Certain precautionary information is of value in shadow-casting:

(1) At the moment of shadowing care must be exercised to avoid damage to the specimen from thermal radiation from the evaporant.

(2) Some metals have the tendency to form granular structure under
the impact of the electron beam in the microscope. Granularity can appear as a result of insufficient vacuum at the moment of shadowing and of
migration of the metal along the surface of the sample.

(3) A baffle placed between evaporant and specimen helps direct the
beam of metal vapor at the specimen and prevent scattered particles from
depositing on the specimen.

(4) Heating the filament by slowly raising the current until the tungsten
wire glows red, then reducing the current, and repeating two or three
times before going to the evaporation temperature is good practice. This
outgasses the filament and the metal and avoids sputtering.

(5) Finally, evaporation should be done rapidly.

c. Replication Technique. The replica technique is used to study surfaces of samples too thick for direct microscopic examination. An impression of the fiber surface is made so that the topography can be explored at a level of resolution somewhat better than that of the scanning microscope [92]. An imprint of the fiber surface is made in a thermoplastic
polymer, the fibers are removed, and a replica of the plastic imprint is
prepared by vacuum evaporation of carbon onto the polymer replica. The
polymer is then dissolved, and the carbon replica shadowed with platinum
to emphasize differences in elevation on the surface. These shadowed replicas in the transmission electron microscope give photographs which are
faithful reproductions of the fiber surface. The objectives of replication
are:

   to compare surface rugosities;
   to detect evidence of mechanical damage;
   to investigate distribution of surface coating;
   to observe phenomena of partial solution or etching.

Such surface studies may be used in cellulose fiber research to follow
progressive effects of purification or modification processes such as scouring, bleaching, coating, impregnation, and derivative formation or to

observe the physical phenomena associated with abrasion, heat damage, microbial degradation, and beating and grinding processes.

Many different replica methods are discussed in textbooks on electron microscopy. Bradley gives a full listing and detailed descriptions in Kay's book on Techniques for Electron Microscopy [89]. Coté et al. [98] summarizes the various replica methods used for wood and paper research. Those most frequently used on cotton fibers are described in detail below.

(1) Polystyrene Replica Technique. A replica solution is made up of 5 g of low molecular weight polystyrene in 91 ml xylene with 1 ml dibutyl phthalate as softener. Most of the polystyrene commercially available must be degraded by heat (175° C for 24 h) in order to make it more easily soluble in the solvents used for replica preparations. To prepare a film of polystyrene (about 0.25 mm thick), a glass slide is coated with the solution and allowed to remain in the air for 1 h until the polystyrene is hardened. This forms a clear substrate. The size and thickness are governed by the textile material to be replicated. The fiber, yarn, or fabric is placed on the film with a glass slide on top of it and three Hoffman screw clamps are applied to press the fibers into the polystyrene film between the slides. After 1/2 h in the oven at 50° C, the preparation is removed and allowed to cool. The fibrous material can then be lifted from the polystyrene film with tweezers.

The polystyrene film with its negative replica of the fiber surface is placed in a shadow-casting unit and carbon is evaporated on it from a vertical position. This forms the positive impression in carbon. (Sufficient thickness of carbon has been deposited when the exposed surface of the glass slide shows a gray color). The replicas are cut in squares to cover about 8 specimen grids, and arranged carbon-side-down on small pieces of filter paper. A drop of xylene is placed at the edge of the filter paper. These small pieces are mounted on a larger piece of filter paper which is kept saturated with xylene in a covered Petri dish for 24 h. This brings about the gradual solution of the polystyrene leaving a very thin positive carbon replica. This replica is then metal shadowed at a 2:1 angle before it is examined in the microscope.

(2) Methacrylate Replica Technique. A viscous solution of partially prepolymerized (semipolymerized) methacrylate (2 parts butyl and 3 parts methyl methacrylate) is put on a glass slide and covered with a glass coverslip. This is allowed to set up for 2 h on a warming table at 65° C. After cooling, the methacrylate sheet is removed from the slide by soaking the assembly in water or by chilling it with a short blast of compressed Freon liquid. The methacrylate sheet is trimmed with a razor blade to remove rough edges and laid on a fresh glass slide. The sample is placed on this sheet, and another slide is clamped on top with Hoffman screw compressor clamps and put in the oven at 100° C for curing. Fabric samples are heated for 1 h; fiber samples for 15 min. When the assembly has cooled, all traces of fibers are removed under a widefield optical microscope.

The methacrylate mold is placed on a slide and carbon coated at 90; then cut into squares to cover several grids. These are arranged with the coated side down on a 1-in. square piece of filter paper. The filter paper sheet containing the grids is now placed in a Petri dish on 10 thicknesses of cotton gauze saturated with methyl ethyl ketone. The dish is covered and the specimens allowed to stand overnight in a vacuum desiccator. This extraction is repeated with fresh gauze and solvent to ensure complete removal of the methacrylate. Removal of methacrylate from the carbon should always be undertaken on the same day that the replicas are initially made, otherwise complete removal is difficult. The individual grids are shadow-cast at a 2:1 angle with platinum and the replica examined and photographed in a transmission electron microscope.

Other variations of the replica technique applied to cellulose are described by Coté et al. [98], Jayme and Hunger [99], Gröbe et al. [100], Kassenbeck [101], and Sloan and Gardner [102].

(3) Replication of Wet Surfaces. To investigate the surfaces of wet materials such as swollen fibers, uncoagulated rayon filaments, undried wood structures, etc., various modifications of the replica process are used. Hall [103] advocated freeze drying to preserve the surface characteristics of wet materials for replication. The specimen is brought into contact with a copper block precooled to liquid air temperature and placed in a bell jar of the vacuum evaporator. When a good vacuum is obtained the temperature of the specimen is raised to about -60 to -90° C by heat from an external lamp to sublime any ice crystals which may have formed and to secure a surface etch sufficient for a good replica. The specimen is then shadowed with chromium and a support film of SiO_2 evaporated onto the specimen. The cold block is removed from the vacuum, the glass plate detached, and the carbon film floated off onto water from whence it is picked up on a specimen grid. Fischbein [104] recommended the use of a warm water bath to replicate surfaces of wet biological specimens with polystyrene sheet. Tripp et al. [105] modified this approach for wet cotton fibers by clamping the fibers and polystyrene sheet between two glass slides, and then immersing the whole assembly in a beaker of hot water for long enough to soften the polystyrene sufficiently for a good imprint of the surface of the wet fiber. The polystyrene mold is coated with evaporated carbon and the polymer subsequently dissolved with appropriate solvent. The carbon replica is then shadowed with platinum and photographed.

Norberg [106] found that to obtain the best structure preservation using the freeze-drying method on wet wood surfaces, cooling must be as fast as possible to produce very low temperature. A minute drop of liquid containing a few fibers is placed on a thin polystyrene plate which is immersed in liquid nitrogen and kept there for about 30 sec. The deep-frozen suspension is then rapidly transferred to a stage in the bell jar kept at about -80° C by streaming liquid nitrogen through it at different rates. As soon as the frozen suspension is placed on the stage, high vacuum is applied and the freeze-drying of the fibers begun. When freeze-drying is complete the fibers are chromium-shadowed from about 30° and finally coated with carbon at 90°. These procedures are executed in one sequence without breaking the vacuum and with the temperature of the cold stage kept at around -60° C. As the vapor pressure of ice is very low at this temperature, complete freeze-drying takes at least 3 h. After carbon evaporation, the temperature of the stage is allowed to attain room temperature slowly and atmospheric pressure is carefully admitted to the bell jar to avoid disturbance of the delicate replica. Further operations are completed according to the techniques developed by Coté et al. [98] in which a sheet of polystyrene is softened by heating to about 150° C on a glass slide and the carbon-coated specimen then gently brought into contact with it. When the sandwich is allowed to cool the polystyrene hardens and the polymer-supported carbon replica can then be freed from the fiber by solution of the cellulose in H_2SO_4. Gröbe et al. [100] have used two approaches to replicate wet surfaces of rayon during stages of manufacture. In one case, the swollen viscose filaments are solvent exchanged through glycerol-water mixtures of 30:70, 50:50, 70:30, 90:10 to pure glycerol. The glycerol-filled filaments are then platinum-shadowed at a suitable angle, followed by direct carbon evaporation. The fragile carbon replica is backed with polystyrene to protect it while the cellulose filaments are being removed by solution in H_2SO_4. Subsequently, the polystyrene is dissolved in chloroform, and the carbon replica observed in the transmission electron microscope. Another method [107] involved freezing in isopentane at -159° C, then subliming in a vacuum of 10^{-4} Torr and finally shadowing and treating with acid to remove the fiber.

Surface topography of fibers, fabrics, and papers are investigated with the transmission electron microscope by means of replicas. The natural rugosities of the cotton fiber surface which are the result of wrinkling of the primary wall are well represented by electron micrographs shown in publications by Tripp et al. [105, 108], and Rollins and Tripp [109], in which the effect of bleaching operations and of tension mercerization on the surface characteristics are shown. The surface of a cotton fiber taken from a green and growing boll is relatively smooth and shows few wrinkles, whereas the surface of a fully dried cotton fiber is typically corrugated (Fig. 9). Bleaching does not remove these wrinkles but does remove the noncellulosic material of the cuticle attached to the primary wall to reveal the woven network typical of primary walls of native fibers from many different plant species (Fig. 15b). Comparison of slack and tension-mercerized fibers by surface replicas indicates that the effect of tension during mercerization is smoothed-out wrinkles and orientation of wrinkles closer to the fiber axis. Starch, still widely used for warp sizing in the weaving of cloth, is shown in Fig. 10a as a featureless deposit superimposed on the corrugated fiber surface. Colloidal dispersions of particulate solids such as silica and alumina used for soil-resistance

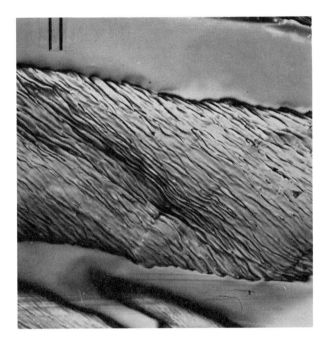

Fig. 9. Replica of cotton fiber surface. U. S. Department of Agriculture photograph.

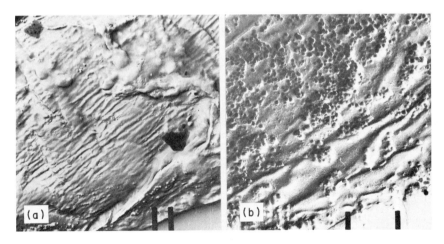

Fig. 10. Replicas of coated cotton fibers: a. starch; b. colloidal silica. U. S. Department of Agriculture photographs.

treatments or for additives to improve fiber friction are shown (Fig. 10b). Replicas also are effectively used in the evaulation of damage to fibers in laboratory abrasion tests and as a result of laundering [110-112]. In studies of soil attachment, "pseudoreplicas" in which particles adhered to the replicating polymer also yield useful information on aggregation of soil on fiber surfaces [113, 114]. Evidence of fungus growth has been recorded by the replica technique, and surface features of rayon filaments demonstrated.

In wood, the patterns of pit formation in various species [22a] (Fig. 11), the fibrillar arrangement in bordered pits [115, 116], the structure of successive layers in a wood cell [117], and the effect of pulping processes for paper manufacture were followed by replicas [118-120].

d. Fragmentation Technique. Because a whole fiber is too thick to examine in the transmission electron microscope, the internal structure may be investigated by comparison of the fibrillate patterns obtained when fibers are disintegrated either by high speed beating in water in a laboratory blender [121, 122], or by ultrasonic vibrations [123].

Cellulose fibers, when beaten in water, fragment in a manner characteristic of the sample. The slurry is allowed to dry on the specimen grid and then metal-shadowed for examination with the transmission electron microscope. Fragmentation patterns and textures of thin splinters resulting from this mechanical disintegration furnish clues to the chemical and physical history of the fiber. Features observed are:

Rupture pattern, whether abrupt or frayed and splintered; Texture of resulting fragments, whether fibrils are discrete or fused, whether debris is characteristically raft-form, spiculate, granular, powder, or amorphous;

Size of fragments, whether dimensions are altered with respect to those observed in fragments of control specimen.

In the study of chemically modified specimens, useful information can be developed by comparing with control samples the shape and size of fragments produced by mechanical fragmentation [124-127]. Thin specimens obtained by mechanical agitation can be subsequently treated on the specimen grid with a variety of reagents for observations of etching, swelling, or dissolution effects indicative of cellulose behavior at the fibrillar level. Sizes of ultimate fibrillar units of cellulose can be determined by measurement of isolated filamentous fragments negatively stained or carefully shadowed [128-134].

A technique particularly fruitful in textile research is the isolation of natural fiber parts in water by mechanical distinegration in a Waring blender or by ultrasonic agitation. The latter method has the advantage that the fiber is likely to break away along natural lines of weakness and so reveal the shape and size of structural units. A typical ultrasonic treatment requires exposure of the fibers in distilled water for several minutes to radiation of 25 kHz frequency. A drop of the slurry is then

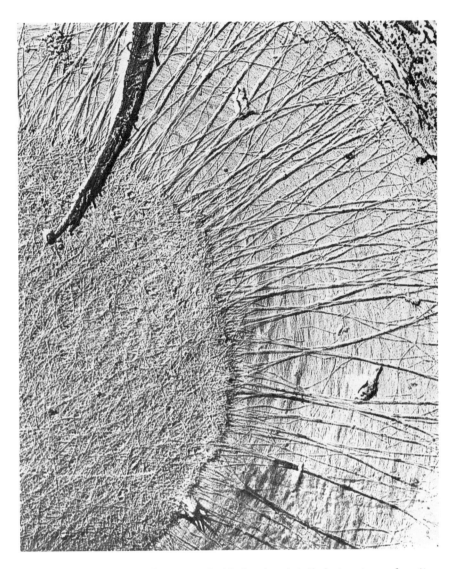

Fig. 11. Replica of pine tracheid showing detailed structure of a pit membrane. Reproduced by courtesy of W. A. Coté.

dried on a filmed grid and the residue shadowed. Untreated cotton normally yields sheets of cellulose in which the parallel arrangement of discrete microfibrils from the secondary wall can be clearly discerned (Fig. 12a). In fragments from mercerized cotton, microfibrils show a certain disorientation, and individual fibrils appear swollen and disarranged

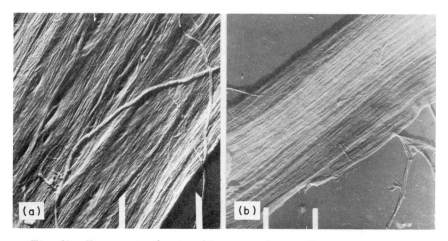

Fig. 12.   Fragments of cotton fiber secondary wall: **a.** untreated; b. scoured, tension mercerized. U. S. Department of Agriculture photographs.

with randomly spaced knots and spicules on individual microfibrils. When photographs are compared, the alignment achieved by tension during mercerization is plainly seen (Fig. 12b). In crosslinked cottons, fragmentation yields few flat sheets of discrete fibrils, but short chunks of fiber occur frequently which, in the electron microscope, appear to consist of several layers tightly bound together (Fig. 13). The pattern of fragmentation of crosslinked cotton reflects the embrittlement that is characteristic of the treated fabric samples. In esterified or etherified cotton, fragments show a progressive loss of fibrillar character as the degree of substitution increases. At low degrees of substitution, discrete fibrils predominate; at the highest substitution, no fibrils are observed and the spongy amorphous material exhibits no features of the original untreated fiber. Figure 14 illustrates a moderate level of acetylation. Drummond [128] showed micrographs of 3 natural and 6 man-made fibers after disintegration in a Waring blender. Morehead [135] and Dlugosz and Michie [123] used ultrasonics to disperse hydrolysis particles from both acid-degraded cotton and rayon. Levavasseur et al. [129] also employed ultrasonics. Rånby 136 was able to demonstrate the amorphous character of the lignin residue derived from unpurified spruce wood by ultrasonic disintegration. Razikov et al. [137] used the fragmentation technique extensively to evaluate the effect of increasing the chemical activity of cellulose preparations by inclusion of different alcohols and glycerine after swelling in amines. Rollins et al. [138] compared the fibrillar nature of cotton fibers after slack and tensioned mercerization.

 e. Embedding Technique. To cut sections to an appropriate thickness for transmission electron microscopy, the specimens must be immobilized in a rigid supporting medium with suitable slicing consistency. Paraffin

Fig. 13. Fragment of cotton from durable press fabric (dimethylolethyleneurea crosslinking process); note sheared end of fibril sheaf and apparent
fusion of individual microfibrils. U. S. Department of Agriculture photograph.

and other waxes traditionally used for optical microscope techniques do not
make sufficiently tough embedments for ultrathin sections. Moreover,
even for light microscopy, few wax preparations are satisfactory for sectioning the tough natural cellulosic fibers.

Polymers widely used for ultrathin sectioning are methacrylates, epoxy
resins, and polyesters [78, 84, 139]. The problem in cellulose work is
that the contrast between cellulose and these polymers is so low that satisfactory observations of structure are difficult, and while low contrast
can be overcome in most biological work by staining, cellulose fibers are
not readily stained with elements heavy enough to scatter electrons in the
electron beam. For this reason it is advantageous in the study of cellulose
materials to remove the embedding material after sectioning but prior to
shadowing.

Methacrylate and polyvinyl alcohol (PVA) are more often used in the
study of cellulose materials than the other common embedding polymers.
Many variations in the application of these resins for embedding cellulose
specimens exist, but typical procedures are described in the following.

(1) Methacrylate Medium. The acrylic resins N-butyl and methyl
methacrylate can be mixed in different proportions to produce blocks of
various hardness and cutting qualities [139-141]. It is desirable to have

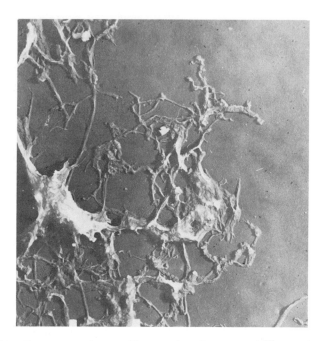

Fig. 14. Fragment of partially acetylated cotton: 19% acetyl. U.S.
Department of Agriculture photograph.

the hardness of the embedding medium matched to the hardness of the
specimen to avoid difficulty in cutting uniform sections. Pure n-butyl
methacrylate polymerizes to a relatively soft block; the addition of methyl
methacrylate increases the hardness to suitable quality for the material under investigation. Methacrylates are supplied as monomers containing
hydroquinone as a polymerization inhibitor. This inhibitor can be removed
by washing with NaOH, but this has been found in practice to be unneces-
sary. The catalyst normally used in methacrylate polymerization is ben-
zoyl peroxide, but for convenience most laboratories prefer to use a combination plasticizer and catalyst in liquid or paste form. One such preparation is "Luperco CDB." It consists of 50% 2, 4-dichlorobenzoyl peroxide in dibutyl phthalate. Both the catalyst and the methacrylate should be
kept refrigerated.

Before embedding cellulose fibers it is convenient to partially polymerize the methacrylate mixture prior to putting the fibers in it. One useful
procedure is as follows:

100 to 150 ml of monomer are made up in the proportions of 3 parts
methyl to two parts n-butyl methacrylate in a 250-ml Erlenmeyer flask
with a ground glass stopper, with 2% of the catalyst Luperco CDB.
(This product is 50% 2, 4-dichlorobenzoyl peroxide, thus requires 2 g
per 100 g methacrylate monomer.) This is well mixed at room

temperature until the catalyst dissolves and is then allowed to stand
overnight in the refrigerator. (All methacrylates should be stored under refrigeration at all times). To prepolymerize this methacrylate
mixture to form a more viscous embedding medium, the flask is suspended in a water bath at a temperature of 60-65° C. (CAUTION: Do
not allow the water temperature to go much higher than 65° ! At high
temperatures the reaction is difficult to control and polymerization may
occur so rapidly as to explode without warning.) The solution is stirred
often to keep the reaction uniform throughout. During polymerization
an ice bath should be immediately available to receive the flask containing methacrylate so that the solution may be cooled quickly if the temperature becomes too high or polymerization proceeds too rapidly. The
time generally required for prepolymerization to a favorable consistency under these conditions is 30 min to 1 h, depending on the final viscosity desired. The polymer becomes considerably thicker on cooling.
Since polymerization continues to occur to some extent over long periods of storage, even under refrigeration, caution should be taken not to
have the initial polymerization mixture so thick as to become unusable
after short storage periods.

(a) Flat Embedding. For small samples the flat embedding method of
Borysko et al. [142-144] is most convenient. A layer of methacrylate approximately the size of a 22-mm coverslip is formed on a glass slide and
the specimen is placed over this layer. Generally, better penetration of
the sample is accomplished if the slide is then placed under vacuum for
several minutes before final polymerization of the methacrylate. Finally,
the embedding area is reflooded with methacrylate and placed on a warming table at 65° C. As the resin becomes viscous more solution is added
until a thick enough block is formed. A coverslip is placed on the embedding and the assembly heated at 65° C until completely hardened. A 2-4
h polymerization period is usually long enough for complete hardening;
however, occasionally, longer periods are required. Thickness of the embedding should be approximately 1 mm (or greater) for good stability in
sectioning. If the embedding is not thick enough after polymerization, the
coverslip can be removed, more methacrylate added and a coverslip replaced, for a second period of polymerization on the warming table. The
embedding is removed from the microscope slide and cut in half to form 2
blocks approximately 10 x 20 mm. An isosceles triangle is formed with a
razor blade by cutting from opposite corners of the 20-mm side to the middle of the other 20-mm side so that the fiber bundle is at the apex. This
triangle is then placed in the flat embedding holder supplied with the microtome and trimmed with a sharp razor blade under a widefield microscope. Generally, smaller surface areas (0.5 mm) are easier to section,
but those of 1 or even 2 mm can be sliced satisfactorily. If serial sections
are desired, ribbons of sections may be cut from a block whose surface

has been trimmed to an isosceles trapezoid and which is mounted with the two sides of the trapezoid horizontal. If serial sections are not necessary, any approximately round or elliptical surface is satisfactory for sectioning.

(b) Capsule Embedding. An alternative method of embedding involves the use of gelatin or polyethylene capsules [78, 122, 139] . A small bundle of fibers is mounted on a nichrome wire frame and clamped to the frame at each end with fragments of lead wire. Fibers and frame are coated with a dilute solution of cellulose nitrate and allowed to dry. The frame is placed in a No. 00 gelatin capsule and covered with embedding solution containing 1 part n-butyl methacrylate, 4 parts methyl methacrylate and 1% Luperco CDB (50% 2, 4-dichlorobenzoyl peroxide in dibutyl phthalate). This mixture is prepared ahead of time and kept in the refrigerator. The open capsule is placed in a desiccator and subjected to a low vacuum for about 0.5 h, during which time the vacuum is released and redrawn several times to remove bubbles of entrapped air from the specimen. The top is sealed on the capsule by painting the seam with cellulose nitrate solution, and the capsule is placed in a 50° C oven overnight to polymerize the embedding resin. When polymerization is complete, the gelatin capsule is dissolved in hot water, leaving a smooth block of methacrylate containing the embedded fiber specimen. On a jeweller's lathe the methacrylate is machined to the approximate size of the specimen chuck of the microtome, and, finally, the face of the block is cut with a razor blade to a truncated pyramid with a face approximately 0.5 mm on a side. This block is now placed in the microtome chuck and sliced with the face of the diamond knife as near vertical as possible. Sections of appropriate thickness show silver gray interference colors as they float on the water-ethanol mixture in the specimen trough behind the knife (Table 3).

TABLE 3

Color-Thickness Correlation For Thin Sections [145]

Thickness range[a] Å	Interference color
600	Gray
600 to 900	Silver
900 to 1500	Gold
1500 to 1900	Purple
1900 to 2400	Blue
2400 to 2800	Green
2800 to 3200	Yellow

[a]Index of refraction, 1.5.

(c) Boat Embedding. Some laboratories embed in aluminum foil boats instead of on a glass slide or in a capsule. The boats can be made in any desired size and depth, and removal of the embedding from the mold is no problem. Evaporation from the surface can cause some nonuniformity in polymerization.

(2) Polyvinyl Alcohol Medium (PVA). Samples such as cellulose acetate or partial esters of cotton or rayon, which may be swollen by methacrylate monomer or by the solvents used in the methacrylate embedding process, require other embedding resins. Moreover, for swollen cellulose samples a water-soluble embedding medium is required. Polyvinyl alcohol (PVA) fulfills this requirement [145a]. The polyvinyl alcohol used is Gelvatol 60-20 from Monsanto [146]. Stock solutions of 2% and of 8% are prepared in 100-ml quantities by wetting 0.5-g amounts of PVA in distilled water at room temperature and mixing thoroughly to ensure complete solution without formation of lumps. After the resin is thoroughly dispersed, heat is applied to speed the dissolving process. It is advisable to add small portions of powdered PVA to the water at room temperature while stirring constantly to break up gel formation and then to heat to the boiling point. The operations are repeated until the required amount of 2 g or 8 g of PVA is dissolved. A clear solution, stable at room temperature, is obtained. A few drops of clear merthiolate (1:1000) are added to prevent bacterial growth. Solutions of PVA in various concentrations can be obtained from Polysciences, Rydal, Pa. Bundles of fibers are mounted on glass slides, coated with the prepared 2% aqueous PVA solution, and allowed to solidfy at room temperature. Layers of the 8% PVA solution are then applied until a thick enough block is formed to be held in a flat specimen chuck on the microtome. The material is allowed to polymerize at room temperature between coatings. Embedding usually requires 3 or 4 applications and takes 2 days to complete. This procedure can be accelerated by heating at 50-60° C after each coating. These flat embeddings are then lifted from the slide and removed to a clean slide. The edges are taped to the slide to prevent curling, and the embeddings are finally cured in an oven at 70° C. The embeddings must be loosened from the glass slide prior to heating because removal is impossible after they have thoroughly hardened. The blocks of embedded fiber pick up moisture after long exposure to the air and should be stored in a desiccator or reheated in the oven at 70° C before sectioning. The embeddings are trimmed with a razor blade into the form of a pyramid, and sections are cut on a microtome with a diamond knife. Sections are floated in the microtome trough on glycerol and picked up on carbon-coated grids directly from the trough. The grids are held with tweezers and the sections picked up by simply touching the ribbon of floating sections to the carbon side, lifting and turning the grid so that sections are face up on the carbon substrate. When these grids are placed on filter paper saturated with water, the sections settle down firmly on the carbon surface.

(3) Glycol Methacrylate Medium (GMA). Water-miscible embedding media are of special importance for studies of certain types of specimens. Gilev [147] used gelatin, but this results in extremely hard, brittle blocks when dry. Scarpelli and Sittler [148], deGruy [149], and Cannizzaro et al. [145a] used polyvinyl alcohol of low viscosity with care not to cure at too high temperatures, so that the polymer remained water soluble even when polymerized to a sliceable hardness.

"Glycol methacrylate" (ethylene glycol monomethacrylate) is widely used because it is miscible with water in all proportions [150, 151]. Polcin and Karhanek [152] used glycol methacrylate for embedding swollen cellulosic fiber from wood in the proportions 95% glycol methacrylate to 5% methyl methacrylate and polymerized it at 45-50° C for 24 h. Betrabet and Rollins [153] exchanged highly swollen cotton fibers through 1-h soaks of 50, 75, and 90% aqueous solutions of glycol methacrylate followed by a final overnight soak in a solution containing glycol methacrylate, methyl, and n-butyl methacrylates in the proportions 90: 6: 4, and containing 1% Luperco CDB. The fiber samples are placed in an aluminum foil boat filled with the methacrylate mixture. Air bubbles are removed in a vacuum desiccator, and the polymer cured at 65° C for 10 h. Blocks are ground to a suitable thickness for the microtome chuck, trimmed to a pyramidal cutting face, and sections cut on the microtome. Floated on 10% aqueous ethanol, the sections are transferred with a nichrome loop to carbon-coated grids and shadowed at approximately 25° C.

(4) Epoxy Resin Medium. Because of the damage which often develops in specimens during methacrylate embedding, other media have been adopted for biological work. Epoxy resins of several types are routinely used. Glauert and Glauert [154] perfected a technique using Araldite M (a trade name for epoxy resins made by Ciba) with a hardener (dodecenyl succinic anhydride) 964B and an amine accelerator 964C. To control the hardness of the final block dibutyl phthalate is added as a plasticizer. Araldite is very viscous and penetration of the specimen is a problem. Epon is an epoxy resin of better penetrating power which is used in a technique developed by Luft [155]. Maraglas is another epoxy resin which slices well [156], but it, too, gives penetration problems.

(5) Polyester Resin Medium. Polyester resins harden uniformly with little shrinkage. Vestopal W [157] is insoluble in ethanol. It is dissolved in acetone, and the specimen is usually infiltrated with styrene or with acetone. Polymerization of Vestopal W is initiated by peroxides as in the methacrylate media. It penetrates well, but its low contrast with cellulose and its difficulty of removal from sections militate against its use in cellulose microscopy.

Because contrast between cellulose and epoxy resins and polyester resins is so poor these embedding media have been very little used in studies of cellulose, especially since these resins cannot be removed from the sections except by destructive means. Details of embedding methods with epoxy and polyester resins are given in the textbooks by Pease [83], Kay [89], and Wischnitzer [3].

f. Sectioning. Details of the technique of ultrathin sectioning are described by Mercer and Birbeck [84], Pease [83], Kay [89], and in the Sorvall pamphlet "Thin Sectioning and Associated Techniques for Electron Microscopy" [78]. The embedded specimen is placed in the microtome head and aligned with the knife so that there is an angle of about 10° between knife and specimen. Thin sections of cellulose fibers require a relatively slow speed of cutting with smoothly deliberate motion; sections cut too fast often show ripples which are difficult to remove and which distort the specimen.

It is necessary to float the sections onto a water suface behind the knife as they are cut, to keep the embedding polymer from sticking to the knife. Most microtome knife holders have a built-in reservoir; for those which do not, satisfactory boats may be made of paper or foil and fastened to the knife with wax. The reservoir is filled with a 10 to 20% solution of alcohol or acetone to lower surface tension and prevent water from creeping over the edge of the knife. It is necessary for the edge of the knife to be wet, but the solution must not spill over the back. The trough should be filled so that the water is level with the knife edge but without a meniscus. If the level is higher than the knife edge the specimen block will pick up water as it passes the knife edge and, if lower, the sections will not float off properly. For PVA embeddings the trough is filled with glycerol rather than water.

A binocular microscope head of 10X to 50X magnification is attached to the microtome to enable close observations of manipulations in cutting and picking up sections. In addition, adequate lighting of the cutting area is necessary; a miniature fluorescent tube placed a few inches behind the knife is optimum. The lighting must be so arranged that a reflection of the light source in the water surface of the reservoir is seen by both eyes in the binocular microscope. Most modern microtomes are supplied with appropriate illumination and magnifier. The thickness of the section is judged by the interference colors observed in sections floating on the surface of the liquid as indicated by Peachey [145] and Satir and Peachey [158] (Table 3). For cellulose sections only those in the silver to gray range are suitable for examination in the electron microscope; sections in the gold range are too thick. Sections may be scooped up from the surface of the liquid in the reservoir by inverting the carbon-coated specimen grid over the section or they may be lifted with a small platinum wire loop and placed in a drop of liquid on the carbon-coated grid.

During cutting, the sections often become compressed along the axis at right angles to the knife. This effect is usually slight, but in very thin sections may be as much as 50%. The stresses produced by the compression cause the section to expand back toward their undistorted shape. The smoothing out can be hastened either by transferring the section to a warm water bath or by holding a piece of filter paper soaked in chloroform, methyl ethyl ketone, trichloroethylene, or other volatile solvent a few millimeters above the water bath in the microtome trough. The vapors expand the sections.

Removal of Embedding Media. The epoxy resins and the polyester resins are not removable by any of the common solvents. Contrast between these embedding media and the embedded specimens is dependent upon staining of the specimen prior to embedment. Glycol methacrylate, though water soluble in the low molecular weight form, becomes insoluble after polymerization. For this reason, the methyl and butyl methacrylate mixtures and polyvinyl alcohol are preferred embedding media for cellulose samples.

The methacrylate embedding medium is removed from the sections by the action of solvent vapors in a closed Petri dish. A group of specimen-loaded grids is placed on clean filter paper saturated with a suitable solvent. The filter paper is laid on about 10 thicknesses of cotton gauze in a Petri dish, and a small cup of solvent is set inside the dish before it is closed. Storage in a vacuum desiccator is suggested. Usually the embedding medium is dissolved by condensing vapors of the solvent in an overnight period. If not, the grids are placed on fresh filter paper supported by fresh gauze, and the process repeated with fresh solvent. Sections may be examined under the optical microscope for evidence of remaining polymer. The clean sections are shadowed with platinum or platinum-carbon at a 25-30° angle (2:1 positioning in the vacuum evaporator) and examined in the transmission electron microscope.

To remove PVA and glycerol, the specimens are washed a number of times with distilled water and heated in an oven at 60°C for 2 h. The water-soaked filter paper is changed twice during the heating. The grids are again thoroughly washed with water and placed on water-saturated paper in Petri dishes overnight.

g. Microsolubility. In investigations of chemically treated cellulose fibers, unmodified areas can be dissolved in appropriate solvents leaving reacted areas intact, and viceversa [159]. Such procedures are used to evaluate grafting reactions and crosslinking phenomena. The method is as follows:

Warp or filling yarns are removed from the fabric under investigation and defibered with care not to break any fibers. Fibers are combed parallel and arranged in small bundles (about 75 fibers). The bundle is tied together with a nylon loop and laid on a glass microscope slide and embedded in methacrylate in the conventional flat embedding procedure described in Section 2e. Sections on the grid are freed of the embedding polymer with appropriate solvents, then the grid is immersed in a drop of 0.5 M cupriethylenediamine hydroxide (cuene) in a spot-test dish for 30 min and washed gently with water until all traces of the blue color of the cuene solution are gone. The sections are then shadowed with platinum or

gold-palladium by vacuum evaporation of the metal at a 2:1 angle and examined with the transmission electron microscope. Observations are made
of the response of the specimen to the solvent to judge uniformity of treatment, location of reacted regions, and apparent texture of the cellulose
residue after solvent extraction.

Alternatively, solvents for noncellulosic additives may be used to dissolve the free material and leave the unreacted cellulose or the insoluble
cellulose-reaction product.

Ultrathin sections of cotton fibers examined with the transmission electron microscope usually show no structural features of the fiber (Fig. 15).
However, by immersing sections free of embedding medium of 0.5 M solution of cupriethylenediamine hydroxide (cuene) it is possible to compare
crosslinking treatments with respect to the amount of swelling the section
will exhibit in the cellulose solvent [159]. Sections of untreated cotton
fibers, when immersed in the cellulose dispersing agent, are dissolved
completely with only noncellulosic materials remaining (Fig. 16). Specimens with extensive crosslinking are immobile in this dispersing agent
(Fig. 17); those with limited crosslinking swell with dissolution of isolated
areas. With various types of crosslinking agents and procedures, varying
degrees of immobilization of the fiber structures exist with respect to cellulose solvents, and comparisons in a series of treatments can be valuable
when the solubilities of cross sections correlate with changes in physical
properties or chemical histories of samples (Figs. 18-20). In impregnation
of the fiber with resin-forming monomers, the cellulose of the fiber is removed by dissolution, but the noncellulosic impregnant remains virtually
intact. Subsequently, it may be dissolved by solvents appropriate to the
specific polymer in question. These microsolubility techniques are useful
in locating reactions with respect to fiber periphery or core in the comparison of different crosslinking treatments and in tracing the position of grafted
copolymer resulting from cellulose activation prior to or during the application of polymer additive [160, 161] (Fig. 21). The distribution of inorganic
inclusions in rayons has also been shown by this method.

h. "Expansion" Technique. A variation of the sectioning technique
takes advantage of the "explosion" effects of methacrylate polymerization
in comparison of samples of certain series of finishing treatments. Bundles of approximately 30 fibers are tied with nylon filament and the fibers
cut at both ends, then brought to a boil in 50% aqueous methanol containing
1% deceresol or other wetting agent. The bundles are allowed to soak
overnight in the solution. The wet bundles are lifted out of the solution,
the excess water is shaken off, and the bundles of water-wet fibers are
laid on a glass slide in semipolymerized but thin rather than viscous n-butyl methacrylate; no fiber ends should protrude from the coverslip. The
slide is placed on a warming table at 65° C. When a milky emulsion begins

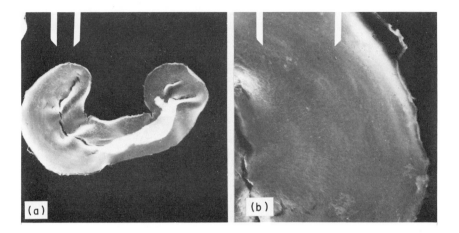

Fig. 15. Transmission electron micrographs of cross sections of untreated cotton: a. whole fiber; b. enlarged portion of fiber. U. S. Department of Agriculture photographs.

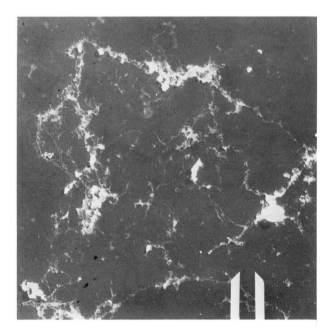

Fig. 16. Transmission electron micrograph of cross section of untreated cotton fiber after 30-min immersion in cupriethylenediamine hydroxide (cuene). U. S. Department of Agriculture photograph.

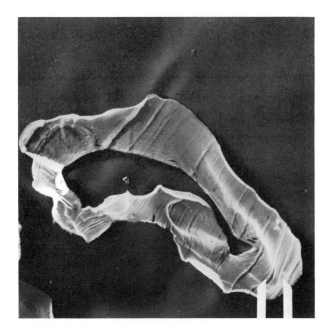

Fig. 17. Transmission electron micrograph of section of fiber cross-linked by episilicate treatment, after 30-min immersion in cuene. U.S. Department of Agriculture photograph.

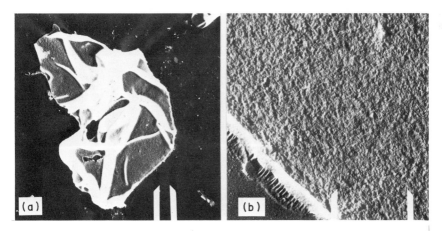

Fig. 18. Transmission electron micrographs of section of fiber cross-linked with dimethyloldihydroxyethyleneurea ($Zn(NO_3)_2$ catalyst), after 30-min immersion in cuene ("extensive" crosslinking): a. whole section residue; b. enlarged portion of section residue showing texture of swollen cellulose. U.S. Department of Agriculture photographs.

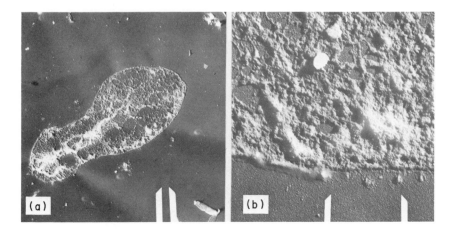

Fig. 19. Transmission electron micrographs of section of fiber cross-linked with dimethyloldihydroxyethyleneurea steam-cure method but with only 4% add-on ("limited" crosslinking): a. whole section residue; b. enlarged portion of section showing texture of swollen cellulose. U. S. Department of Agriculture photographs.

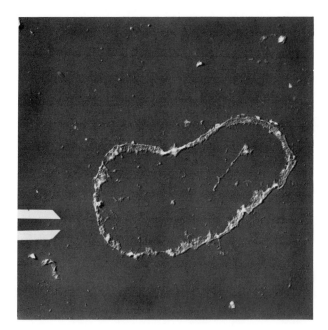

Fig. 20. Transmission electron micrograph of section of fiber cross-linked by the "wet-fix" method prior to curing steps: (note peripheral reaction). U. S. Department of Agriculture photographs.

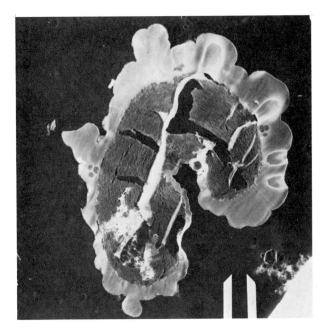

Fig. 21. Transmission electron micrograph of cross section of cotton fiber grafted with 1-3 butylene dimethacrylate in methanol by gamma radiation at 0.5×10^6 R. After 30-min immersion in cuene, and overnight extraction in methyl ethyl ketone, a heavy deposit remains on outside of fiber. U. S. Department of Agriculture photograph.

to form around the bundles (usually about 25 to 30 min after placing on the warming table) a needle is used to make a tiny opening near the edge of the coverslip, and a sturdy toothpick is run across the fiber bundles to remove the milky substance from under the coverslip by pressure. The polymer is then allowed to set up completely on the slide (about 2 h at 65° C). The slide is then cooled to room temperature and the embedding removed. This is trimmed with scissors until a small rectangle containing the fibers is left. The trimmed sample is placed on a clean slide, and the more viscous prepolymerized mixture of methyl and n-butyl methacrylates normally used for dry embeddings described in (e) is used. When the block begins to fuse into the syrupy polymer mixture, the slide is put on the warming table to polymerize at 65° C for 2 to 3 h. Cooled to room temperature, the assembly is separated as previously described by cooling the slide for release of the coverslip and of the embedded block. The block is trimmed with a razor blade under a widefield microscope, and the cutting face shaped into a pyramid, with the block clamped into the jaw of the microtome. Sections are cut as described in Section III Af and freed of their embedding polymers by slow solvent extraction with condensing vapors of a volatile solvent. The sections are then shadowed

and examined in the conventional way. In untreated cotton, or purified
fibers from other natural sources, the layering pattern is typically concentric (Fig. 22), and when the embedding polymer is dissolved out, the
unsupported layers often collapse sufficiently that they present a lateral
surface to view (Fig. 23). In these "tipped-over" layers one may observe
the microfibrillar pattern of the cellulose that composes each layer. In
fibers from fabrics which have a high dry wrinkle recovery angle, no layering results (Fig. 24); in fibers from mercerized fabrics the layering
is not concentric and the sections appear to have a spongy texture (Fig.
25). Variations of the layering pattern are expected from any prehistory
of drying or swelling. Still, in a series of treatments it is advantageous
to observe the effect of this "expansion" embedding to compare the extent
of resistance to separation exhibited by fibers from different treatments.
This expansion phenomenon is not completely understood, but an explana-
tion of the polymerization mechanism which governs it was offered by
Dlugosz [162]. Calling the phenomenon "polymerization swelling" he
stated that it occurs rather rapidly after a period of induction of several
hours during which no appreciable swelling can be detected.

The appearance of polymerization swelling coincided with a sharp
rise in the viscosity of the external medium. This suggests that the
polymerization swelling may be connected with the well-known auto-
acceleration effect [163]. It is well known that the bulk polymerization of methyl methacrylate proceeds initially as a first-order reaction; at about 25% conversion, when the viscosity of the system increases markedly, the reaction accelerates. The auto-acceleration
is believed to be brought about by a decrease in the termination rate
caused by the rise in viscosity of the polymerization medium. This
is supported by the finding of Tromsdorff [164] that the polymerization
is hastened if the viscosity of the monomer is increased by the dissolution in it of its polymer. Similar conditions exist in the medium
permeating the cotton fiber; the presence of cellulose, while not appreciably affecting the reactions of initiation and propagation impedes
the chain termination. Movement and growth of macroradicals are
restricted to the spaces that were originally accessible to water and
are now filled with monomer. A macroradical in the fiber is, to some
extent, isolated from other macroradicals of cellulose; the probability
of colliding and reacting with another macroradical is reduced and
hence its lifetime is increased, during which it continues to add on
monomer molecules and so grows longer. Thus the rate of polymerization is higher and auto-acceleration occurs earlier than in the external medium. In the system comprising a cotton fiber impregnated
with catalyzed methyl methacrylate monomer and immersed in it the
external medium is in communication with the medium in the fiber;
during polymerization therefore, a gradient of concentration of monomer

is developed across the boundary of the fiber. Monomer tends to diffuse through the outermost lamella into the medium in the first inter-
lameller space and to dilute it. This results in a swelling pressure
which causes the outermost lamella to expand and to separate from the
rest of the fiber. The tension in the expanding lamella splits it up.
The now open structure of the outermost lamella facilitates the diffusion of monomer. Consequently, a concentration gradient is set up
across the second lamella, which, in its turn expands and splits up.
In this way the dispersion of cellulose is propagated layer by layer to-
wards the lumen. The following factors are thought to contribute to
this state of affairs:

1. The medium within the expanding part of the fiber, containing
polymer of higher molecular weight and hence being more viscous,
polymerizes faster than the external medium.

2. The restricted mobility of the polymer molecules confined to
grow in the spaces that had been opened by swelling the fiber in water
may result in the existence of long-lived macroradicals. On being
brought into contact with the diffusing monomer, these trapped radicals
would resume their polymerization.

3. The viscoelastic flow of cellulose during polymerization swelling involves first the breaking of van der Waals and hydrogen bonds,
and as stress concentrations exceed chemical bond strength, probably
the scission of cellulose chains.

This disruptive tendency of the methacrylate embedding medium
is nevertheless useful in structural studies of materials lacking inherent contrast. The methacrylate embedding technique disperses
the structural building units of cellulose to a considerable extent, without altering their relative positions and thus provides an exploded view
of the fiber so that its architecture can be studied. In particular, the
technique sheds light on the problem of water accessibility; it shows
where in the fiber imbibed water goes. At the same time it reveals
the shape and size of cellulose crystallites since they do not swell with
water, water swelling being intercrystalline.

This expansion phenomenon is useful for comparing fibers from different textile finishing treatments in any given series, and it is construed to
reflect relative bond strengths in treated cellulose.

In the expanding technique, as a result of rapid polymerization of methacrylate which may have penetrated into natural cleavage areas of the fiber enlarged by swelling in water, the layers of cellulose within the fiber
wall are forced apart. In untreated cotton, the layering is typically extensive and uniform (Fig. 37). When the embedding polymer is dissolved
out, the unsupported layers often collapse sufficiently that they present
a lateral surface view; in these "tilted over" layers one can observe the
microfibrillar pattern of the cellulose that composes each layer (Fig. 23).

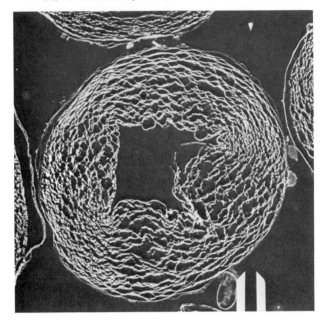

Fig. 22. Transmission electron micrograph of section of untreated cotton fiber by layer expansion technique. U. S. Department of Agriculture photograph.

Figure 25 shows the expansion pattern of fibers from cotton yarns mercerized under tension. Although expansion of the fiber occurred by the preparation technique, the lacunae are not so nearly concentric as in the untreated fiber of Fig. 22, and the general effect is of a spongy and randomly honeycombed structure. Individual fibrils are less easily seen in the enlarged view than in the sections of untreated fiber, and the partitions between capillary spaces appear several layers of fibrils thick. Figure 24 shows a fiber section from a sample treated with the crosslinking agent dimethylolethyleneurea (DMEU) to produce a permanent press fabric. It is obvious that crosslinking was effective here, for the fiber section was not separated into layers by the expansion technique. The method was extensively used by Rollins and co-workers [124-126, 138, 165-167] and by Nikonovich et al. [168] in the study of crosslinked cellulose; and by Razikov et al. [169] in the study of structure of cotton fibers of different maturities. Betrabet et al. [170-171] employed this technique for comparison of inter- and intracrystalline swelling of cotton fibers.

i. Staining

(1) Positive Stains. A major problem of examining organic matter in the electron microscope is that it consists mainly of light weight atoms

that scatter electrons weakly and so cannot be distinguished from the atoms that cause scattering in the support films or embedding media. This problem can be solved if the organic materials react with chemicals that contain heavy atoms; the heavy atoms enhance the visibility of organic materials in the electron microscope (positive staining). The most widely used electron stains are osmium tetroxide, uranyl acetate, and compounds containing lead, tungsten, or molybdenum. Unfortunately cellulose is not reactive with these compounds and positive staining is limited to cases in which reducing groups can attach metal ions. Positive staining was used by Müthlethaler who stained cellulose fibers with potassium dichromate $(K_2Cr_2O_7)$

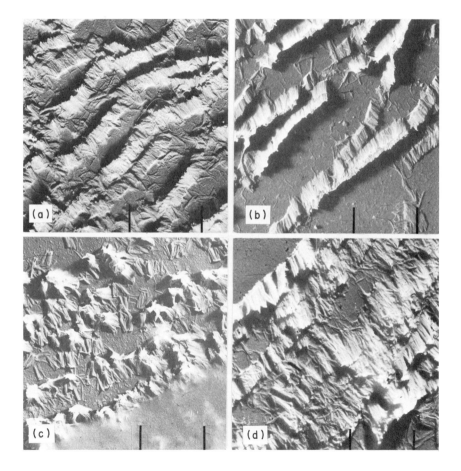

Fig. 23. Transmission electron micrographs of expansion patterns of native cellulose fibers: a. cotton; b. flax; c. hemp; d. ramie, U.S. Department of Agriculture photographs.

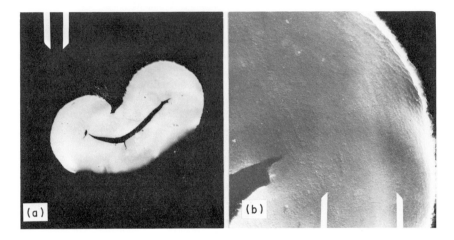

Fig. 24. Transmission electron micrographs of expansion pattern of cotton fiber crosslinked by conventional dimethylolethyleneurea process ($Zn(NO_3)_2$ catalyst) after commercial mercerization: a. whole fiber; b. enlarged portion of "a." U. S. Department of Agriculture photographs.

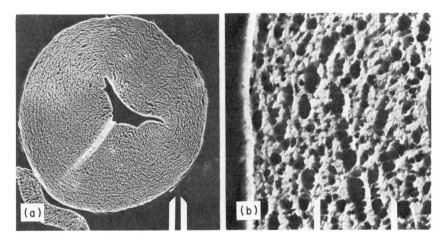

Fig. 25. Transmission electron micrographs of expansion pattern of tension-mercerized cotton: a. whole fiber; b. enlarged portion of "a." U. S. Department of Agriculture photographs.

32, 172 . Tripp et al. 173 stained cotton with Congo red (Direct Red 28; Color Index No. 22120) which does attach to cellulose, and subsequently reacted the Congo red with lead salts to indicate by electron microscopy the location of the stained area. More recently, Broughton et al. 174 reacted cotton with an acrylamide and subsequently stained the reaction product with osmium, indicating that the amide lay along the layers within the

cell wall structure. Hebert [175] reacted the cotton fiber with acrylic acid
to form an unsaturated ester which was subsequently osmium stained to
show dimensions of fibrils. Bredereck [176] stained with NaI longitudinal
sections of cellulose fibers crosslinked with n-methylochloroacetamide
and chloroacetylchloride.

(2) Negative Stains. Another way of enhancing visibility is the technique
of negative staining in which the specimen is surrounded by a deposit dried
down from a solution of an electron-dense chemical, such as phosphotungstic
acid, with which it does not react. The heavy metal cannot penetrate the
specimen but accentuates the outline of specimens of low scattering power
by entering into pores or openings, sharply delineating any surface irregularities [177-179]. The most commonly used negative stains are salts of
phosphotungstic acid (PTA) or of uranium. For successful staining of
cellulose, PTA is used at a pH of 7.0. Negative staining was extensively
used in determination of the width of elementary fibrils of cellulose [131,
172, 180-184]. Care should be taken not to have too thick a deposit around
the specimen or fine details will be masked out entirely. Negative staining
often provides information not obtainable by shadowing; the two methods
should be regarded as complementary. Boylston's review [185] of electron
microscopical stains for textiles summarizes the literature on staining of
fibrous materials to 1970.

Because of their crystalline nature, cellulose fibrils are not readily
stained. The problem is circumvented by modification of the cellulose and
staining of the reaction product. Kassenbeck and Hagege [186] oxidized
cotton and ramie fibers and stained them with lead citrate and uranyl acetate to investigate fibrillar characteristics; he stained soda cellulose by ex-
changing the sodium with cesium from cesium butyrate. He also modified
cotton and ramie by periodate oxidation followed by "fixation" in formaldehyde and treatment with thiosemicarbazide subsequently reduced with potassium borohydride; the final product was stained with silver nitrate to show
fibrils of 40 Å diameter unstained between stained portions of similar
dimensions. Sauer et al. [187] used samples crosslinked with 4,4'bis (beta-
sulfatoethylsulfonyl) diphenyldisulfide which he converted to a lead mercap-
tide to render it visible in the electron microscope. Mühlethaler [32] treated cotton fibers and bacterial cellulose with $K_2Cr_2O_7$ to demonstrate reactive spots along the faces of microfibrils. Silver nitrate was used by
Colvin [134] on oxidized fibers of ramie, bacterial cellulose and Valonia,
and by Kassenbeck and Hagege [186] on oxidized cotton. Hess et al. [188,
189] stained rayon fibers and purified wood fibers with iodine and concluded
that the regularly spaced stained spots along the length of the fibril are
regions of noncrystalline cellulose alternating at regular intervals with
crystalline cellulose; other workers attribute this periodicity to artifacts
created by degradation in the purification procedures used. To stain wood
fibers Asunmaa [190] used thallous ethylate, Heyn [18] uranyl acetate, and

Preston [24] silver nitrate; presumably the heavy metals attached to residual hemicelluloses.

Staining cellulose materials for electron microscopical evaluation can also be effected by depositing within the cellulose an alien substance which reacts with a heavy metal for contrast with the cellulose. Hagege et al. [191] included an unsaturated polymer (butadiene or isoprene) which was subsequently polymerized within the cellulose of both cotton and flax fibers. This was stained by immersion of the fibers in osmium textroxide solution to fix the heavy metal at the double bond of the included polymer. Tripp et al. [173] used osmium tetroxide to stain fibers modified with dimethylolethyleneurea to show that this reaction occurred within as well as between lamellae of the fiber wall. Broughton et al. [174] deposited acrylamide within the cellulose and stained it with osmium; Hebert [175], to measure fibril diameter in cotton fibers, reacted the cotton with acrylic acid which then reacted with osmium tetroxide at its double bond. Bates and Lyness [192] treated purified wood fibers with dimethylsulfoxide which they converted to the sulfone, a derivative that they thought would remain immobile through staining and embedding procedures and could be stained with osmium.

Negative staining is extensively used to determine dimensions of cellulose fibrils [130-132, 172, 180, 181, 184, 193-199]. Phosphotungstic acid, lead acetate, osmium tetroxide, and silver nitrate are employed. Phosphotungstic acid is normally used at pH 7.0 for untreated cellulose. Manley [183] used sodium phosphotungstate. Considerable discussion has been carried on with respect to accuracy of measurements on negatively stained fibrils, especially in preparations in which the solution creeps up the edges of the specimen. Comparisons with shadow-cast specimens are appropriate in all cases.

j. Freeze Etching. In the freeze etching technique the tissue sample is flash frozen and either cut or fractured to expose an internal surface [200] . Then it is warmed slightly in a vacuum so that water vapor sublimes from the ice in the tissues to produce an etching effect on the exposed surface that reveals details of structure. For reproducibility, this step must be done under highly controlled conditions. A replica is made of the surface by evaporating carbon in vacuum onto the frozen or slightly thawed fractured surface. The resulting layer of carbon, a few hundred angstroms thick, is a faithful model of the original fractured surface and can be detached by dissolving the tissue chemically or thawing it until it collapses. The replica is supported on a mesh grid and examined in the transmission miscoscope. Freeze etching, introduced by Steere [200] , and used by Bullivant and Ames [201] and Moor et al. [202, 203, 204] has mostly been applied in the study of unicellular organisms, but may have some applicability to cellulose samples, especially in connection with scanning microscopy. McAlear et al. [205, 206] and Kay [89] have excellent descriptions of the method.

k. Ionic Etching. Gas discharge etching is a possible technique for differentiation of components of polymer mixtures not otherwise readily distinguishable. This technique (also called "ion bombardment") makes it possible to reveal fine details which might be obliterated by conventional chemical deep-etching techniques. Sections, or whole fibers, are subjected to direct current discharge of 3 mA for 3 min in oxygen. The specimen is then shadowed with platinum and examined in the transmission microscope. Anderson and Holland [207] in etching synthetic fibers found a clear etch was obtained by using a constantly changing atmosphere of argon at pressures of 1 and 2 cm mercury. Spit [208-210] has used ion etching on cotton fibers and Kassenbeck [211, 212] on rayon. Their experiments indicate that the most accessible regions are affected first, and, after etching, the homogeneous regions appear discrete, as if composed of separate systems. Other reports of ionic etching of cellulose fibers are by Jakopic [213], Dlugosz [214], Nikonovich and co-workers [215], and Peck and Carter[216]. The results have been difficult to interpret.

l. Autoradiography. Autoradiography involves incorporation of a radioactive material into a specimen and its subsequent detection by exposure of a photographic emulsion to the radioactive specimen [217]. The radioactive specimen is coated with photographic emulsion and kept in the dark during a suitable exposure interval, during which time the radioactive material decays and the resulting emission of ionizing radiation produces a latent image in the emulsion. After photographic processing the developed image appears as an accumulation of silver grains in those areas where the radioactivity was localized. By means of electron autoradiography, the functions of various cell organelles have been determined. In connection with cellulose synthesis in living cells, some experiments in autoradiography combined with lead staining of ultrathin sections of tissue have been enlightening.
Methods for electron autoradiography have been described by Saltpeter and Bachmann [218], Caro and Van Tubergin [219, 220], Peters and Ashley [221], and Hay and Revel [222]. Paul et al. [223] used autoradiography in the investigation of sizing on cotton fiber, but applications to chemical modifications of cellulose have not been reported, although this approach might well prove instructive. Northcote and Pickett-Heaps [224], by incorporation of tritiated glucose into polysaccharides of actively growing cells of plant tissue and subsequent examination by autoradiography of sections of the cells, determined that polysaccharide materials were formed within the Golgi apparatus and were stored in vesicles for transporation. Wooding [225] by chemical and autoradiographic studies on stems of sycamore seedlings confirmed these findings and indicated that incorporation of materials into the cell walls was simultaneous at all points along the wall.

B. Scanning Electron Microscope

1. Microscope Operation

Operation of the scanning electron microscope is perhaps less taxing
than that of the transmission instrument, but for optimum results considerable attention must be paid to such details as angle of specimen to beam,
beam intensity, and critical focusing. Handbooks supplied by the manufacturer should be consulted for details of instrument operation. In the most
advanced scanning microscopes pump-down cycle is automatic, and electronics for visual inspection almost so. X and Y translation, X-Y-Z tilt,
and rotate motions of the specimen stage are commercially available, and
specimen stages specially wired for heating and cooling and for stretching
the specimen during examination can be made or bought. Selection of operating parameters, such as accelerating voltage and scanning rate, are
dictated by the type of specimen under observation. Magnification is simply
the ratio of the size of the display area on the cathode-ray tube to the distance the probe (electron beam) is scanned. The minimum magnification
obtainable is between 10X and 15X, which conveniently overlaps the range
of low magnification of the light microscope. The limit of maximum magnification is dependent on the resolving power of the particular instrument
used and is of the order of 200,000X. Most instruments have a specimen
air lock to permit rapid specimen exchange and to maintain cleanliness of
the column. When the probe is properly focused on the specimen, changing
magnifications does not require any change in focus. Specimens can be as
large as 1 cm in each dimension, and specimen preparation is relatively
simple. Choice of arrangement of specimen with respect to the beam is
dependent on the type of specimen and the information desired. Most SEM
pictures of cellulosic materials have been taken with the sample mounted at
45° to the electron beam, but other angles may prove advantageous for
specific purposes.

2. Specimen Preparation

Preparation of the sample for SEM, though not as complicated as preparation of ultrathin sections or replicas for the transmission microscope,
still presents some difficulty. For textiles, the yarn, fiber, or fabric is
attached to the specimen stub, usually with double-coated adhesive cellophane tape, and then vacuum coated with gold-palladium alloy to a thickness
of 100-500 Å. In vacuum, cellulose becomes a very good insulator. The
specimen must thus be provided with a conducting surface to bleed off the
static charge which builds up when the dry cellulose is bombarded with
electrons. This must be a continuous surface; if there is a crack in the

conducting surface static charging prevents focus on that portion of the specimen. The gold coating must be grounded to drain the electric charge; unfortunately most cellulosic samples are multistructured and incapable of being grounded entirely. Since the specimen normally used in the scanning microscope can be half an inch square, the necessity of having an unbroken conducting surface creates certain problems, especially with textile fabrics in which weave or knitting pattern is especially irregular. If the sample is not completely coated with a conducting material it builds up a charge either immediately or within a very short time of the electron bombardment used for SEM examination. The image as seen on the viewing screen may then appear white, blank, or streaked; fine detail is absent or fades out during viewing. A prime difficulty with cellulose samples is that not only do they become insulators in a vacuum, but also that fibrous materials are flexible so that almost any movement of the specimen can crack the conducting film.

It is customary in evaporating gold or gold-palladium to rotate the specimen stage, and, at the same time, to operate it on a tilt of about 30° so that all surfaces of the specimen become adequately and uniformly coated. If this feature of a rotating, tilting state is not available, various devices can be made for achieving a uniform coating. One way to overcome the cracking of the conducting film is to reduce the primary electron beam so that undue charging is prevented. In practice, the lower the atomic number of the elements in the sample, the lower the kilovolts which must be used. Also, if the sample has rough edges, it may be necessary to use a low kilovoltage. Cotton fabrics are usually examined at a low voltage of not more than 5 kV. However, in reducing voltage the resolution is compromised.

In general, antistat coatings are partially successful in supplanting metal coatings as charge drainers. Commercial aerosol antistat materials which yield acceptable results at relatively low magnifications are "Duron," "Statikil," "Neutro-stat," and Triton X-100 [226]. They are useful for specific power levels. A 2% solution of the household detergent "Joy" in isopropanol is especially recommended [227]. The useful life of coatings in the SEM is limited, depending on the particular combination of metal and plastic. The antistat coating may be effective only to about 5- or 10-kV power levels for a 1- to 10-min scan time. Beyond this approximate level, the coating fails to perform its function fully. Antistat coatings are easy to remove; metal coatings are impossible.

Typical applications of scanning electron microscopy to textile research are illustrated in photographs of cotton. Figure 26 shows characteristic positions of cotton fibers in the yarns of a print cloth. Figure 27 is a detailed view of a typical convolution along the length of a cotton fiber. Figure 28 compares the effects of washer damage and dryer damage in laundering. The penetration of coatings into yarns is illustrated in Fig. 29, a photograph of the edge of a typical awning fabric. Important to the strength of coated fabrics is freedom of movement of fibers within the yarns.

Fig. 26. Scanning electron micrograph of surface of a scoured and bleached print cloth at 64° tilt, showing arrangement of cotton fibers in both warp and filling yarns. U. S. Department of Agriculture photograph.

IV. APPLICATION TO CELLULOSE

A. Bacterial Cellulose

The bacterium Acetobacter xylinum produces cellulose in a form especially easy to investigate. The mats of cells which form in a culture of this organism are examined by merely drying them onto a coated specimen grid, followed by appropriate staining or shadowing for contrast. Usually, the specimen is washed to remove free bacterial cells, and the mats are dispersed in a laboratory blender or by ultrasonic radiation to obtain specimens thin enough for optimum viewing. Diameters of individual filaments of bacterial cellulose have been measured by Mühlethaler [121, 228], Rånby [229], Frey-Wyssling and Mühlethaler [230], Hestrin and Schramm [231], Glaser [232], Colvin et al. [233], Jayme and Hunger [234], Ohad et al. [130, 131], Colvin [134], and Manley [183].

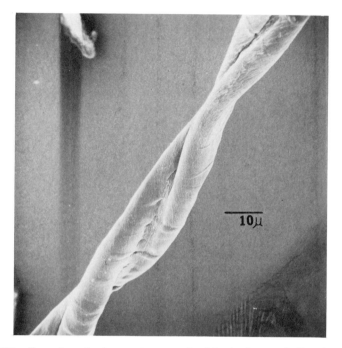

Fig. 27.   Scanning electron micrograph of typical convolution in cotton
fiber.   U. S. Department of Agriculture photograph.

B.   Algal Cellulose

Valonia ventricosa, a species of alga consisting of relatively large spherical cells, was much studied by Preston et al. [235, 236]; Rånby [237], and
Steward and Mühlethaler [238].   According to findings of Preston the elementary fibrils are large, ranging from 200 to 300 Å and averaging about 250 Å
in width.   However, Frey-Wyssling et al. [35] found that by ultrasonic degradation, Valonia fibrils develop slip planes in which elementary fibrils
of 35 Å diameter could be measured, and Wardrop and Jutte [239] found that
upon enzyme degradation of Valonia the microfibrils tend to separate into
elementary fibrils and to become oblique pointed at their ends.   The effect
of wet heat and of mercerization on the fibrillar structure of Valonia was
described by Okajima and Kai [240].
    Other algae in which cellulose has been studied microscopically are
Ulva lactuca and Enteromorpha [241], Nitella axilaris [242], Griffithsia
flosculosa [243] .

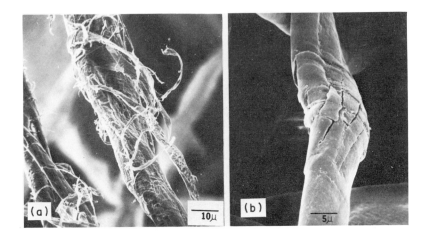

Fig. 28.   Scanning electron micrographs of cotton fibers from a laundering experiment: a. typically damaged fiber from repeated washing without drying; b. typically damaged fiber from repeated drying without washing. U. S. Department of Agriculture photographs.

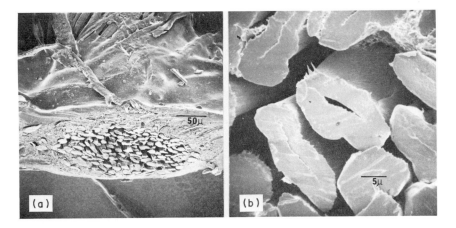

Fig. 29.   Scanning electron micrographs of a vinyl-coated awning fabric: a. edge view showing penetration of polymer coating into yarns; b. enlarged view of yarn cross section showing freedom of individual fibers essential to fabric flexibility. U. S. Department of Agriculture photographs.

C. Cotton Fibers

Early electron micrographs of cotton fibers by Ruska and Kretschmer [244], Eisenhut and Kuhn [245], Barnes and Burton [246], Kinsinger and Hock [247], and Rånby [229, 237] showed fibrillar structure in thin fragments obtained by beating or ultrasonic degradation and in surface replicas of thick specimens. It was apparent that cellulose fibers break into microfibrils only in the presence of polar liquids. A series of papers by Kling and Mahl [248-254] explored the organization of the cotton fiber and the effects upon it of laundering practices. Rollins, Tripp, and co-authors [108, 109, 255, 256] investigated cell wall organization and showed the effects of purification processes such as alcohol extraction, scouring, kiering, and bleaching on the fiber cell wall. Both groups made extensive use of replication and sectioning techniques.

Surface topography of the cotton fiber by the replication technique reveals a system of roughly parallel ridges and grooves spiraling around the fiber at an angle of about 20-30° to the fiber axis. The average height and distance between ridges is approximately 0.5μ, and many of the ridges are 10 or more microns in length. The distribution is so nonuniform, however, that this feature cannot be considered a criterion for identification of variety or species. Electron micrographs of surface replicas of cotton fibers from fresh bolls showed [105] that the primary wall offers a smooth, structureless appearance on the undried fiber but becomes wrinkled, fluted, and corrugated on the mature fiber in the dried state, giving rise to the characteristic rugosities noted above. When freed from the fiber by beating in water, the primary wall unfolds to form a smooth, flat sleeve much greater in width than the fiber itself, presumably approximately the size and shape of the primary wall in the growing state before initial drying and shrinkage of the fiber.

Microchemical analyses of primary wall sheets stripped from the fiber and separated from secondary wall [256] showed that most of the noncellulosic constituents of the fiber are concentrated in the primary wall. Electron microscopical observation of purified wall specimens revealed that the cellulosic portion is a network of microfibrils interlaced in a fabric in which the general system of orientation is axial on the outer face and transverse on the inner face [108]. The same scheme of fibrillar arrangement was observed in primary wall from other plant sources such as oat coleoptile [257], eucalyptus wood [258], root hairs of corn [259], moss protonema [260]. Both size and number of microfibrils appear to increase with increased age of the primary wall. This was observed in cotton by Tripp et al. [256] and O'Kelley [261], Frey-Wyssling and Stecher [262], and Roelofsen and Houwink [263] and in moss by Gunther [260].

The effects on the primary wall of commercial purification treatments such as scouring in dilute sodium hydroxide, bleaching in hypochlorite or

Fig. 30. Surface replicas of cotton fiber: a. after alcohol extraction; b. "a" after alkali extraction. Reproduced by courtesy of H. Dolmetsch.

in hydrogen peroxide were investigated by Tripp et al. [108] and by Kling and Mahl [249] for cotton. Similar electron micrographs illustrating the appearance of the primary wall network on the surface of wood fibers after digestion in simulated paper mill operations were published by Jurbergs [116] and by Jayme and Hunger [264] and Dolmetsch and Dolmetsch [265]. A detailed study of the effects of each step of purification of the cotton fiber primary wall by alcohol extraction and by kiering was shown by Dolmetsch and Dolmetsch [266] (Fig. 30).

The separation of the cotton fiber cell wall into concentric layers was demonstrated by Castiaux et al.[267], Kling [251, 269], Tripp et al. [173, 255], Dlugosz [162], Rollins et al. [138, 165], Dolmetsch and Dolmetsch [268], Nikonovich et al. [270], Razikov et al. [169], and Betrabet et al. [170, 171]. The layers are not normally seen in cross section of fibers embedded dry. The expansion into layers is a reflection of the amount of pressure exerted by the polymerization as the polymer builds up in spaces within the fiber initially enlarged by water or other polar liquid.

Individual layers are composed of bundles of microfibrils in parallel array. The outermost layer is clearly the multinet structure of the primary wall. Just beneath the primary wall is a transition layer commonly referred to as the first layer (S-1 layer) of the secondary wall. Hock [271] called it the "winding layer," and this terminology has persisted. It is frequently observed in specimens beaten in water and is distinctly different in arrangement from either the fibrillar network of the primary wall or the parallelism of the fibrillar cellulose of the secondary wall. The S-1 layer appears to consist of alternating bands of coarse transverse fibrils loosely arranged, and closely packed parallel fibrils (Fig. 31). It has been clearly pictured in other cell walls by Bosshard [272], Hodge and Wardrop [273], and Meier [274].

D. Bast and Leaf Fibers

Of the natural fibers, cotton and wood claim the major attention of investigators; bast fibers have been studied comparatively little. However, flax, ramie, jute, and hemp also played parts in the development of the electron microscopy of cellulose.

1. Flax

The fine structure of the flax examined in 1940 [275] revealed microfibrils comparable in appearance to those of cotton and wood. Mühlethaler [121, 228] and Wardrop [276] also examined the morphology of flax fibers. Wardrop stained thin fragments with gold chloride and found microfibrils of 70 to 100 Å. More recently, Nettelnstroth [277] studied details of the morphology of the linen fiber, and Lambtinou [278] considered changes induced by finishing. Shekhterman and co-workers [279] investigated the effects of structural changes induced by purification techniques on the rate of copolymerization of the cellulose with acrylonitrile.

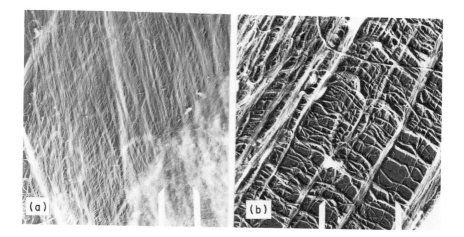

Fig. 31. Fragments of cotton fiber S-1 layer: a. with overlying primary wall; b. detail of S-1 fibrillar structure. U.S. Department of Agriculture photographs.

2. Ramie

Ramie has received more attention than the other bast fibers. Wergin [280] measured unshadowed microfibrils from ramie fibers at 80 to 100 Å. Wuhrman et al. [281] shadowed microfibrils obtained by sonication and obtained a 60-Å width. Eisenhut and Kuhn [245], without shadowing, reported widths of 100 to 200 Å, but Hermans and Weidinger [282] reported a range of 50 to 700 Å. Kinsinger and Hock [247] shadowed specimens obtained by beating and measured microfibrils of 370 Å, but Mühlethaler [121, 228] obtained 250 Å and Vogel [283] 173 to 203 Å, which evidently reflected the extent of fibrillation in beating. Heyn [180] more recently obtained values of 48 and 36 Å for ramie elementary fibrils in wet and dry states.

3. Jute

Mukherjee and Woods [284] found jute fibrils of 130 Å, but Heyn recently [180] measured elementary fibrils of jute at 28 Å. Mukherjee et al. [285] determined micelle dimensions after acid hydrolysis for jute, hemp, ramie, cotton, and Fortisan.

4. Sisal

Sisal fibers disintegrated in a Waring blender by Mühlethaler [121] were measured as early as 1949. Generally, the more commercially valuable fibers received the most attention in microscopical studies.

E. Wood Fibers

The electron microscopy of wood during the period 1950 to 1960 was ably reviewed by Liese and Coté [22a]. For the most part, replica techniques were used, and the investigations dealt with cell wall pits and pit membranes (Fig. 21). The submicroscopic sturcture of cell walls was also studied by fibrillation and investigations of the effects of paper beating processes and of chemical procedures are numerous. Early work included studies of wood morphology by Bosshard [272], Hodge and Wardrop [273], Wardrop and Dadswell [286], Ribi [287], Isenberg [288], Asunmaa et al. [289, 290], Asplund [291], Coté [292], and Liese [293]. More recently Coté and co-workers studied with the transmission electron microscope the pit structures of coniferous wood [294], vestured pits in dicotyledonous tissues [295], morphology of heartwood in white birch in which the gums appear to correspond to tyloses in other species of wood [296], and the structure of gelatinous fiber [297]. The anatomy of reaction wood as revealed by transmission microscopy was discussed by Casperson [298], and scanning microscope pictures of reaction wood were provided by Scurfield and Silva [299]. Heyn [181] used ultrathin sections of wood fibers negatively stained "in the natural swelling state" to show elementary fibrils of 35-Å width and assumed that interfaces between elementary fibrils are formed by less-ordered cellulose chains and by polyose chains. Sullivan and Sacks [300] and Sullivan [182] presented new measurements of dimensions of elementary fibrils ("protofibrils") with evidence from negatively stained preparations that fibril widths vary considerably from species to species with a range of from 38 Å for Ponderosa pine to 26 Å for red oak. Dunning [117] by an ingenious peeling technique and a remarkable exhibition of dexterity investigated the architecture of a pine fiber by hand-stripping the single fiber layer by layer and replicating the peeled area to show fibrillar relationships in different structural regions of the fiber. He examined the late wood of longleaf pine and its intertracheid and pit membranes. Thomas and Nicholas [301] also investigated the ultrastructure of bordered-pit membranes in southern pine. A spectacularly beautiful album of electron micrographs is presented in the Coté-Day brochure, "Wood Ultrastructure of the Southern Yellow Pine" [302]. Sacks and co-workers [303] investigated lignin distribution in cell walls, and Wooding [225] used autoradiographic techniques to investigate incorporation of chemicals into sycamore vascular tissue. In studies of microbial attack in wood decay, Cowling [304] discussed features of cellulose that influence susceptibility to enzymatic hydrolysis and proposed the use of microbial systems as selective tools in wood anatomy [305]. Koran [306] presented photographs of tensile failures in black spruce wood at various temperatures by replicas of tangential tracheid surfaces. Bates and Lyness [192] in studies of the morphology of purified fibers made cross sections of fibers swollen in dimethylsulfoxide prior to staining with osmium tetroxide; they suggested that the DMSO was converted to sulfone and thus became an insoluble inclusion in the structure indicating permeability sites. Their interpretation of the photographs is probably open to question. Norberg investigated by freeze–drying the details of wet fiber surfaces in paper wood [119]; by ultrathin sections morphological changes in fibers from sulfite and sulfate cooks [118], and, by both sectioning and freeze replication techniques, the effects of drying on cellulose fibers [120].

F. Paper Pulp

Perhaps the earliest instance of electron microscopy of paper was reported in 1942 by Sears and Kregel [307] who discussed applications of this new tool to problems of the pulp and paperboard industry. The use of reflection electron microscopy on surfaces of paper and of individual fibers was described in a preliminary paper by Emerton et al. in 1954 [68] and a series of other papers on their method of examination followed [69, 72-75, 308, 309]. However, the distortion of perspective and foreshortening of the image made these pictures difficult to interpret, and the advent of the scanning microscope better fulfilled the need in surface studies of this type.

The hornification of paper pulp under different conditions of drying was investigated by Jayme and Hunger [264] by transmission electron microscope studies of replicas. Asunmaa and Steenberg [289] discussed the characteristics of beaten pulp, and the microscopy of fiber-to-fiber bonds in paper. Page and Emerton [310] described the general microscopic examination of paper, board, and wood, and Emerton [71] further elaborated on fundamentals of the beating process. McIntosh [311] studied the effect of refining on the structure of the fiber wall, and Casperson et al. [312] discussed difference between pulps from alkaline and acid cooks of reaction wood from poplar. Wardrop [313, 314] described morphological features of wood involved in pulping and beating for paper manufacture. Degradation of the fiber from the lumen toward the outside wall was described by Lantican et al. [315] in exploration of the possibility of using ozone gas as a pulping medium. Jayme and Bergh [316] determined that in sulfite wood pulp the middle lamella shows evidence of chemical attack by the sulfite cooking, but that in sulfate pulp the attack, though of the same type, is only slight. Polcin and Karhanek [317] used both cross sections and surface replicas to investigate the effect on spruce sulfite and sulfate fibers, as well as on native cotton, of swelling in such agents as water, sodium hydroxide cuprammonium hydroxide, cupriethylenediamine hydroxide, cadoxen, iron tartrate, and concentrated phosphoric acid. They found marked differences between cotton and wood, but also considerable difference in the response of sulfate and sulfite fibers. The swelling agents were extracted from swollen samples by solvent exchange through methanol, benzene, and ether. Sections were made in aqueous glycol methacrylate containing 3 to 5% methyl methacrylate monomer.

G. Regenerated Cellulose (Rayon)

The application of electron microscopy to the study of rayon is almost as extensive as that on cotton. Studies of Guthrie [318] and Sikorski [319] report the most significant earlier work. An excellent review of Jayme and Balser [320] discussed the effect of drawing speed, and of concentration of spinning bath as ably demonstrated by Kassenbeck [321-325] and by Morehead [135]. The rayon process was investigated more recently and completely by Gröbe and co-workers [107, 326-339].

Some of the earliest investigations of textiles with the electron microscope showed the fibrillation of viscose and cuprammonium rayon fibers by mechanical agitation in water [340, 341] and demonstrated with replicas the network patterns on filament surfaces. Morehead [135], Mühlethaler[342], Frey-Wyssling [343], and Kassenbeck [321], exploring the morphology of rayon by fragmentation and replica techniques, demonstrated the effect of solution characteristics and of processing variables on fibrillation and surface striations. In 1955 Kobayashi and Utsumi [344] described the characteristics of regenerated cellulose micelles and microfibrils. Kassenbeck [322-325] published electron micrographs of ultrathin sections of viscose rayon indicating that the highest packing density lay in the cuticle region of the fiber. Rånby et al. [345] reported a thorough investigation of the viscose process; and Samuelson et al. [346] showed the fibrillar structure of viscose model fibers derived from saponified cellulose acetate. Spit [347] studied the voids in viscose rayon using Araldite instead of methacrylate for embedding to avoid the effects of "explosion polymerization." He also used replicas to study filament surfaces. Dlugosz and Michie [123] used electron microscopy to study the effect of acid swelling on the fine structure of regenerated cellulose. In the same year Sisson [348] published on the skin and core structure of cellulose films, describing crystalline and amorphous regions about 104 Å in width in the skin as compared to 206 Å for those in the core, and Sperling and Easterwood [349] investigated the chemical and physical nature of insolubles in acetone solutions of cellulose acetate and found many to be identifiable fragments of fiber walls. The year 1962 saw the first paper [326] in the extensive and excellent series on the electron microscopy of rayons by Gröbe and co-workers at the Institute of Fiber, Research of the German Academy of Science, Berlin. Drisch and Priou [350] published electron micrographs of hydrolyzed polynosic yarns. Morehead [351] described the various steps in the viscose reaction, and Kling et al. [352] showed surface replicas, fibrillation patterns, and cross sections of modified fibers including high tenacity, high wet-modulus, and polynosic fibers. Kaeppner [353] published a comparison of the electron microscopical characteristics of Fortisan, a high wet-modulus rayon, and a high tenacity tire yarn with especial emphasis on differences in surface characteristics as evidence of an underlying fibrillar structure difference. Kaeppner [354] also studied voids in viscose tire yards.

The year 1964 saw another paper in the series by Maron et al. [328] , and in 1967 Jayme and Balser [355] reported on electron optical studies of viscose films and on comparative properties of high wet-modulus rayon and other regenerated cellulose fibers [320]. In the same year Balcerzyk and Wloderski [356] investigated viscose fiber structure by hydrolysis and by replicas of fractured surfaces. Classical viscose, polynosic, and cord fiber were photographed. They reported that high modulus fibers had no distinct skin region, and apparently contained perfectly oriented fibrils. A paper by Nikonovich and co-workers in 1968 [357] reported studies on supramolecular structure of hydrate cellulose cord. The year 1968 also saw 6 papers from the group at the Textile Research Institute in Berlin on experimental viscose rayon filaments [107, 329-333]. Of particular interest are two papers [107, 329] describing a method for the preparation of surface replicas of wet fibers.

The structure of cellulose gels has recently come into consideration. Coalson et al. [358] discussed their structure by both light scattering and thin section microscopy, and Boman and Treiber [359, 360] used density gradient and freeze-drying techniques to investigate gel particle size. Gröbe and Gensrich [335] and Gröbe [334] reported on ultrathin sectioning of regenerated cellulose fibers and summarized findings on the formation of surface and internal structure during the coagulation process. This latter paper reported model experiments to explore sol/gel transformations of viscose and described electron microscopical techniques used, including scanning electron microscopy, ultrathin sectioning, replication, and freeze-etching.

H. The Architecture of Cell Walls

The electron microscope permits elucidation of the organization of naturally occurring cellulose. The gross morphology of plant cell walls appears to follow a general architectural plan in cells from a variety of tissues of widely separated families of plants (algae, wood, bast fibers, seed hairs). In general, the wall consists of:

Cuticle
Primary wall
Secondary wall
Tertiary wall
Lumen

1. Cuticle

The cuticle is the outermost layer of the cell wall of independent cells like seed hairs and contains wax and pectinaceous materials. In a stricter sense, the cuticle is the thin continuous sheet of wax which covers the outside of the fiber, the other noncellulosic materials being considered a part of the primary wall. On hairs such as cotton, kapok, and milkweed floss, the noncellulosic materials correspond to the middle lamellae of cells of meristematic or xylem tissues.

2. Primary Wall

An interlaced network of cellulose microfibrils is universally characteristic of the structure of primary walls from any source. The literature contains almost identical photographs of open "sleeves" of primary wall from such diverse sources as oat coleoptile [257, 273], cotton seed hairs [108], kapok and milkweed floss [263], pine tracheids [258], root tip and parenchyma cells of onion [259], corn roots and stems of the marsh reed, Juncus [259], sisal leaf fibers [121], and moss protonema [260]. With increase in age of primary wall, both size and number of microfibrils increase. The microfibrils on the inner face of the primary wall are transverse to those on the outer face of this sheet of cellulose.

3. Secondary Wall

In the secondary wall, which constitutes 90% or more of the weight of cell walls, the microfibrillar arrangement is closely appressed and highly parallel, occurring in sheets of tightly packed cellulose filaments which appear to exist in more or less concentric layers. The forces bonding microfibrils together in a layer are greater than the forces holding adjacent layers together. Individual layers are not cleanly discrete, but appear to coalesce at random. In an unswollen sample no evidence of layering can be detected, so tightly are the microfibrils bonded one to the other, but upon mechanical agitation after swelling in water, the lines of weakness appear localized between layers of sheets of microfibrils. Microfibrils are threadlike units 50 to 250 Å wide and of indeterminate length, reaching perhaps 50,000 Å or more. A widely held concept is that these microfibrils are fasciates of elementary fibrils whose dimensions are 35–40 Å on each side. The "S-1 layer" of the secondary wall is a transition layer just beneath the primary wall, but entirely different from either the fine-textured primary wall network (Fig. 32) or the parallel fibrillar arrangement of the subsequent "S-2 layer" of the secondary wall (Fig. 33). The "S-1 layer"

(Fig. 31) appears to be a continuous sheet less than 5000 Å thick made up of alternating bands of fibril bundles which spiral about the fiber at an angle of 45 to 70° to the fiber axis. The wider bands have a closely packed parallel arrangement of fibrils while the narrower bands which connect them consist of two systems of fibrils at right angles to each other in an open weave pattern. The electron microscope picture of the "S-1 layer" isolated by Rollins and Tripp [109] from a 17-day old cotton fiber is virtually identical with the pictures of the "S-1 layer" shown by Hodge and Wardrop [273] for eucalyptus wood and for oat coleoptile; by Emerton [361] for pine, and by Hodge and Wardrop [362] for Douglas fir.

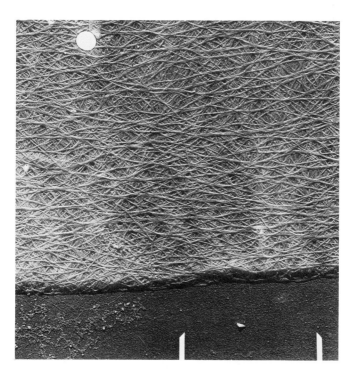

Fig. 32. Isolated fragment of cotton fiber primary wall purified to show three-directional network of microfibrils. U. S. Department of Agriculture photograph.

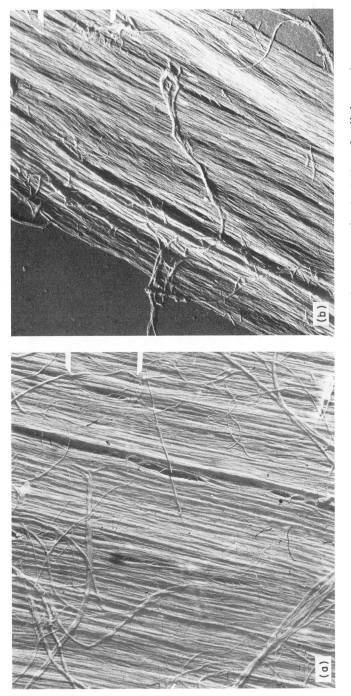

Fig. 33. Enlarged view of secondary wall fragments showing character and orientation of cellulose micro-fibrils: a. flax; b. cotton. U. S. Department of Agriculture photographs.

4. Tertiary Wall

The tertiary wall which lines the lumen apparently has a winding pattern similar to that of the "S-1 layer," but this phenomenon is often not visible in cotton. Bucher [363] showed it in both conifers and hardwoods. Meier [364] noted that a conspicuous helical tertiary wall is prominent in tension wood, but missing from compression wood.

5. Lumen

The lumen or central canal is the hollow space within the collapsed tube of the dried fiber; it often contains dried protoplasmic residue.

A consideration of this architectural pattern of the plant cell wall is of interest to the paper and paperboard industries and to both the cordage and the textile industries. Chemical treatments to enhance the performance of commercial products depend upon the response of the cell wall to the individual reagents and prevailing conditions of the process. The contribution of the techniques of electron microscopy toward an appreciation of the problem involved is recognizable.

I. Current Concepts of the Cellulose Fibril

Much effort is expended in attempts to characterize the primary unit of native cellulose. Early electron microscopy of cellulose consisted of fibrillation studies of fibers disintegrated by ultrasonic agitation or by beating in a Waring blender [235, 246, 247, 365]. At that time there was question as to whether the fine threads observed were natural entities or merely the results of mechanical force. When bacterial cellulose from Acetobacter xylinum was studied [121] it became apparent that the cellulose microfibril is a unique building element produced by biological growth.

In the period 1967 to 1970 a plethora of papers appeared discussing the fibrillar structure of cellulose: Mühlethaler [32, 33], Frey-Wyssling [36], Preston et al. [31, 366], Rånby and Rydolm [23], Jayme and Balser [29], Jeffries et al. [367], Fengel [34], and Peterlin and Ingram [30]. Unfortunately through the past 30 years, the terminology used to describe the macromolecular aggregates in plant cell walls was rather loosely applied, creating confusion in the literature. The following general meanings have evolved:

The plant cell wall is made up of thread-like macrofibrils which often can be seen with the light microscope, especially in highly swollen specimens; these range from 2000 to 4000 Å in thickness. Subunits of these macrofibrils are microfibrils, 100 to 250 Å wide, 50 to 125 Å thick, and apparently endless in length. These are the units most frequently seen in electron micrographs of mechanically fibrillated cellulose materials which have been shadow-cast or negatively stained to enhance contrast in the instrument. Subunits of the microfibrils were identified by Frey-Wyssling and Mühlethaler [172] as elementary fibrils seen both by negative staining methods and by shadow-casting and measured at a statistically determined average diameter of 30 to 40 Å. These elementary fibrils apparently readily fasciate by hydrogen bonding to form strap-shaped microfibrils.

In the literature on cell walls, values ranging from 60 Å to 600 Å in width were reported for microfibrils (see tables in reviews by Warwicker et al. [27] and Usmanov and Nikonovich [16]), but it is generally accepted at present that microfibrils are about 100 to 300 Å in width, half as thick, several microns in length, and can be regarded as morphological units of plant cell walls. Ohad et al. [132] corrected measurements of shadow-cast fibrils for shadow height by plotting the measured image width against the sine of the angle between the long axis of the fibril and the direction of shadow and extrapolated to zero angle. He found that the elementary fibrils of cotton and of bacterial cellulose have a real width of 30 Å. The fibrils measured by either this corrected shadow technique or by negative staining have a height of 20 Å. In the negatively stained material the 30-Å dimension is predominant, and a few fibrils are up to 40 Å wide. It follows that the elementary fibril is of rectangular cross section about 20 x 30 Å. Frey-Wyssling and Mühlethaler [172] reported that bacterial cellulose and primary wall cellulose from oat coleoptile consist of 35-Å elementary fibrils which behave as rather brittle crystalline needles. They confirmed the size of the elementary fibril both by positive staining with $K_2Cr_2O_7$ and by negative staining with phosphotungstic acid. Heyn [194] found that lateral width of microcrystallites ranges from 30 to 60 Å for elementary fibrils in ultrathin sections of cotton by negative staining with phosphotungstic acid at pH 7.0. He commented on a beadlike appearance which he considered suggested a twisted state in microfibrils. Heyn [180] also measured fibrils of cotton at 35 Å, jute 28 Å, and ramie 36 Å dry, 48 Å wet, in negatively stained ultrathin sections. In 1969 he reported a 35-Å width for the negatively stained fibrils in sections of wood "in their native swollen state" [181]. Sullivan [182] had determined the range of fibrillar widths in various woods and reported statistical means of 28 Å for red oak and 38 Å for pine, with characteristic widths for various pines. Hebert [175] reacting cotton with acrylaminde, and staining with osmium tetroxide, confirmed the 35-Å width of the fibrils. Morosoff and Ingram [368], however, determined a 50-Å diameter by negative staining, and Jeffries and co-workers [367] indicated that fibrils of cotton are "less than 75 Å." Nieduszynski and Preston [366] still maintain that in Valonia the native minimum dimension is in excess of 200 Å.

In the assessment of fibril dimensions consideration of the following limitations of electron microscopy are appropriate:

Calibration of the transmission electron microscope is precise within only $\pm$ 10%, due to fluctuation of electrical currents and column contamination during even relatively short periods of use.

Preparation of specimens thin enough for transmission microscopy requires rather drastic mechanical agitation which may alter the natural state, or, conversely, may not completely separate all of the fasciated fibrils.

Inevitable strong dehydration occurs under the high vacuum necessary for electron microscopy.

Over-shadowing or over-staining may introduce some error in measurement.

During positioning, focusing, and photographing, the specimen undoubtedly suffers degradation due to exposure to the intense electron beam.

As pointed out by Johnson and Tyson [369], variations in apparent size may be due to phase contrast effects during focusing. He found that fibrils of Fortisan varied from 18 Å to 55 Å in width depending on the phase enhancement due to over- or underfocusing.

Serious potential inaccuracies are present in degree of photographic enlargement of the electron micrograph negative prior to measurement of fibrils.

Potential inaccuracies occur due to selection of areas of the photograph to be measured, and to biased selection of individual fibrils in the field.

It is thus reassuring to find that determinations by different scientists on totally different plant materials in laboratories in different countries are in reasonable agreement with respect to dimensions of the cellulose fibrils derived from native sources by mechanical agitation.

There is some discussion still about apparent regularity of spacing of stained and unstained areas along the length of individual microfibrils. Hess and Mahl [370] using iodine and thallium as stains demonstrated apparent periodicities of approximately 650 Å along the length of microfibrils from rayon and in synthetic fibers (perlon). Later papers [188, 189] further elaborated on the periodicities of distribution of stained areas along the lengths of microfibrils indicating a periodic disorder in ramie, cotton, and Fortisan. Cross striae were observed by Manley [183] in negatively stained ramie fibrils, and other investigators have occasionally commented on this apparent regularity. However, the periodic patterns vary from 100 Å to 2000 Å, depending on the treatment, so that justification of these as being the result of natural morphological features is difficult, especially since no indications of such irregularities are observed in shadowed specimens. Nevertheless, since by other physical methods native cellulose is found to be 15 to 30% paracrystalline, it is tempting to allocate the disordered areas to the observed discontinuities. Moreover, in hydrolysis, oxidation and enzymolysis, the breakdown of long fibrils into short segments follows the changes in degree of polymerization accompanying degradation. Peterlin and Ingram [30] proposed that the fibril is entirely crystalline and that the discontinuities occasionally observed in negative staining are the result of crystal faults, tilts, dislocations, and other growth imperfections.

Because synthetic polymers form folded chains, attempts were made to interpret the cellulose fibril as consisting of folded cellulose chains. Manley [183], having observed by negative staining a regular spacing between stained and unstained regions along the length of the fibril, proposed a novel scheme of chain folding to account for this arrangement. He conceived of a ribbon composed of a folded chain which in turn is curled so as to give a spiral in which the general chain direction is parallel to the fibril length. Experimental evidence offered to support this model was meager. It has since been pointed out that the center of such a fibril should be hollow and thus density would be much less than that experimentally observed for cotton. Other models of chainfolding within microfibrils were proposed by Tonnesen and Ellefsen [371], Dolmetsch and Dolmetsch [372], and Marx-Figini and Schulz [373]. However, in 1969 a thesis by Muggli [35, 374, 375] reported experiments and data to prove that in native cellulose the chains exist in an extended rather than a folded state. By comparing the molecular weight distribution of cut and uncut ramie fibers it was shown that a folding of chains in elementary fibrils is not supported by the evidence.

Frey-Wyssling and Mühlethaler [172] considered the elementary fibril as entirely crystalline, but Dolmetsch and Brederek [195] suggested a 35-Å crystalline core embedded in a matrix of paracrystalline cellulose, or of noncellulose polyoses. Nieduszynski and Preston [366] and Colvin [134] considered the 100-250 Å crystalline core as the original native fibril dimension, and Preston [24] suggested that this may be surrounded with various noncellulosic paracrystalline materials. Methods of preparation, purification, and, most importantly, drying after purification, undoubtedly have marked influence on the amount of coalescence of fibrils into larger units. In addition, some attention must be paid to the fact that the algae produce their cell walls immersed in water, whereas the cell walls of land plants may be produced under slightly less aquatic situations.

The mechanism by which cellulose is formed is still unclear, but the electron microscope has made possible considerable progress toward solution of the mystery of cellulose synthesis. Current thinking is that elementary fibrils are produced enzymatically, outside the cell protoplasm, from precursors which arise in a matrix of pectin and hemicellulose produced in the vesicles of Golgi bodies. The cellulose fibril apparently arises from clusters of glucoprotein particles which accumulate on the cell membrane. The association of granular protein bodies with sheets of fibrils was first noted in Cladophorales by Preston [376] who suggested that the arrangement of these granules directs the multidirectional orientation of fibril alignment in primary walls. Moor and Mühlethaler [204] used the freeze etch technique to study yeast cell walls and found that the protoplasmic surface is covered by numerous particles having a diameter of 60 to 140 Å, and that groups of these particles are concentrated in a hexagonal arrangement containing 20 to 50 units and have a lattice period of 180 Å. With higher resolution it was found that the hexagonal arrangements are penetrated by

fibrils perpendicular to the wall; these are about 50 Å in diameter corresponding in size to glucan strands. Tomlinson and Jones [377] studying the
spores of Enteroamoeba by freeze fracture concluded that the clusters of
granules observed on the cell membrane are the manufacturing sites for
cellulose fibrils. In both yeast and Acantamoeba, Mühlethaler [32] found a
striking correlation between the occurrence of the membrane particles and
wall formation. He indicated that a single elementary fibril of cellulose
may grow from the center of a "spinneret" consisting of 8 protein granules
in a cluster. The particles have been analyzed and found to contain glyco-
protein, and may well be the sites of formation of elementary fibrils. Hopefully, the sequence of enzyme systems which results in the cellulose I configuration in native cell walls will become clearly defined in the near future.
Reviews of the literature on cellulose synthesis were published by Rollins
[17], Preston and Goodman [31], Mühlethaler [32] and Frey-Wyssling [36].

J. Chemical Modification of Cellulose

Assessment of morphological changes which accompany chemical modifications designed to improve physical properties of cotton materials was
covered in detail by Rollins and co-workers in several papers in the period
1962 to 1970. Methods were described in the American Dyestuff Reporter
for 1963 [125], 1965 [126], and 1968 [145a]; ASTM Special Publication No.
257 [378]; Norelco Reporter for 1966 [124]; and in Methods in Carbohydrate
Chemistry, Vol. III [122]. The effect of mercerization on morphology were
described in 1965 [138] and the effects of esterification were described by
Tripp et al. in 1957 [127] and by Cannizzaro et al. in 1968 [379]. Variations
in crosslinking treatments with ethyleneurea derivatives using different
catalysts, under a range of time and temperature conditions, were investigated by Rollins in 1966 [165], and a report on the morphology of cellulose-
polymer grafts in cotton was published by her in 1968 [160]. Techniques
included fragmentation, surface replicas, ultrathin sections with subsequent
solution by cellulose solvents or polymer solvents, and the layer expansion
technique of sectioning. Results of specific investigations are reviewed
below.

1. Additive Finishes

Desirable modifications of the properties of cotton yarns and fabrics are
often achieved by the application of additive finishes in the form of surface
coatings, such as vinyl resins, starch, or rubber, which give little evidence
of penetrating the fiber. Particulate deposits such as carboxymethylcellulose
powder, or poly (methyl methacrylate) particles are special cases of surface
coating effective in soil resistance. Colloidal alumina and colloidal silica

increase friction of fibers in yarns; emulsions of such polymers as acrylics, silicones, polyurethanes, and polyethylenes are used as coatings for water repellency or as softeners to improve the handle of the fabric. Demonstrations of the distribution of soil on fiber surfaces were made by Kling and Mahl [252, 253], Tripp et al. [114] , and Reumuth [113] .

The effect of antisoiling additives such as colloidal silica and particulate methacrylate were shown by Tripp et al. [380]. The relationships of vinyl coatings to the yarn constructions of the fabric were illustrated by scanning electron micrographs of Rollins et al. [67], and the failure of such additives to penetrate were illustrated by scanning micrographs of cross sections (Fig. 29). Position of penetrants within the fiber wall was investigated by transmission electron micrographs of sections after appropriate solvent extraction of the section to remove the cellulose and thus reveal the location of the noncellulosic additive (Fig. 21). The scanning electron microscope is best suited for evaluating large areas of such treated fabrics or yarns for uniformity and, in edge views, of depth of penetration and distribution of the polymer within the substrate. The transmission microscope can reveal details of the character of the additive on individual fibers.

2. Purification

In the preparation of cotton for textile finishing, purification steps include scouring by an alkaline pressure boil of several hours duration followed by an oxidative bleach, or alternatively, continuous bleaching in an alkaline peroxide system. The main purpose is the removal of noncellulosic materials, such as waxes, pectic compounds, and nitrogenous matter occurring naturally in the fiber, to achieve more rapid wetting out of the fabric with subsequent treating liquids, and thus a more uniform treatment. Dolmetsch and Dolmetsch [266], Kling and Mahl [249], and Tripp et al. [108] followed the effects of sodium hydroxide scouring, and of hypochlorite and hydrogen peroxide bleaching by treating isolated primary wall fragments which were examined in the electron microscope at different stages of purification. Kling and Mahl [252] used surface replicas for their studies of the effect of laundering practices on cotton fiber structure. Dolmetsch and Dolmetsch [266] made replicas of the fiber surfaces which had been purified by extraction at various time intervals in the extraction procedure. In all cases, the noncellulosic materials were gradually removed, and the primary wall structure revealed as a netlike fabric with fibrils of many different diameters interwoven in at least three orientations to the axis of the fiber.

Purification of wood pulp fibers for paper making had similar scrutiny, and notable papers are those of Jurbergs [116] and McIntosh [311], who studied the effect of refining on the structure of the fiber wall, and of Casperson and co-workers [312] who discussed differences between pulps from alkaline and acid cooks of poplar wood.

3. Swelling

The effect of various swelling agents on the morphology of cellulose fibers was examined often [27]. Razikov et al. [169] investigated changes in the supermolecular structure of cellulose after treatment with activating agents, which involved solvent occlusion of isoamyl and heptyl alcohols after initial swelling of the fiber in alkali, and studied interfibrillar swelling in aqueous solutions of glycerin. Work [381] investigated the microfibrillar morphology of normal and swollen cotton cellulose with special emphasis on the effect on fibrillar bonding in inter- and intracrystalline swelling agents; a later paper showed that discreteness of fibrillar outline is progressively lost as decrystallization occurs [171]. The effects of trimethylbenzylammonium hydroxide on cotton fiber morphology was described by Betrabet et al. in 1970 [170].

4. Mercerization

Mercerization is a special application of intracrystalline swelling employed commercially to enhance dyeability and luster of cotton materials. It consists of immersion of the fabric in from 18 to 25% sodium hydroxide at ambient conditions and under tension to prevent loss of yardage, or under slack conditions to induce stretch in the yarns. The cellulose of the fiber is swollen rapidly and its cross section becomes less oval but more nearly circular (Figs. 22 and 25). The sodium hydroxide is washed out and the fabric subsequently neutralized, but the effect on the fiber is irreversible. The cellulose is so modified that the luster and absorption properties of the fibers are improved greatly and when dyed, the resulting colors are more brilliant than in the case of unmercerized cotton. X-ray diffraction measurements indicate that the orientation of the cellulose crystallites is altered, and the crystal lattice changed from that of cellulose I to that of hydrate cellulose, cellulose II. Refractive indices are also affected. The refractive index parallel to the fiber axis is 1.578 for native cotton (cellulose I) but 1.566 for mercerized cotton (cellulose II). A microscopical study of mercerization was published by Rollins and co-workers in 1965 [138], and similar information on fibrillar characteristics was given in the 1963 paper in the American Dyestuff Reporter [125].

5. Substitution

Substitution involves chemical modification of the cellulose molecule by such reactions as esterification (i.e., acetylation) and etherification (i.e., carboxymethylation, cyanoethylation, benzylation). The new materials retain the physical form of the fiber, but have entirely different chemical and physical properties. In acetylation, for example, acetyl groups are introduced into the cellulose chain to produce a new fiber resistant to heat and microbial attack and with the dyeing properties of cellulose acetate, but with the original fibrous form of the cotton fiber and its useful properties. Electron microscopical examinations of esterified cottons were undertaken by Tripp et al. [127] and by Cannizzaro et al. [379]. When esters of even low degree of substitution (D. S. 0.2) are embedded in methacrylate for sectioning, the fiber cell walls tend to separate into concentric layers due to partial solution in the methacrylate. Notable exceptions are the ricino-leate, linoleate, and the 12-hydroxystearate esters, but no valid explana-tion for this anomalous behavior was offered. Tripp et al. [127] found that, upon fragmentation in water, the cellulose partial esters break into groups of microfibrils and masses of amorphous material; at higher degrees of substitution more amorphous material and less fibrillar material are ob-tained (Fig. 14). Work [381] reported similar observations on cellulose acetate fibers. The same phenomenon was observed in the case of ethers, especially the cyanoethyl derivatives.

6. Crosslinking

To improve resilience and thus the wrinkle resistance of cotton fabrics, crosslinking treatments are applied in which the reactions form covalent bonds between cellulose chains to immobilize the grosser structural ele-ments of the fiber. Results are an increase in the Monsanto crease-recov-ery angle [382] accompanied by decreases in moisture regain, extensibility, elongation at break, and breaking strength. Crosslinking is accomplished by the use of formaldehyde in any of several forms, followed by an after wash and a cure at elevated temperature; by the use of aminoplast resins such as methylolmelamine and dimethylol urea; by the use of ethyleneurea derivatives, and by the use of such reagents as butadienediepoxide [383] or tris-aziridinyl phosphine oxide [384]. The fibrillar layers and the micro-fibrils within them are immobilized by bonding the cellulose structure to-gether more firmly than it exists in the natural state. Tripp et al. [173] showed by cross sections and fiber fragments of cotton modified by cross-linking treatments the location of reacted regions and discussed the effects of specific chemical treatments on fiber extensibility, swelling capacity, and solubility. Rollins et al. [165] showed the effects of crosslinking by ethyleneurea derivatives on the morphology of the cotton fiber, exploring by ultrathin sectioning methods the efficacies of different catalysts and differ-ent curing conditions (Figs. 19, 20, 24, 34). Nikonovich and co-workers [270] showed electron micrographs of cellulose fibers modified by treatments with acrolein, and Rowland et al. [385] demonstrated various degrees of heterogeneity in the distribution of formaldehyde crosslinks in cotton cellu-lose. Dolmetsch and Bredereck [195] investigated crease-resistant cottons finished by conventional wet and dry processes. Nikonovich et al. [168] studied the structures of cottons crosslinked with dimethylolurea, formaldehyde, acrolein, epichlorohydrin, and propylene oxide. Lauchenauer et al. [386] and Meier and Zollinger [387] showed the effects of core-crosslinking on cotton fabric. Ward et al. [388] investigated the reactions of cotton and of diethylaminoethyl cotton with 1-chloro-2, 3-epithiopropane and showed differences in ultrathin cross sections associated with wet and dry wrinkle recovery properties. deGruy et al. [166] investigated the effects of various solvents on reactions of diethylaminoethyl cellulose with epichlorohydrin, and Guidry et al. [167] presented microscopical features of reaction products of halogenated, 1, 2-epoxides with diethylaminoethyl cotton and aminized cotton.

7. Grafting

The interaction of high-energy radiation with polymers and the production of graft copolymers by radiation-induced reactions of cellulose and vinyl polymers were investigated. Chemical initiation by ceric ion, ferrous ion, or persulfate ion was also promoted. The property changes are dependent not only on the nature and amount of graft polymer formed but on its location within the fiber structure. The distribution of graft copolymer within the cell wall was evaluated by observations of ultrathin sections of fibers. Arthur and Demint [389] showed localized distribution of acrylonitrile-cellulose copolymer after simultaneous irradiation treatment. Blouin and Arthur [390] and Rollins et al. [391] showed effects of activation by irradiation of the cotton at different levels prior to treatment with the polymer (Fig. 21). Kaeppner and Huang [161] made similar observations on rayon filaments treated with styrene, and Kaeppner investigated, by surface replications and ultrathin sections, the distribution of styrene-cellulose graft copolymers in Fortisan, high tenacity tire yarn, and a modifier-type high wet modulus rayon yarn. Changes in the structure of cross sections by alternate extraction with solvents for polystyrene and for cellulose showed that, from a macroscopical point of view, the styrene-grafted rayons are homogeneous, but at the molecular level they consist of a gross heterogeneous mixture of polystyrene homopolymer, cellulose, and true cellulose-styrene graft copolymer. Razikov and Usmanov [392] studied the effect of preparation treatments on the microstructure of cellulose grafted with methacrylates. Liggett et al. [393] investigated grafting of cotton with acrylonitrile by chemical initiation with ceric, persulfate, and ferrous ions. Rollins et al. [165] and Blouin et al. [394] discussed vinyl-cellulose copolymers prepared by radiation-initiated reactions. Grafts investigated included polyacrylonitrile, polystyrene, poly (methyl methacrylate), and polyvinyl acetate onto cotton. Although more precise methodology exists for the chemical analysis of reacted cellulose, electron microscopy offers the best approach for locating within the fiber structural alterations that bear on physical behavior of fabrics (Fig. 35).

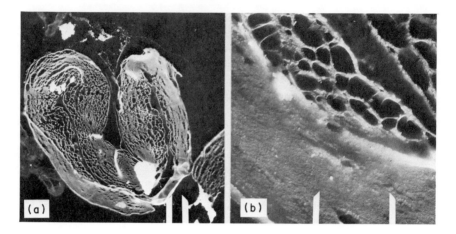

Fig. 34. Transmission electron micrographs of expansion pattern of cotton fiber crosslinked by dihydroxyethyleneurea (Zn(NO3)2 catalyst) at low temperature: a. whole fiber; b. enlarged portion of "a." U. S. Department of Agriculture photograph.

8. Degradation

The deterioration of cellulose fibers under various environments was also studied with the electron microscope. The microfibril is susceptible to attack by dilute mineral acids and by microbial cellulases. Under either of these conditions deterioration occurs by chain scission, and results in transverse breakdown of the microfibrils into elongated tabular particles (Fig. 36). The similarity of appearance of specimens of lamellae degraded by these two agents is striking. Hock [395] described degradation by oxidation as well as by hydrolysis. The breakdown of cellulose by mineral acid hydrolysis was used in chemical analysis as a technique for estimating the degree of crystallinity of the cellulose. Initially, degradation occurs very rapidly but soon levels off to an almost imperceptible rate. At "leveling off d. p." the undissolved residue of cotton is approximately 80 to 85% of the original weight; in delignified wood it is about 70%. X-ray diffraction patterns of this "hydrocellulose" residue indicate that it is more nearly crystalline than the original substance. For cotton, flax, ramie, and other textile fibers the representative length of the tabular fragments of the hydrocellulose residue is of the order of 1000 Å; for wood pulp it is about 600 Å. The work of Rånby and Ribi [396], Morehead [135], and Immergut and Rånby [397] indicated that attack by hydrolyzing agents reduces the length of the particles in the fibrillar residue, but not their width. At a limiting length the fragments suddenly disappear. Krassig and Kitchen [398] showed the existence of primary morphological units in hydrolyzed native cotton and in Fortisan regenerated cellulose ester, and discussed accessibility of interlinkages between them with respect to tensile properties of the bulk material. Dlugosz and Michie [123], applying acid swelling and ultrasonic

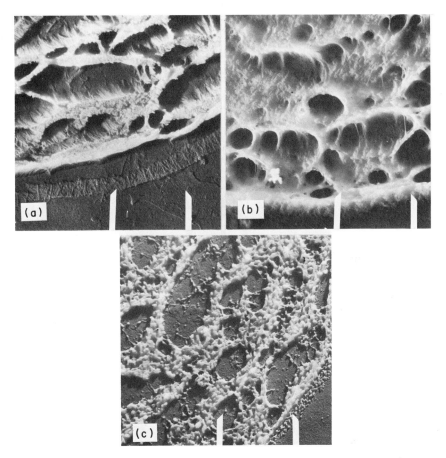

Fig. 35. Transmission electron micrographs of cross sections of cotton fibers grafted with polystyrene by simultaneous irradiation: a. whole section by conventional methacrylate embedding technique (polystyrene evidently swells in methacrylate); b. parallel specimen of "a" after a 30-min immersion in cuene to remove unreacted cellulose; c. parallel specimen of "b" after a 30-min immersion in xylene to remove unreacted styrene. U. S. Department of Agriculture photographs.

disintegration to several samples of regenerated celluloses, revealed elongated, threadlike particles longer than the crystallites of cellulose but having approximately the same width. Immergut and Rånby [397] found, in a comparison of dewaxed cotton and a chemical grade sulfite pulp from hemlock, that the initial fast hydrolysis to a yield of 80% involves the decomposition of elementary cellulose fibrils into shorter fragments. Other studies of hydrocellulose particles appear in publications by Rånby [229], Vogel [283], Krassig and Käppner [399] and Battista et al. [400, 401, 402].

Fig. 36.   Hydrocellulose particles from H_2SO_4 digestion of kiered cotton. U. S. Department of Agriculture photograph.

Polcin and Karhanek [403] reported the effects of gamma rays on the chemical, physical–chemical, and morphological properties of wood pulp and cotton linters by examination of the irradiated fibers after vibration mill or ultrasonic treatment.   Porter et al. [404] followed the effect of increasing gamma-ray exposure on purified cotton fibers.   Changes in fibrillate structure of cotton on hydrolysis and oxidation were found by Porter and Goynes [405] to be parallel to changes in moisture regain, crystallinity index, copper number, carboxyl content, and degree of polymerization. Bobeth and co-workers [406] investigated the response of fibers to axial compression, and Brestkin [407] discussed degradation of cellulose in hot alkali.

Reese et al. [408] and Porter et al. [409] showed various stages of cotton breakdown by microbial action.   Liu and King [410] described the treatment of hydrocellulose with the cellulase from Trichoderma viride and suggested that the C–1 enzyme component of cellulase may not be an enzyme in the usual sense, but a protein which hydrogen bonds to cellulose more tightly than cellulose hydrogen bonds to itself, thus causing gradual collapse of the initial particle and ultimate release of micelles.

ACKNOWLEDGMENTS

The authors wish to express their thanks to Miss Rosalie Babin for preparation of the photographic enlargements, and to the various members of the staff of the Southern Regional Research Laboratory for their helpful suggestions in revision of the manuscript.

REFERENCES

[1] G. L. Clark, ed. , The Encyclopedia of Microscopy, Reinhold, New York, 1961.

[2] S. Bradbury, The Microscope, Past and Present, Pergamon, New York, 1968.

[3] S. Wischnitzer, Introduction to Electron Microscopy, 2nd ed. , Pergamon, New York, 1970.

[3a] R. O. Partsch, Carl Zeiss, New York, 1970.

[4] R. D. Heidenreich, Fundamentals of Transmission Electron Microscopy, Wiley-Interscience, New York, 1964.

[5] V. K. Zworykin, G. A. Morton, E. G. Ramberg, J. Hillier, and A. W. Vance, Electron Optics and the Electron Microscope, Wiley, New York, 1945.

[6] R. W. G. Wyckoff, Electron Microscopy, Wiley-Interscience, New York, 1949.

[7] R. W. G. Wyckoff, The World of the Electron Microscope, Yale Univ. Press, New Haven, Connecticut, 1958.

[8] C. E. Hall, Introduction to Electron Microscopy, 2nd ed. , McGraw-Hill, New York, 1966.

[9] B. M. Siegel, ed. , Modern Developments in Electron Microscopy, Academic, New York, 1964.

[10] R. B. Fischer, Applied Electron Microscopy, Indiana Univ. Press, Bloomington, 1953.

[11] D. G. Drummond, ed. , The Practice of Electron Microscopy, J. Royal Microscopical Society, Ser. 3, 70, 56 (1950).

[12] V. E. Cosslett, Practical Electron Microscopy, Butterworth, London, 1951; Academic, New York, 1952.

[13] V. E. Cosslett, Modern Microscopy, or Seeing the Very Small, Cornell Univ. Press, Ithaca, New York, 1966.

[14] M. E. Haine and V. E. Cosslett, The Electron Microscope, E. and F. Spon, London, 1961.

[15] R. Barer and V. E. Cosslett, Advances in Optical and Electron Microscopy, Academic, London and New York, 1968.

[16] Kh. U. Usmanov and G. V. Nikonovich, Electron Microscopy of Cellulose, Academy of Sciences, Uzbek SSR, Tashkent, 1962. English translation by Joint Publication Research Service, Office of Technical Services, U. S. Dept. of Commerce, Washington, D. C. , 1963.

[17] M. L. Rollins, Forest Prod. J., 11, 493 (1961).

[18] M. L. Rollins, Forest Prod. J., 18, 91 (1968).

[19] A. Frey-Wyssling, Die Pflanzliche Zellwand, Springer-Verlag, Berlin, 1959.

[20] P. A. Roelofsen, The Plant Cell Wall, Gebruder Borntraeger, Berlin, 1959.

[21] E. Treiber, Die Chemie der Pflanzenzellwand, Springer-Verlag, Berlin, 1957.

[22] R. D. Preston, Molecular Architecture of Plant Cell Walls, Wiley, New York, 1952.

[22a] W. Liese and W. A. Coté, Proc. Fifth World Forest. Cong., Seattle, September 1960, World Forestry Congr. Proc., 5 (2), 1288 (1962).

[23] B. Rånby and S. A. Rydolm, in Polymer Processes in High Polymers, (C. E. Schildknecht, ed.), Vol. 10, Chap. 9, Wiley-Interscience, New York, 1956.

[24] R. D. Preston, in The Interpretation of Ultrastructure (R. J. C. Harris, ed.), Academic, New York, 1962.

[25] K. Mühlethaler, Festschr. Frey-Wyssling, Beih. Z. Schweiz. Forstverein, 30, 55 (1960).

[26] A. Frey-Wyssling, in The Interpretation of Ultrastructure (R. J. C. Harris, ed.), Academic, New York, 1962.

[27] J. O. Warwicker, R. Jeffries, R. L. Colbran, and E. N. Robinson, A Review of the Literature on the Effect of Caustic Soda and Other Swelling Agents on the Fine Structure of Cotton, Shirley Institute Pamphlet No. 93, Cotton Silk and Man-Made Fibres Research Association, Manchester, 1966.

[28] S. M. Betrabet, J. Sci. Ind. Res., 26 (5), 102 (1967).

[29] G. Jayme and K. Balser, Melliand Textilber., 51, 3 (1970).

[30] A. Peterlin and P. Ingram, Textile Res. J., 40, 345 (1970).

[31] R. D. Preston and R. N. Goodman, J. Roy. Microscop. Soc., 88 (4), 513 (1967).

[32] K. Mühlethaler, Ann. Rev. Plant Physiol., 18, 1 (1967).

[33] K. Mühlethaler, J. Polymer Sci., Pt. C, 28, 305 (1969).

[34] D. Fengel, Tappi, 53, 497 (1970).

[35] A. Frey-Wyssling, K. Mühlethaler, and R. Muggli, Holz Roh-Werkstoff, 24, 443 (1966).

[36] A. Frey-Wyssling, Progr. Chem. Org. Natur. Prod., 27, 1 (1969).

[37] K. C. A. Smith, Pulp Paper Mag., Can., 60 (12), 3 (1959).

[38] P. R. Thornton, Scanning Electron Microscopy, Chapman and Hall, London, 1968.

[39] C. W. Oatley, W. C. Nixon, and R. F. Pease, in Advances in Electronics and Electron Physics, Vol. 21, Academic, New York, 1965, p. 181.

[40] K. C. A. Smith and C. W. Oatley, Brit. J. Appl. Phys., 6, 391 (1955).

[41] S. Kimoto and J. C. Russ, Mater. Res. Std., 9 (1), 8 (1969).

[42] S. Kimoto and J. C. Russ, Am. Sci., 57 (1), 112 (1969).

[43] T. E. Everhart, Proc. Symp. Scanning Electron Microscop., The Instrument and Its Applications, Chicago, Ill., April 30-May 1, 1966. Ill. Inst. Technol. Res. Inst., Chicago, 1966.

[44] J. T. Black, Phot. Appl. Sci., Technol. and Med. 4 (3), 29 (1970).

[45] A. V. Crewe, J. Wall, and L. Welter, Rev. Sci. Instr., 39, 576 (1968).

[46] A. V. Crewe in Proceedings 27th Annual Meeting Electron Microsc. Soc. Amer. (C. J. Arceneaux, ed.), Claitor's Publishing Division, Baton Rouge, 1969.

[47] A. V. Crewe and J. Wall, J. Mol. Biol., 48, 375 (1970).

[48] A. V. Crewe, Quart. Rev. Biophys., 3, 137 (1970).

[49] Anon., Ind. Res., 12 (7), 19 (1970).

[50] Anon., Phot. Appl. Sci., Technol. and Med., 4 (18), 38 (1970).

[51] J. G. Buchanan and O. V. Washburn, The Surface and Tensile Fractures of Chemical Fibre Handsheets as Observed with the Scanning Electron Microscope, Tech. Rept. 294, Pulp and Paper Research Institute of Canada, Montreal, 1962.

[52] J. Sikorski, J. S. Moss, P. H. Newman, and T. Buckley, J. Sci. Instr., Ser. 2, 1, 29 (1968).

[53] J. Sikorski, and A. Hepworth, Proc. First Ann. Symp. Scanning Electron Microsc. (O. Johari, ed.), Ill. Inst. Technol. Res. Inst., Chicago, Ill., 1968.

[54] J. W. S. Hearle, Man-Made Textiles, Skinner's Rec., 44 (7), 44; (8), 95 (1967).

[55] H. Reumuth, Z. Ges. Textil-Ind., 69, 227 (1967).

[56] H. Reumuth, Melliand Textilber., 48, 489 (1967).

[57] H. R. Billica, Mod. Textiles Mag., 49 (12), 69 (1968).

[58] A. Rezanowich, Proc. Symp. Scanning Electron Microsc., Ill. Inst. Technol. Res. Inst., Chicago, 1968.

[59] H. Arai and I. Maruta, J. Am. Oil Chemists, Soc., 45, 448 (1968).

[60] H. P. Sennett, R. J. Diexel, and H. H. Morris, Tappi, 51, 567 (1968).

[61] E. J. Heiser and H. M. Baker, Tappi, 51, 528 (1968).

[62] A. Resch and R. Blaschke, Planta, 78, 85 (1968).

[63] K. G. Norberg and E. L. Back, Svensk Papperstid., 72, 605 (1969).

[64] I. U. Frölander and N. Harter, Svensk Papperstid., 73, 197 (1970).

[65] W. R. Goynes, Proc. Second Ann. Stereoscan Colloq., Engis Equip. Co., Chicago (1969).

[66] W. R. Goynes, Textile Res. J., 41, 226 (1971).

[67] M. L. Rollins, W. R. Goynes, and R. J. Brysson, Polymer Symp., (In press) (1971).

[68] H. W. Emerton, J. Watts, K. Amboss, and A. Simpson, Proc. Tech. Sect. Brit. Paper and Board Makers' Ass'n. Inc., 35, 487 (1954).

[69] H. W. Emerton, D. Page, and J. Watts, Proc. Tech. Sect. Brit. Paper and Board Makers' Assoc. Inst., 37, Part i, 105 (1956).

[70] J. A. Chapman and J. W. Menter, Proc. Roy. Soc. (London), Ser. A, 226, 400 (1954).

[71] H. W. Emerton, Fundamentals of the Beating Process, The British Paper and Board Makers Association, Kenley, England, 1957.

[72] K. Amboss, H. W. Emerton, and J. Watts, Proc. Intern Conf. Electron Microscop., London, 1954, Royal Microscopical Society, London, 1956, p. 560.

[73] H. W. Emerton, D. H. Page, and J. Watts, Proc. Stockholm Conf. Electron Microscopy, 1956, Almquist and Wiksell, Stockholm, 1957, p. 287.

[74] J. W. Menter, J. Phot. Sci., 1, 12 (1953).

[75] H. S. Bennett, H. P. Jenerick, and I. Warshaw, Science 168, 506 (1970).

[76] V. E. Cosslett, Proc. 28th Meet. Electron Microscop. Soc. Am. (C. J. Arceneaux, ed.), Claitor's Publishing Division, Baton Rouge, La., 1970.

[77] G. Thomas, Proc. 28th Meet. Electron Microscop. Soc. Am. (C. J. Arceneaux, ed.), Claitor's Publishing Division, Baton Rouge, 1970.

[78] Anon., Thin Sectioning and Associated Techniques for Electron Microscopy, 2nd ed., Ivan Sorvall, Norwalk, Conn., 1965.

[79] A. M. Porter and J. Blum, Anat. Record, 117, 685 (1953).

[80] F. S. Sjostrand, Electron Microscopy of Cells and Tissues, Vol. I: Instrumentation and Techniques, Academic, New York and London, 1967.

[81] F. S. Sjostrand, Experientia, 9, 114 (1953).

[82] R. F. Baker and D. C. Pease, J. Appl. Phys., 20, 480 (1949).

[83] D. C. Pease, Histological Techniques for Electron Microscopy, 2nd ed., Academic, New York and London, 1964.

[84] E. H. Mercer and M. S. C. Birbeck, Electron Microscopy, a Handbook for Biologists, 2nd ed., Blackwell Scientific Publications, Oxford, 1966.

[85] H. Latta and J. F. Hartman, Proc. Soc. Exptl. Biol. Med., 74, 436 (1950).

[86] H. Fernandez-Moran, Exptl. Cell Res., 5, 255 (1953).

[87] H. Fernandez-Moran, J. Biophys. Biochem. Cytol., 2 (4), Suppl. 29 (1956).

[88] J. A. Horner, Sci. Instr. News, 12, 10 (1967).

[89] D. H. Kay, ed., Techniques for Electron Microscopy, 2nd ed., Blackwell Scientific Publications, Oxford, 1967.

[90] A. Engstrom and J. B. Finean, Biological Ultrastructure, Academic, New York, 1958.

[91] M. L. Watson, J. Biophys. Biochem. Cytol., 1, 183 (1955),

[92] D. E. Bradley, J. Roy. Microscop. Soc. , 79, 101 (1960).
[93] D. E. Bradley, Brit. J. Appl. Phys. , 10, 198 (1959).
[94] R. C. Williams and R. W. G. Wyckoff, J. Appl. Phys. , 15, 712 (1944).
[95] R. C. Williams and R. W. G. Wyckoff, J. Appl. Phys. , 17 (1),
       23 (1946).
[96] L. E. Preuss, RCA Sci. Instr. News, 4 (1), 13 (1959).
[97] D. E. Bradley, in Techniques for Electron Microscopy (D. Kay, ed.),
       Blackwell Scientific Publications, Oxford, 1961.
[98] W. A. Coté, Z. Koran, and A. C. Day, Tappi, 47, 477 (1964).
[99] G. Jayme and G. Hunger, Monatsh. Chem. , 87, 8 (1956).
[100] A. Gröbe, R. Maron, and K. P. Rose, Faserforsch. Textiltech. ,
       19, 253 (1968).
[101] P. Kassenbeck, Melliand Textilber. , 39, 55 (1958).
[102] R. D. Sloan and W. F. Gardner, Program Ann. Meet. Electron
       Microscop. Soc. Am. , J. Appl. Phys. , 37, 3919 (1966).
[103] C. E. Hall, J. Appl. Phys. , 21, 61 (1950).
[104] I. W. Fischbein, J. Appl. Phys. , 21, 1199 (1950).
[105] V. W. Tripp, A. T. Moore, and M. L. Rollins, Textile Res. J. ,
       27, 419 (1957).
[106] P. H. Norberg, Svensk Papperstid. , 17, 869 (1968).
[107] A. Gröbe, Svensk Papperstid. , 71, 646 (1968).
[108] V. W. Tripp, A. T. Moore, and M. L. Rollins, Textile Res. J. ,
       24, 956 (1954).
[109] M. L. Rollins and V. W. Tripp, Textile Res. J. , 24, 345 (1954).
[110] I. V. deGruy, J. H. Carra, V. W. Tripp, and M. L. Rollins,
       Textile Res. J. , 32, 873 (1962).
[111] M. L. Rollins, I. V. deGruy, J. H. Carra, and T. P. Hensarling,
       Textile Res. J. , 40, 903 (1970).
[112] A. R. Markezich, Textile Bull. , 94 (8), 26 (1968).
[113] H. Reumuth, S. V. F. Fachorg. Textilveredlung., 6 (7) 245; (8), 285
       (1951); 7 (1), 85 (1952).
[114] V. W. Tripp, A. T. Moore, B. R. Porter, and M. L. Rollins,
       Textile Res. J. , 28, 447 (1958).
[115] W. A. Coté, Forest Prod. J. , 8, 1296 (1958).
[116] K. A. Jurbergs, Tappi, 43, 865 (1960).
[117] C. E. Dunning, Tappi, 52, 1326 (1969).
[118] P. H. Norberg, Svensk Papperstid. , 71, 869 (1968).
[119] P. H. Norberg, Svensk Papperstid. , 72, 575 (1969).
[120] P. H. Norberg, Svensk Papperstid. , 73, 208 (1970).
[121] K. Mühlethaler, Biochim. Biophys. Acta, 3, 15 (1949).
[122] M. L. Rollins and V. W. Tripp, in Methods for Carbohydrate
       Chemistry, Vol. III: Cellulose (R. W. Whistler, ed.), Academic,
       New York, 1963.
[123] J. Dlugosz and R. I. C. Michie, Polymer, 1, 41 (1960).
[124] M. L. Rollins, I. V. deGruy, A. M. Cannizzaro, and J. H. Carra,
       Norelco Reptr. , 13, 119 (1966).

[125] M. L. Rollins, A. T. Moore, and I. V. deGruy, Am. Dyestuff Reptr., 52, 479 (1963).

[126] M. L. Rollins, A. T. Moore, W. R. Goynes, J. H. Carra, and I. V. deGruy, Am. Dyestuff Reptr., 54 (14), 36 (1965).

[127] V. W. Tripp, R. Giuffria, and I. V. deGruy, Textile Res. J., 27, 14 (1957).

[128] D. G. Drummond, Shirley Inst. Mem., 26, 53 (1952).

[129] G. Levavasseur, H. Troitzsch, and G. Chaume, Mem. Serv. Chim. Etat (Paris), 39, 287 (1954).

[130] I. Ohad, D. Danon, and S. Hestrin, J. Cell. Biol., 12, 131 (1962).

[131] I. Ohad and D. Danon, J. Cell. Biol., 22, 302 (1963).

[132] I. Ohad, D. Danon, and S. Hestrin, J. Cell. Biol., 17, 321 (1963).

[133] A. W. Khan and J. R. Colvin, Science, 133, 2014 (1961).

[134] J. R. Colvin, J. Appl. Polymer Sci., 8, 2763 (1964).

[135] F. F. Morehead, Textile Res. J., 10, 549 (1950).

[136] B. G. Rånby, Svensk Papperstid., 57, 9 (1954).

[137] K. K. R. Razikov, E. D. Tyagai, P. P. Larin, and K. L. U. Usmanov, Polymer Sci. (USSR), 9 (2), 442 (1967).

[138] M. L. Rollins, A. T. Moore, I. V. deGruy, and W. R. Goynes, J. Roy. Microscop. Soc., 84, Pt. 1, p. 1 (1965).

[139] S. B. Newman, E. Borysko, and M. Swerdlow, Science, 110, 66 (1949).

[140] W. E. Yolland, Textile Mfr., 65, 300 (1939).

[141] E. I. duPont deNemours & Co., Tech. Service Rept. No. 4, Plastics Department, E. I. duPont deNemours & Co., Arlington, N. J., 1945.

[142] E. Borysko and P. Sopramaulka, Bull. Johns Hopkins Hosp., 95, 68 (1954).

[143] E. Borysko, Biophys. Biochem. Cytol., 2 (4) Suppl. (1956).

[144] E. Borysko and J. Roslanski, Ann. N.Y. Acad. Sci., 78, 432 (1958).

[145] L. D. Peachey, J. Biophys. Biochem. Cytol., 4, 233 (1958).

[145a] A. M. Cannizzaro, W. R. Goynes, and M. L. Rollins, Am. Dyestuff Reptr., 57 (2), 23 (1968).

[146] R. L. Davidson and M. Sitting, Water Soluble Resins, Reinhold, New York; Chapman and Hall, London, 1963.

[147] V. P. Gilev, J. Ultrastruct. Res., 1, 349 (1958).

[148] D. G. Scarpelli and C. V. Sittler, Proc. EMSA, J. Appl. Phys., 30, 2039 (1959).

[149] I. V. deGruy, Textile Res. J., 25, 887 (1955).

[150] M. Rosenberg, P. Bartl, and J. Leska, J. Ultrastruct. Res., 4, 298 (1960).

[151] E. Leduc, V. Marinozzi, and W. Bernhard, J. Roy. Microscop. Soc., 81, 119 (1963).

[152] J. Polcin and M. Karhanek, Naturwissenschaften, 35, 224 (1966).

[153] S. M. Betrabet and M. L. Rollins, The Microscope, 18, 193 (1970).

[154] A. M. Glauert and R. H. Glauert, J. Biophys. Biochem. Cytol.,
4, 409 (1955).
[155] J. Luft, J. Biophys. Biochem. Cytol., 9, 409 (1961).
[156] B. O. Spurlock, V. C. Cattine, and J. A. Freeman, J. Cell. Biol.,
17, 203 (1963).
[157] A. Ryter and E. Kellenberger, J. Ultrastruct. Res., 2, 200 (1958).
[158] P. S. Satir and L. D. Peachey, J. Biophys. Biochem. Cytol.,
4, 345 (1958).
[159] M. L. Rollins, A. T. Moore, and V. W. Tripp, Proc. Fourth
Intern. Congr. Elec. Microscop., Berlin, 1958, Springer-Verlag,
Berlin, (1960), p. 712.
[160] M. L. Rollins, A. M. Cannizzaro, F. A. Blouin, and J. C. Arthur,
J. Appl. Polymer Sci., 12, 71 (1968).
[161] W. M. Kaeppner and R. Y-M. Huang, Textile Res. J., 35, 504
(1965).
[162] J. Dlugosz, Polymer, 6, 427 (1965).
[163] C. H. Bamford, W. G. Barb, A. D. Jenkins, and P. F. Onyon,
Kinetics of Vinyl Polymerization by Radical Mechanism,
Butterworths, London, 1958.
[164] E. Trommsdorff, E. Koehle, and P. Lagelly, Makromol. Chem.,
1, 769 (1948).
[165] M. L. Rollins, J. H. Carra, W. J. Gonzales, and R. J. Berni,
Textile Res. J., 36, 185 (1966).
[166] I. V. deGruy, J. C. Guidry, J. H. Carra, and M. L. Rollins,
Textile Res. J., 39, 223 (1969).
[167] J. C. Guidry, I. V. deGruy, J. H. Carra, and M. L. Rollins,
Textile Res. J., 39, 866 (1969).
[168] G. V. Nikonovich, S. A. Leont'eva, and K. U. Usmanov,
J. Polymer Sci., Pt. C, 16, 877 (1967).
[169] K. Kh. Razikov, E. D. Tyagai, V. I. Sadovnikova, and
Kh. U. Usmanov, Polymer Sci., USSR, 11A (8), 1948 (1969).
[170] S. M. Betrabet, L. L. Muller, V. W. Tripp, and M. L. Rollins,
J. Appl. Polymer Sci., 14, 2905 (1970).
[171] S. M. Betrabet and M. L. Rollins, Textile Res. J., 40, 917 (1970).
[172] A. Frey-Wyssling and K. Mühlethaler, Makromol. Chem.,
62, 25 (1963).
[173] V. W. Tripp, A. T. Moore, I. V. deGruy, and M. L. Rollins,
Textile Res. J., 30, 140 (1960).
[174] E. M. Broughton, M. L. Rollins, and W. K. Walsh, Textile Res.
J., 40, 672 (1970).
[175] J. J. Hebert, private communication, 1970.
[176] K. Bredereck, Textilveredlung., 5, 368 (1970).
[177] S. Brenner and R. W. Horne, Biochim. Biophys. Acta, 34, 103
(1959).

[178] R. W. Horne, in Techniques for Electron Microscopy (D. H. Kay, ed.), F. A. Davis Co., Philadelphia, 1965.
[179] H. Dolmetsch, Papier (Darmstadt), 22, 196 (1968).
[180] A. N. J. Heyn, J. Cell. Biol., 29, 181 (1966).
[181] A. N. J. Heyn, J. Ultrastruct. Res., 26, 52 (1969).
[182] R. D. Sullivan, Tappi, 51, 501 (1968).
[183] R. St. J. Manley, Nature, 204, 1155 (1964).
[184] J. R. Colvin, J. Cell. Biol., 17, 105 (1963).
[185] E. K. Boylston and M. L. Rollins, The Microscope. 19, 255 (1971).
[186] P. Kassenbeck and R. Hagege, Textile Res. J., 38, 196 (1968).
[187] W. Sauer, O. A. Stamm, and H. Zollinger, Textilveredlung, 1, 4 (1966).
[188] K. Hess, H. Mahl, and E. Gutter, Kolloid-Z., 155, 1 (1957).
[189] K. Hess, E. Gutter, and H. Mahl, Kolloid-Z., 158, 115 (1958).
[190] S. Asunmaa, Svensk Papperstid., 57, 367 (1954).
[191] R. Hagege, P. Kassenbeck, D. Meimoun, and A. Parisot, Textile Res. J., 39, 1015 (1969).
[192] N. A. Bates and W. I. Lyness, Tappi, 51, 405 (1968).
[193] I. Ohad and D. Mejzler, J. Polymer Sci., Pt. A 3, 399 (1965).
[194] A. N. J. Heyn, J. Appl. Phys., 36, 2088 (1965).
[195] H. Dolmetsch and K. Bredereck, Melliand Textilber., 48, 561 (1967).
[196] K. Bredereck and H. Dolmetsch, Melliand Textilber., 48, 699 (1967).
[197] H. and H. Dolmetsch, Melliand Textilber., 48, 1449 (1967).
[198] K. A. Jurbergs, J. Polymer Sci., Pt. C, 28, 169 (1969).
[199] H. Maeda, H. Kawada, and T. Kawai, Makromol. Chem., 131, 169 (1970).
[200] R. L. Steere, J. Biophys. Biochem. Cytol., 3, 45 (1957).
[201] S. Bullivant and J. Ames, J. Cell. Biol., 29, 435 (1966).
[202] H. Moor, K. Mühlethaler, H. Waldner, and A. Frey-Wyssling, J. Biophys. Biochem. Cytol., 10, 1 (1961).
[203] H. Moor, Intern. Rev. Exptl. Pathol., 5, 179 (1966).
[204] H. Moor and K. Mühlethaler, J. Cell. Biol., 17, 609 (1963).
[205] J. H. McAlear and G. O. Kreutziger, Proc. 25th Ann. Meet. EMSA, 1967, Claitor Publishers, Baton Rouge, 1967.
[206] J. H. McAlear, G. O. Kreutziger, and R. F. W. Pease, J. Cell Biol., 35, 89A (1967).
[207] F. Anderson and V. F. Holland, J. Appl. Phys., 31, 1516 (1960).
[208] B. J. Spit, Proc. Second Eur. Reg. Conf. Elec. Microscop. De Nederlandse Vereniging Voor Electronenmicroscopie, Delft, 1960, Vol. 1, 564, Delft, 1961.
[209] B. J. Spit, Polymer, 4, 109 (1963).
[210] B. J. Spit, Faserforsch. Textiltech., 18, 161 (1967).
[211] P. Kassenbeck, Bull, Inst. Textile, France, 43, 44 (1953).
[212] P. Kassenbeck, Melliand Textilber., 39, 55 (1958).

[213] E. Jakopic, Proc. Eur. Reg. Conf. Electron Microscop., Delft, 1960, Vol. I, 559, Delft, 1961.

[214] J. Dlugosz, Proc. Fifth Intern. Congr. Electron Microscop., Philadelphia, 1962, Vol. I, Academic, New York, 1962, p. BB11.

[215] G. V. Nikonovich, N. D. Burkanova, S. A. Leont'eva, and Kh. U. Usmanov, Cell. Chem. Tech., 1, 171 (1967).

[216] V. Peck and W. L. Carter, Proc. 26th Ann. Meet. EMSA, Claitor Publishing Co., Baton Rouge, 1968, p. 408.

[217] A. W. Roger, Techniques of Autoradiography, Elsevier, Amsterdam, 1967.

[218] M. M. Saltpeter and L. Bachmann, J. Cell Biol., 22, 469 (1964).

[219] L. C. Caro, in Methods in Cell Physiology (D. M. Prescott, ed.), Vol. I, Academic, New York, 1964, p. 327.

[220] L. C. Caro and R. P. Van Tubergin, J. Cell Biol., 15, 173 (1962).

[221] T. Peters and C. A. Ashley, J. Cell Biol., 33, 53 (1967).

[222] E. D. Hay and J. P. Revel, J. Cell Biol., 16, 29 (1963).

[223] D. Paul, A. Gröbe, and F. Zimmer, Nature, 227, 488 (1970).

[224] D. H. Northcote and J. D. Pickett-Heaps, Biochem. J., 98, 159 (1966).

[225] F. B. P. Wooding, J. Cell. Sci., 3, 71 (1968).

[226] J. Sikorski, J. S. Moss, P. H. Newman, and T. Buckley, J. Sci. Instr., 1, 29 (1968).

[227] C. J. Owen and W. R. Merwarth, Res. Develop., 21 (3), 66 (1970).

[228] K. Mühlethaler, Biochim. Biophys. Acta, 5, 1 (1950).

[229] B. G. Rånby, Diss. Uni. Uppsala, Sweden, 1952.

[230] A. Frey-Wyssling and K. Mühlethaler, Fortschr. Chem. Org. Naturstoffe, 8, 1 (1959).

[231] S. Hestrin and M. Schramm, Biochem. J., 58, 345 (1954).

[232] L. Glaser, J. Biol. Chem., 232, 627 (1958).

[233] J. R. Colvin, S. T. Bayley, and M. Beer, Biochim. Biophys. Acta, 23, 652 (1957).

[234] G. Jayme and G. Hunger, Mikroskopie, 13, 24 (1958).

[235] R. D. Preston, R. Reed, and A. Millard, Nature, 162, 665 (1948).

[236] R. D. Preston and B. Kuyper, J. Exptl. Botany, 2, 247 (1951).

[237] B. G. Rånby, Tappi, 35, 53 (1952).

[238] F. C. Steward and K. Mühlethaler, Ann. Botany (London), 17, 295 (1953).

[239] A. B. Wardrop and S. M. Jutte, Wood Sci. Technol., 2, 105 (1968).

[240] S. Okajima and A. Kai, J. Polymer Sci., Pt. A-1, 6, 910 (1968).

[241] R. D. Preston, E. Nicolai, and B. Kuyper, J. Exptl. Botany, 4 (10), 40 (1953).

[242] P. B. Green, J. Biophys. Cytol., 7, 289 (1960).

[243] A. Myers, R. D. Preston, and G. W. Ripley, Proc. Roy. Soc. (London), Ser. B., 144, 450 (1956).

[244] H. Ruska and M. Kretschmer, Kolloid-Z., 93, 163 (1940).

[245] O. Eisenhut and E. Kuhn, Angew. Chem., 55, 198 (1942).
[246] R. B. Barnes and C. J. Burton, Ind. Eng. Chem., 35, 120 (1943).
[247] W. G. Kinsinger and C. W. Hock, Ind. Eng. Chem., 40, 1711 (1948).
[248] W. Kling and H. Mahl, Melliand Textilber., 31, 407 (1950).
[249] W. Kling and H. Mahl, Melliand Textilber., 32, 1 (1951).
[250] W. Kling and H. Mahl, Melliand Textilber., 33, 32 (1952).
[251] W. Kling and H. Mahl, Melliand Textilber., 33, 328 (1952).
[252] W. Kling and H. Mahl, Melliand Textilber., 33, 829 (1952).
[253] W. Kling and H. Mahl, Melliand Textilber., 35, 640 (1954).
[254] W. Kling and H. Mahl, Melliand Textilber., 35, 1252 (1954).
[255] V. W. Tripp and R. Giuffria, Textile Res. J., 24, 345 (1954).
[256] V. W. Tripp, A. T. Moore, and M. L. Rollins, Textile Res. J., 21, 886 (1951).
[257] A. B. Wardrop, Biochim. Biophys. Acta, 21, 200 (1956).
[258] A. B. Wardrop, Australian J. Botany, 4, 193 (1956).
[259] A. L. Houwink and P. A. Roelofsen, Acta Botan. Neerl., 3, 385 (1954).
[260] I. Gunther, J. Ultrastruct. Res., 4, 304 (1960).
[261] J. C. O'Kelley, Plant Physiol., 28, 281 (1953).
[262] A. Frey-Wyssling and H. Stecher, Experientia, 7, 420 (1951).
[263] P. A. Roelofsen and A. L. Houwink, Acta Botan. Neerl., 2, 218 (1953).
[264] G. Jayme and G. Hunger, Monatsch., 87, 8 (1956).
[265] H. Dolmetsch and H. Dolmetsch, Papier, 22, 196 (1968).
[266] H. Dolmetsch and H. Dolmetsch, Textile Res. J., 39, 568 (1969).
[267] P. Castiaux, G. Raes, and G. Vandermeersche, Ann. Sci. Textiles Belges, 2, 29 (1956).
[268] H. Dolmetsch and H. Dolmetsch, Melliand Textilber., 48, 1449 (1967).
[269] W. Kling, C. Langner-Irle, and T. Nemetschek, Melliand Textilber., 39, 879 (1958).
[270] G. V. Nikonovich, S. A. Leont'eva, Kh. Dustmukhamedoy, Kh. U. Usmanov, and T. G. Gafurov, Uzbeksk. Khim. Zh., 10, 30 (1966).
[271] C. W. Hock, in Cellulose and Cellulose Derivatives (E. Ott, H. M. Spurlin, and M. W. Griffin, eds.), Vol. I, Wiley-Interscience, New York, 1954.
[272] H. Bosshard, Ber. Schweiz. Botan. Ges., 62, 482 (1952).
[273] A. J. Hodge and A. B. Wardrop, Australian J. Sci. Res., Ser. Biol. Sci., 3, 265 (1950).
[274] A. H. Meier, Svensk Papperstid., 61, 633 (1958).
[275] B. Erikson and S. Saverborn, Acta Agr. Suec., 2, 233 (1940).
[276] A. B. Wardrop, Biochim. Biophys. Acta, 13, 306 (1954).
[277] K. Nettelnstroth, Melliand Textilber., 49, 565 (1968).

[278] I. Lambtinou, Melliand Textilber., 49, 251 (1968).
[279] E. J. Shekhterman et al., Zh. Prikl. Khim. (Leningrad) 42 (6), 1425 (1969) (in Russian, through CA, 71, 92528 (1969)).
[280] W. Wergin, Kolloid-Z., 98, 131 (1942).
[281] K. Wuhrman, A. Heuberger, and K. Mühlethaler, Experientia, 2, 105 (1946).
[282] P. Hermans and A. Weidinger, Makromol. Chem., 13, 30 (1954).
[283] A. Vogel, Makromol. Chem., 11, 111 (1953).
[284] S. Mukherjee and H. Woods, Biochim. Biophys. Acta, 10, 499 (1953).
[285] S. Mukherjee, J. Sikorski, and H. Woods, J. Textile Inst., 43, T196 (1952).
[286] A. B. Wardrop and H. E. Dadswell, Australian Pulp & Paper Ind., Tech. Assoc. Proc., 4, 198 (1950).
[287] E. Ribi, Exptl. Cell. Res., 5, 161 (1953).
[288] I. H. Isenberg, Tappi, 39, 882 (1956).
[289] S. Asunmaa and B. Steenberg, Svensk Papperstid., 61, 686 (1958).
[290] S. Asunmaa and P. Lange, Svensk Papperstid., 60, 751 (1957).
[291] A. Asplund, Svensk Papperstid., 61, 701 (1958).
[292] W. A. Coté, Forest Prod. J., 8, 296 (1958).
[293] W. Liese, Proc. Second Symp. Cellulose Res., Dehra-Dun, India, Feb.-Mar., 1958, Council of Sci. Ind. Res., New Delhi, 1960.
[294] W. A. Coté and R. L. Kramer, Tappi, 45, 119 (1962).
[295] W. A. Coté and A. C. Day, Tappi, 45, 906 (1962).
[296] W. A. Coté and R. Marton, Tappi, 45, 46 (1962).
[297] W. A. Coté and A. C. Day, Forest Prod. J., 12, 333 (1962).
[298] G. Casperson, Svensk Papperstid., 68, 534 (1965).
[299] G. Scurfield and S. Silva, Australian J. Botany, 17, 391 (1969).
[300] J. D. Sullivan and I. B. Sacks, Forest Prod. J., 16, 83 (1966).
[301] R. J. Thomas and D. D. Nicholas, Tappi, 52, 2160 (1970).
[302] W. A. Coté and A. C. Day, Tech. Publ. 95, State University College of Forestry at Syracuse University, Syracuse, New York, 1969.
[303] I. B. Sachs, I. T. Clark, and J. C. Pew, J. Polymer Sci., Pt. C, 2, 203 (1963).
[304] E. B. Cowling, in Advances in Enzymic Hydrolysis of Cellulose and Related Materials (E. T. Reese, ed.), Pergamon, London, 1963.
[305] E. B. Cowling, in Cellular Ultrastructure of Woody Plants (W. A. Coté, ed.), Syracuse Univ. Press, Syracuse, 1965.
[306] Z. Koran, Svensk Papperstid., 71, 567 (1968).
[307] G. R. Sears and E. A. Kregel, Paper Trade J., 114, 43 (1942).
[308] H. W. Emerton and V. Goldsmith, Holzforschung, 10, 108 (1956).
[309] H. W. Emerton, Tappi, 40, 542 (1957).

[310] D. H. Page and H. W. Emerton, Svensk Papperstid., 62, 318 (1959).
[311] D. C. McIntosh, Tappi, 50, 482 (1967).
[312] G. Casperson, V. Jacopian, and B. Philipp, Svensk Papperstid., 17, 482 (1968).
[313] A. B. Wardrop, Svensk Papperstid., 66, 231 (1963).
[314] A. B. Wardrop, Tappi, 52, 396 (1969).
[315] D. M. Lantican, W. Coté, and C. Skaar, Ind. Eng. Chem., Prod. Res. Develop., 4(2), 66 (1965).
[316] G. Jayme and N. O. Bergh, Holz Roh- Werkstoff, 26, 427 (1968).
[317] J. Polcin and M. Karhanek, Svensk Papperstid., 17, 441 (1968).
[318] J. C. Guthrie, J. Textile Inst., 47, P248 (1956).
[319] J. Sikorski, Fourth Intern. Conf. Electron Microscop., Berlin, 1958, Springer-Verlag, Berlin, 1960, p. 686.
[320] G. Jayme and K. Balser, Pure Appl. Chem., 14, 395 (1967).
[321] P. Kassenbeck, Bull. Inst. Textile France, 43, 43 (1953).
[322] P. Kassenbeck, Bull. Inst. Textile France, 61, 7 (1956).
[323] P. Kassenbeck, Bull. Inst. Textile France, 61, 15 (1956).
[324] P. Kassenbeck, Ann. Sci. Textiles Belges, 4, 176 (1956).
[325] P. Kassenbeck, Bull. Inst. Textile France, 59, 7 (1956).
[326] A. Gröbe, H. Klare, R. Maron, H. Jost, and G. Casperson, Faserforsch. Textiltech., 13, 1 (1962).
[327] R. Maron, A. Gröbe, H. Klare, and G. Casperson, Faserforsch. Textiltech., 14, 313 (1963).
[328] R. Maron, A. Gröbe, and G. Casperson, Faserforsch. Textiltech., 15, 202 (1964).
[329] A. Gröbe, R. Maron, and K. P. Rose, Faserforsch. Textiltech., 19, 253 (1968).
[330] R. Maron, K. P. Rose, and A. Gröbe, Faserforsch. Textiltech., 19, 319 (1968).
[331] K. P. Rose, A. Gröbe, and R. Maron, Faserforsch. Textiltech., 19, 358 (1968).
[332] H. J. Purz and A. Gröbe, Faserforsch. Textiltech., 19, 460 (1968).
[333] K. P. Rose, R. Maron, and A. Gröbe, Faserforsch. Textiltech., 19, 499 (1968).
[334] A. Gröbe, Chemiefasern, 20, 134 (1970).
[335] A. Gröbe and H. J. Gensrich, Faserforsch. Textiltech., 21 (2), 67 (1970).
[336] A. Gröbe and H. J. Gensrich, Faserforsch. Textiltech., 20, 118 (1969).
[337] H. J. Gensrich and A. Gröbe, Faserforsch. Textiltech., 20, 425 (1969).
[338] H. J. Gensrich and A. Gröbe, Faserforsch. Textiltech., 21, 163 (1970).
[339] H. J. Gensrich and A. Gröbe, Faserforsch. Textiltech., 21, 163 (1970).

[340] L. M. Welch, W. E. Roseveare, and H. Mark, Ind. Eng. Chem., 38, 580 (1946).

[341] C. W. Hock, Textile Res. J., 18, 366 (1948).

[342] K. Mühlethaler, Experientia, 6, 226 (1950).

[343] A. Frey-Wyssling, Makromol. Chem., 7, 163 (1951).

[344] K. Kobayashi and N. Utsumi, Kyoto Daigaku Nippon Kagakuseni Kenkyusho Koenshu, 12, 159 (1955).

[345] B. G. Rånby, H. W. Giertz, and E. Treiber, Svensk Papperstid., 59, 117 (1956).

[346] O. Samuelson, F. Alvang, and A. Svensson, Svensk Papperstid., 59, 712 (1956).

[347] B. Spit, J. Textile Inst., 50, T553 (1959).

[348] W. A. Sisson, Textile Res. J., 30, 153 (1960).

[349] L. H. Sperling and M. Easterwood, J. Appl. Polymer Sci., 4 (10), 25 (1960).

[350] N. Drisch and R. Priou, Bull. Inst. Textile France, 101, 667 (1962).

[351] F. F. Morehead, Tappi, 46, 524 (1963).

[352] W. Kling, H. Mahl, and W. Heumann, Melliand Textilber., 44, 35 (1963).

[353] W. M. Kaeppner, Tappi, 46, 637 (1963).

[354] W. M. Kaeppner, Textile Res. J., 38, 662 (1968).

[355] G. Jayme and K. Balser, Svensk Papperstid., 70, 655 (1967).

[356] E. Balcerzyk and G. Wloderski, Ind. Chim. Belge, 32, 653 (1967).

[357] G. V. Nikonovich, N. D. Burkhanova, S. A. Leont'eva, and Kh. U. Usmanov, Cell. Chem. Technol., 2, 231 (1968).

[358] R. L. Coalson, A. Day, and R. H. Marchessault, Holzforschung, 22 (6), 190 (1968).

[359] N. Boman and E. Treiber, Svensk Papperstid., 71, 1 (1968).

[360] N. Boman and E. Treiber, Svensk Papperstid., 71, 267 (1968).

[361] H. W. Emerton, Tappi, 40, 542 (1957).

[362] A. J. Hodge and A. B. Wardrop, Nature, 165, 272 (1950).

[363] H. Bucher, Attisholz/Solothurn, Switzerland, Cellulose Fabrik Attisholz, A. G. Derendingen, Habegger (1953).

[364] H. Meier, Holz Roh- Werkstoff, 13, 323 (1955).

[365] A. Frey-Wyssling, K. Mühlethaler, and R. W. G. Wyckoff, Experientia, 4, 475 (1948).

[366] I. Nieduszynski and R. D. Preston, Nature, 225, 273 (1970).

[367] R. Jeffries, D. M. Jones, J. G. Roberts, K. Selby, S. C. Simmens, and J. O. Warwicker, Cell. Chem. Technol., 3, 255 (1969).

[368] N. Morosoff and P. Ingram, Textile Res. J., 40, 250 (1970).

[369] D. J. Johnson and C. N. Tyson, J. Roy. Microscop. Soc., 88, Pt. I, 39 (1968).

[370] K. Hess and H. Mahl, Naturwissenschaften, 41, 86 (1954).

[371] B. A. Tonnesen and Ø. Ellefsen, Nor. Skogind., 14, 266 (1960).

270   M. L. ROLLINS, A. M. CANNIZZARO, and W. R. GOYNES

[372] H. Dolmetsch and H. Dolmetsch, Kolloid-Z., 185, 106 (1962).
[373] M. Marx-Figini and G. V. Schulz, Naturwissenschaften, 53, 466 (1966).
[374] R. Muggli, Cell. Chem. Technol., 2, 549 (1968).
[375] R. Muggli, H. G. Elias, and K. Mühlethaler, Makromol. Chem., 121, 290 (1969).
[376] R. D. Preston, Advances in Botanical Research, Vol. 2, Academic, London and New York, 1969.
[377] G. Tomlinson and E. A. Jones, Biochim. Biophys. Acta, 63, 194 (1962).
[378] M. L. Rollins, I. V. deGruy, V. W. Tripp, and A. T. Moore, Am. Soc. Testing Materials Spec. Tech. Publ. No. 257, 153 (1949).
[379] A. M. Cannizzaro, W. R. Goynes, M. L. Rollins, and R. J. Berni, Textile Res. J., 38, 842 (1968).
[380] V. W. Tripp, A. T. Moore, and M. L. Rollins, Textile Res. J., 27, 427 (1957).
[381] R. W. Work, Textile Res. J., 19, 381 (1949).
[382] Am. Soc. Test. Mater. Committee D-13, "ASTM Standards in Textile Materials," Pt. 24, Philadelphia, 1964, ASTM Designation D-1295-67.
[383] R. R. Benerito, R. J. Berni, J. B. McKelvey, and B. G. Burgis, J. Polymer Sci., Pt. A-I, 3407 (1963).
[384] G. L. Drake and J. D. Guthrie, Textile Res. J., 29, 155 (1959).
[385] S. P. Rowland, M. L. Rollins, and I. V. deGruy, J. Appl. Polymer Sci., 10, 1763 (1966).
[386] A. E. Lauchenauer, H. H. Bauer, P. F. Matzner, G. W. Toma, and J. B. Zürcher, Textile Res. J., 39, 585 (1969).
[387] P. Meier and H. Zollinger, Textilveredlung., 5, 709 (1970).
[388] T. L. Ward, R. R. Benerito, and J. B. McKelvey, J. Appl. Polymer Sci., 13, 607 (1969).
[389] J. C. Arthur and R. J. Demint, Textile Res. J., 30, 505 (1960).
[390] F. A. Blouin and J. C. Arthur, Textile Res. J., 33, 727 (1963).
[391] M. L. Rollins, A. T. Moore, and V. W. Tripp, Textile Res. J., 33, 117 (1963).
[392] K. Kh. Razikov and Kh. U. Usmanov, Polymer Sci., USSR, 8, 421 (1966).
[393] R. W. Liggett, H. L. Hoffman, A. C. Tanquary, and C. Hamalainen, Am. Dyestuff Reptr., 58 (94), 25 (1969).
[394] F. A. Blouin, A. M. Cannizzaro, J. C. Arthur, and M. L. Rollins, Textile Res. J., 38, 811 (1968).
[395] C. W. Hock, Textile Res. J., 20, 141 (1950).
[396] B. G. Rånby and E. Ribi, Experientia, 6, 12 (1950).
[397] E. H. Immergut and B. G. Rånby, Ind. Eng. Chem., 48, 1183 (1956).
[398] H. Krassig and W. Kitchen, J. Polymer Sci., 51, 123 (1961).

[399] H. Krassig and W. Käppner, Makromol. Chem., 44/46, 1 (1961).
[400] O. A. Battista and P. A. Smith, Ind. Eng. Chem., 54, 20 (1962).
[401] O. A. Battista, Am. Sci., 53, 151 (1965).
[402] O. A. Battista, N. Z. Erdi, C. F. Ferrare, and F. J. Karosinski, J. Appl. Polymer Sci., 11, 481 (1967).
[403] J. Polcin and M. Karhanek, Faserforsch. Textiltech., 14, 357 (1963).
[404] B. R. Porter, V. W. Tripp, I. V. deGruy, and M. L. Rollins, Textile Res. J., 30, 510 (1960).
[405] B. R. Porter and W. R. Goynes, Textile Res. J., 34, 467 (1964).
[406] W. Bobeth, H. Luczak, H. Poessler, and H. Tausch-Marton, Faserforsch. Textiltech., 18, 547 (1968).
[407] Y. V. Brestkin, J. Anal. Chem. USSR, 40, 2449 (1967).
[408] E. T. Reese, L. Segal, and V. W. Tripp, Textile Res. J., 27, 628 (1957).
[409] B. R. Porter, J. H. Carra, V. W. Tripp, and M. L. Rollins, Textile Res. J., 30, 249, 259 (1960).
[410] T. H. Liu and K. W. King, Arch. Biochem. Bioplys., 120, 462 (1967).

Chapter 5

INSTRUMENTAL METHODS IN THE STUDY OF OXIDATION,
DEGRADATION, AND PYROLYSIS OF CELLULOSE

Pronoy K. Chatterjee and Robert F. Schwenker, Jr.

Personal Products Co.
Subsidiary of Johnson & Johnson
Milltown, New Jersey

I. MECHANISM OF THERMAL DEGRADATION

A. Introduction

The thermal degradation of cellulosic materials proceeds through a series of complex chemical reactions. The reactions are highly influenced by the nature and period of heating, the evironmental atomsphere, the inorganic impurities, the noncellulosic ingredients, etc. The reaction pathways can be broadly classified into two categories. One involves fragmentation and formation of combustible volatiles and another involves dehydration and the formation of carbonaceous material or char. The former could feed the flames whereas the latter could lead to glowing ignition or slow burning. Heating at lower temperatures favors dehydration and the subsequent char formation. Heating at higher temperatures favors the formation of flammable volatile products. Of course, at higher temperature also, dehydration and the formation of char is evident. Perhaps, both reactions compete with each other at all temperatures, and one becomes predominant over the other, depending upon the condition of the substrate and temperature. An outline of the general reactions that could lead to the flaming combustion or glowing ignition is shown in Fig. 1 as presented in a review article by Shafizadeh [1].

In the following discussion information pertinent to the mechanism of thermal degradation of cellulose has been summarized into two categories: (a) low temperature degradation and (b) high temperature degradation.

B. Low Temperature Degradation

Perhaps the most important factor influencing cellulose properties at low temperatures (below 100-140°C) is not the temperature but the duration of heating under a given set of conditions. According to Richter [2], when

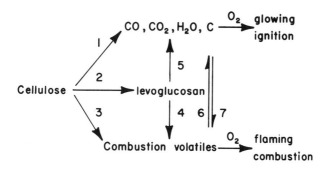

Fig. 1. Major pathways for cellulose thermal decomposition reactions (Shafizadeh [1]).

a rag paper is heated at 38°C for six months, the folding strength decreases to about 19%. Farquhar and co-workers [3] heated raw cotton at temperatures below 100°C for 4-24 h and found considerable degradation. It is also believed that moisture and oxygen play important roles in the thermal degradation at or below 100°C [3-11]. For example, the breaking strength and fluidity of heat-treated purified cotton and viscose appeared to be functions of the relative humidity and the initial moisture content of the materials. Other factors that influence the degradation mechanism are the surrounding atmosphere, impurities (such as salts, resin, and waxes), average chain length, and possibly the supramolecular structure. The initial degradation reactions include depolymerization, hydrolysis, oxidation, dehydration, and decarboxylation.

The first noticeable result of the effect of heat on pure cellulose is the loss of water. On the other hand cellulose triacetate evolves acetic acid instead of water [12]. Therefore, dehydration occurs with cellulose and deacetylation occurs with cellulose triacetate, presumably at random along the chain of glucose units [12, 13]. The decomposition products of several types of chemically modified celluloses are also consistent with this general mechanism [13]. The proportions of carboxyl and carbonyl groups formed on heating cellulose at 170°C in oxygen and nitrogen atmosphere, in terms of milliequivalents per mole of D-glucose residue, have been reported [1]. It is evident that the formation of carboxyl and carbonyl groups is very much influenced by the type of atmosphere. Evolution of water shows a small initial change in the nitrogen atmosphere and rapid progress in the oxygen atmosphere.

Thermal degradation was invariably found slower in the absence of oxygen, which, according to Rutherford [14], indicates that the initial stages of breakdown, at least, include oxidative degradation. Later work [13, 15, 16] characterized the mechanism as essentially nonoxidative in the higher temperature range. However, it is not unlikely that the degradation mechanism at lower temperatures is substantially different from that at higher temperatures. After the initial period of heating in oxygen, the degree of polymerization of cellulose drops to a constant value which corresponds to the postulated length of the crystalline micelles [17, 18]. This observation apparently supports the theory that initial thermal degradation of cellulose proceeds primarily through the oxidation process, with oxygen acting as a nonspecific oxidant.

The formation of ether crosslinkages has been proposed [19] in order to explain certain properties of cellulose after exposure to heat. More recently this autocrosslinking mechanism has been supported and further discussed in connection with high temperature degradation [20]. Back and co-workers [21-24] suggested that crosslinking occurs during thermal treatments of lignocellulosic and cellulosic papers and board materials. They discussed the effect of autocrosslinking [22] on the thermal softening of cellulose, which has been defined as the melting transformation of cellulose crystallites [25-27].

C. High Temperature Degradation

1. Degradation Products

The study on the high temperature decomposition of cellulose (above 250° C) was initiated more than seventy years ago [28]. In 1909, Klason et al [29] reported the reaction products from the destructive distillation of cotton cellulose as water, acetic acid, acetone, tars, CO_2, CO, CH_4, C_2H_4, and coke (char). The next and perhaps the most pioneering effort in the study of pyrolytic degradation of cellulose was the isolation and identification of levoglucosan (1, 6-anhydro-B-D-glucopyranose), in the products obtained from the destructive distillation of cellulose in vacuum by Pictet and Sarasin [30] in 1918.

The literature up to 1947 was reviewed by Coppick [31], with the products of decomposition divided into three main categories as follows [32]:

(i) Gas phase products, which consist of carbon monoxide, carbon dioxide, methane, and hydrogen, usually constituting 20 to 25% of the total products of pyrolysis.

(ii) Liquid phase products, which can be collected as an aqueous distillate (the pyrolozate), consist of tars, water, and volatile condensables at 0 °C in approximately 65% of the products.

(iii) Solid phase products, which consist of a carbonaceous residue of char in approximately 15% of the products.

Subsequent investigations established the fact that levoglucosan is the major product among the numerous volatile products obtained from the pyrolysis of cellulose. In addition to levoglucosan, the cellulose pyrolyzate was also found to contain water plus a number of organic acids, aldehydes, and ketones. An extensive tabulation of these volatile products from cellulose decomposition is available from the work of Schwenker and Pacsu [32] and Schwenker and Beck [16], as shown in Table 1. Using gas-liquid chromotography involving the pyrolysis of cotton, they isolated thirty-seven gas chromatographic peaks. The total number of volatile compounds isolated and identified to date is well over fifty. The larger volatile molecules generated from cellulose above 250°C are levoglucosan (over 80%), 1, 6-anhydro-D-glucofuranose (1.5%), 5-hydroxymethylfurfuraldehyde (1.3%), and 1,4: 3.6-dianhydro-D-glucopyranose (0.7%). Miscellaneous other products such as glyoxal, formic acid, and at least ten polynuclear hydrocarbons such as acetonaphthylene, coronene, etc., are also reported [15, 33].

The characterization and mode of formation of volatile products from cellulose nitrate was extensively studied by Wolfrom and co-workers [34-37]. In addition to standard isolation and characterization they also pyrolyzed cellulose nitrate labeled with ^{14}C at positions C-1 and C-6. C-1 was predominantly converted to CO_2 and lesser amounts of formic and glyoxal, whereas C-6 yields mainly formaldehyde and C-2 and C-5 contribute

TABLE 1

Pyrolytic Degradation Products of Cellulose[a]

Compound	Method of analysis
Carbon monoxide	GLC[b]
Carbon dioxide	GLC
Formaldehyde	Crystalline deriv. GLC
Acetaldehyde	GLC
Acrolein	GLC
Propionaldehyde	GLC
n-Butyraldehyde	GLC
Glyoxal	Crystalline deriv. GLC
Furfural	GLC
5-Hydroxymethyl furfural	GLC
Acetone	GLC
Methyl ethyl ketone	GLC
Methanol	GLC
Formic acid	Paper chrom. GLC
Acetic acid	Paper chrom. GLC
Lactic acid	Paper chrom. GLC
Dilactic acid	Paper chrom.
Glycolic acid	Paper chrom.
Water	GLC
Levoglucosan	Paper chrom. GLC

[a]R. F. Schwenker and E. Pacsu, Chem. Eng. Data Series, 2, 83 (1957); R.F. Schwenker, and L.R. Beck, J. Polymer Sci., C2, 331 (1963).

[b]GLC - gas liquid chromatography.

mainly formic acid and carbon dioxide. Gaelernter et al. [38] also studied the decomposition of cellulose nitrate by radioactive tracer techniques with ^{15}N and concluded that the nitrate at C-6 was more stable than the nitrates at C-2 and C-3. The mechanism and pyrolytic products of cellulose acetate are available in a review by DePuy and Kings [39]. More recently, the decomposition of partially substituted benzhydrylated cotton cellulose was studied [40], and it was proposed that the ether groups undergo a combination of elimination and disproportionation reactions resulting in a cellulosic residue plus a mixture of diphenyl methane and benzophenone.

Modern instrumental techniques such as mass spectrometry, gas chromatography, and infared spectroscopy have helped to analyze the decomposition products extensively. These techniques and their specific applications in cellulose decomposition are discussed later in this chapter.

2. Reaction Mechanism

In 1921 Irvine and Oldham [41] proposed that the first step in the high temperature (pyrolytic) degradation of cellulose was decomposition by hydrolysis to glucose followed by dehydration of the glucose to form levoglucosan. Thus, they believed that the cellulose decomposition mechanism was based on the formation of glucose as an intermediate compound.

Parks and co-workers [42] and Ivanov et al. [43] independently proposed instead a levoglucosan intermediate mechanism. The essential features of the levoglucosan intermediate theory were first suggested by Tamaru [44], although the intermediate was not identified. Depolymerization of cellulose to levoglucosan involving scission of 1, 4 glucosidic linkages with intramolecular rearrangement to the glucosan, proposed in this mechanism, was considered as the first and rate-determining step, as shown in Fig. 2. Levoglucosan was then said to participate in two competing reactions: (a) molecular rearrangement to from 5-hydroxymethyl furfural and polymerization with aromatization to form char and (b) destructive distillation to form flammable tars, volatile gases, and water.

Support of the levoglucosan theory emerged when it was found that on pyrolyzing cellulose samples, to which glucose had been added, the levoglucosan content was depressed [45-48], as shown in Table 2. A decrease would not be expected if levoglucosan was formed by the simple dehydration of the glucose. The alteration of physical structure and the packing density of cellulose has a pronounced effect on the nature of the pyrolysis reaction and the yield of levoglucosan [49]. Pacsu and Schwenker [50] reported that no evidence of glucose was found by paper chromatographic analysis in any of the pyrolyzates obtained by pyrolysis of cotton in air.

Madorsky [13] suggested that cellulose undergoes simultaneous dehydration and random scission at C-O bonds, since C-O bonds are thermally less stable than the C-O bonds. Some cleavage of glucosidic bonds (C1-O, i. e., bond d in Fig. 3) yield hexose units that rearrange to form levoglucosan.

TABLE 2

Dependence of Levoglucosan Yield on
the Experimental Conditions [49, 50]

Cellulose compound	Environment	Levoglucosan yield (%)
Cotton	Vacuum	60
Cotton	Vac in presence of water	58-60
Cotton	Vac in presence of glucose	25-30
Hydrocellulose	Vacuum	5-14
Cotton	Air	10-15
Oxycellulose	Air	4-6

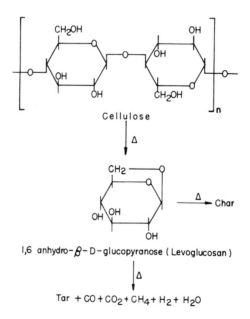

Fig. 2. Formation of levoglucosan and char during the thermal
decomposition of cellulose (Pacsu and Schwenker [50]).

Other cleavages of C–O bonds (C5–O, C1–O involved in the oxygen ring and C4–O bond connecting the d-glucose residue, i.e., a, b, c in Fig. 3) cause breakdown of part of the chain, yielding carbon monoxide and carbonaceous residue. Schwenker and Beck [16] observed that the pyrolysis of cellulose at different rates of heating and in different atmospheres gave very similar products. This observation further supports a nonoxidative mechanism.

Gardiner [51] discussed the pyrolysis mechanism, along the line suggested by Madorsky, as two modes of reaction: the glucose units are (i) preserved as hexose units, e.g., in the form of 1, 6-anhydro-β-D glucopyranose or 1, 6-anhydro-β-D-glucofuranose or (ii) destroyed by fragmentation, or by condensation to a char. The change in conformation of the glucose unit from the normal C1 to the form B1, as shown in Fig. 4, governs the rate of formation of levoglucosan. According to Gardiner, the formation of levoglucosan requires the change from C1 to 1C or 3B, and in the pyrolysis process ample energy is available for these changes. For the mode (ii) reaction they preferred the idea of the carbonium ion intermediate mechanism, originally proposed by Schuyten et al [52]. However, the possibility of a free radical mechanism had not been ruled out. They also suggested some possible mechanisms for the formation of aldehydes, ketones, and acids. It is also reported in the literature [53] that simultaneous oxidation-reduction of adjacent aldehydes and alcohol groups results in the formation of hydroxy acids or hydroxy ketones. On further pyrolysis at higher temperatures numerous secondary conversions could take place, resulting in hundreds of volatile products and fixed gases [12].

An intermediate free radical peeling mechanism has been advanced [54-56]. In one theory [54], mono- and difunctional radicals are formed by the cleavage of glucosidic linkages, which radicals in turn give rise to volatile products including levoglucosan, as shown in Fig. 5. The mechanism suggested is in general conformity with that summarized by Bull [57]. There is, however, some difference between the mechanism stated by Bull and that proposed by Pakhomov [54]. Bull stated that during the initial stage of degradation, cellulose undergoes random chain scission, the excited anhydroglucose group at one end of the break rearranges, transmits the free radical to the next anhydroglucose unit in line, and a levoglucosan molecule splits off. Pakhomov stated that the radicals first split off from the chain and then interact with each other to form levoglucosan.

The proposal of a free radical mechanism is not unreasonable, but vague due to lack of experimental evidence. No one has yet conclusively proved the participation of a free radical in the pyrolysis reaction. However, it must be noted that the appearance of free radicals during the thermal decomposition of cellulose has been recently observed [58, 59] using electron spin resonance. It was concluded that at temperatures below the decomposition temperature, the relative concentration of free radicals in cellulose reaches some equilibrium concentration. Again at 225°C, the concentration of free radicals appears to increase linearly with time. Esteve and Wright

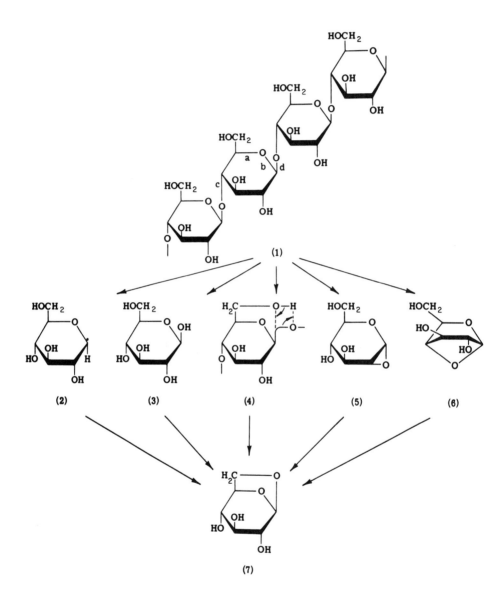

Fig. 3. Intermediates that lead to levoglucosan during the pyrolysis of cellulose (Shafizadeh [1]).

[60] disagreed with the free radical reaction concept on the basis that the energy of activation for the pyrolysis of cellulose was found to be substantially lower (33 ± 3 kcal) than that expected by a free radical mechanism. The argument is not sound since many secondary reaction processes could substantially lower the overall apparent energy of activation.

On the other hand, Kato [61], who supported the intermediate free radical mechanism, rejected the levoglucosan mechanism proposed by Parks et al. [42] based on their observation on the relative amounts of furfural and furan in the cellulose pyrolyszate and levoglucosan pyrolyzate. His argument is that, if most of the pyrolytic degradation proceeds through levoglucosan, the amounts of the volatiles from levoglucosan should be higher than those from cellulose, which was contrary to his experimental observations. However, this observation, though important, is not sufficient to pin down the conclusion drawn. It is known that levoglucosan polymerizes on heating [62], so that the char formation would depend significantly upon the amount of polymer present which, in turn, is dependent upon the initial concentration of its monomer, levoglucosan.

In the case of cellulose, suppose the levoglucosan is the intermediate step. The major portion of it, as soon as it forms, would then immediately undergo volatilization and/or secondary transformation, whereas if pure levoglucosan is taken as the starting material it is likely that the polymerization rate would increase. Consequently, the proportion of the total amount of volatile products will be relatively small and the amount of char will be greater.

Kato [61] finally suggests that the volatile products are produced through two primary reactions: (a) the initial scission of glucosidic linkages to produce radicals and (b) chemical changes in anhydroglucose units of cellulose before scission of the glucosidic linkages.

The possibility of the formation of intermolecular crosslinks between any of the free hydroxyl groups of one glucose unit with another unit due to dehydration during the pyrolysis of cellulose was proposed [20], but other work [63] contradicts the idea of crosslinking in proposing the dehydration mechanism as an intramolecular phenomenon.

It is evident from the literature that the mechanism of the pyrolysis reaction is not yet fully resolved as to whether it is ionic or free radical. However, the preponderance of the experimental data strongly favor an ionic mechanism. It is possible that the reaction mechanism could be significantly influenced and altered by changing the pyrolysis temperature such that it is also possible that the reaction comprises a combined ionic and free radical mechanism, both acting simultaneously. The literature reveals an enormous quantity of data on the pyrolytic products, but much remains to be learned about the actual mechanism of this very complex reaction.

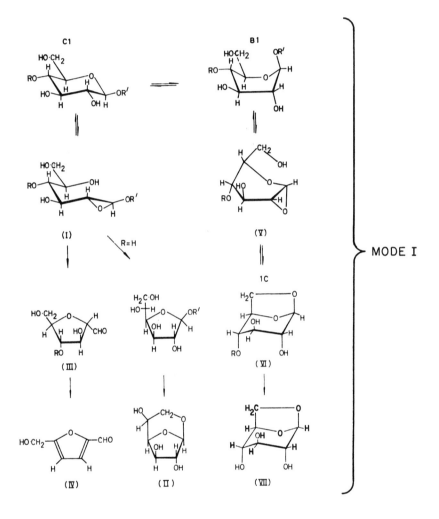

Fig. 4. Two modes of cellulose pyrolysis reactions. Mode 1: mechanism of formation of anhydroglucose and furfurals. Mode 2: mechanism of formation of carbonyl compounds (Byrne et al. [15]).

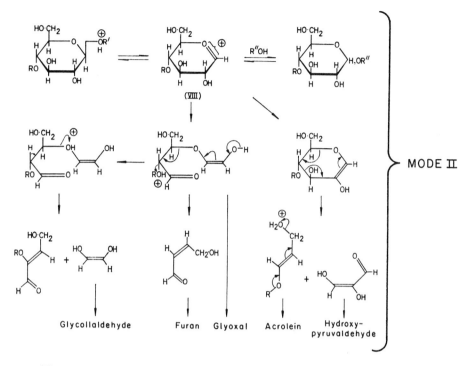

Fig. 4, continued.

3. Kinetics of Reaction

Since so many products result from the thermal degradation and de-composition of cellulose, a complete understanding of kinetics of the process is difficult, if not impossible, with the present state of knowledge. The major portion of the kinetic study has been mainly concerned with the determination of overall kinetic parameters (order, energy of activation, and Arrhenius frequency factors). The results of kinetic measurements merely furnished certain facts, and the mechanisms proposed, in some instances, are the models devised to explain those facts.

Parks et al. [42] proposed that the depolymerization of cellulose to levoglucosan was the first and rate-determining step in its thermal degradation. Madorsky [13] summarized his work on kinetics with the conclusion that after a period of initial decomposition, a first-order law was obeyed. The kinetics of the initial decomposition were found to be more complex than the latter part. A first-order reaction during the latter part of the reaction was reported by many workers [42, 64-68]. Tang and Neill [68] obtained a pseudo-zero-

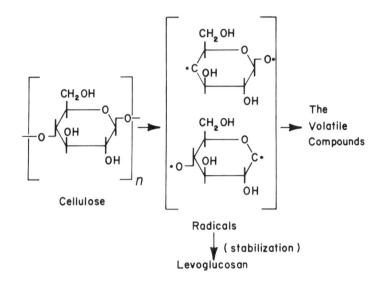

Fig. 5. Intermediate free radical mechanism for cellulose
pyrolysis (Pakhomov [54]).

order type of reaction for the initial part of the reaction from a temperature
programmed kinetic study. Their observation was supported by Lipska and
Parker [64] who studied isothermal kinetics in a fluidized bath. In all of
the above cases, attempts were made to determine the Arrhenius parameters
only by empirical approaches. It is, however, difficult to understand the
real meaning of the order of reaction applied to the thermal decomposition
of long chain polymer molecules, other than its value as an empirical correlation parameter of the relationship between weight loss and rate of decomposition.

Gizycki [69] reported that the weight loss of cellulose nitrate during
heating at 100°C was retarded by diphenylamine and other free radical scav-
engers. This indicates that the reaction, at least in part, is a free radical
type. However, he also proposed that an initial step in this decomposition
was depolymerization, which is not free radical in nature.

From the weight loss measurements and analysis of gaseous products,
Andraev and Samsonov [70] proposed that cellulose decomposition comprised
a combination of simultaneous and consecutive reactions. Andraev [71]
found that at lower temperatures the degradation reaction was first order
but not first order at higher temperatures. The chemical and physical
state of the samples have been shown to affect the rates and mechanism of

reaction [72]. Occurrence of two distinctly different reactions at two temperature levels: (i) 175-250°C and (ii) over 300°C was also reported [73].

Golova and co-workers [45, 46] claimed levoglucosan yields up to about 60%, with chain degradation of the cellulose macromolecule. However, they did not treat the data according to chain reaction theory and therefore their statement on chain decomposition was merely conjecture.

More recently, the major reaction of cellulose pyrolysis has been defined by Chatterjee and Conrad [74] as a chain reaction type consisting of a two-step process. The first step is chain initiation involving glucosidic bond scission. Once fragmented chains are produced due to scission, they undergo further depolymerization from one end due to thermal instability to yield levoglucosan molecules. This reaction should continue either until the molecular chains are unzipped or until other processes such as repolymerization and carbonization become controlling. The unzipping or depolymerization reaction, which is the second step of the process, could be considered as chain propagation.

In the early stage, both processes occur concurrently and continue as long as glucosidic bond scissions are initiated. But, gradually, all the chains which were initiated will transform to degraded molecules consisting of thermally unstable end groups. At this stage the major reaction would be exclusively the propagation type (end group cleavage). Two separate kinetic expressions for the initial and the latter portion of the pyrolysis, respectively, have been derived to analyze the fundamental steps of the major reaction. The occurrence of a pseudo-zero-order type of rate process at the beginning of the reaction and a pseudo-first-order type at the later stage of the reaction could be explained by the chain reaction mechanism [74-76].

Recently, a rigorous kinetic treatment was proposed [77] and applied to the temperature-programmed thermogravimetric analysis of cellulose to further elucidate the chain reaction mechanism of cellulose pyrolysis. Results of the thermal degradation of trityl cellulose derivatives were also consistent with the same mechanism up to the stage of trityl scission [78]. Trityl scission takes place at a very low energy of activtaion, 2.5 kcal/mole, and such a low activation energy is itself indicative of a free radical mechanism [78].

The chain reaction process, though discussed on the basis of the levoglucosan intermediate mechanism, is not contradictory to the free radical mechanism. The process, in general, is based upon the criteria of random scission, the formation of thermally unstable degraded product, and the unzipping reaction. These criteria could as well be applied in a reaction system where free radicals are participating. An interesting feature of the pyrolysis of cellulose is the residue which remains as the char. The phenomenon of char formation has not been kinetically treated, it being merely hypothesized that such a step occurs.

4. Structure of Char

The mechanism of char formation during cellulose pyrolysis is more obscure than the mechanism of volatile compound formation. Such information would be extremely important from the point of view of the flame retardancy mechanism. In general, a flame-retardent causes less tar and more char residue and gases [31]. The charred residue exhibits further weight loss and passes through advanced stages of carbonization on being heated at higher temperature. The yield and properties of the char depend upon the rate and other conditions of heating.

Otani and co-workers [79, 80] reported that the carbonization of cellulose proceeded with the reduction of hydroxyl groups and the formation of carbonyl groups. Higgins [81, 82] indicated the presence of stable unsaturated rings containing carboxyl groups in the residue when cellulose is pyrolyzed.

Hofman et al. [83] indicated that when cellulose was heated above $300°C$, the cellulose structure was gradually replaced by an aromatic condensed system and eventually by a charry material. It has been noted that "cellulose coke" has a composition similar to coal except for a somewhat lower hydrogen content [7, 53, 84]. Cobb [85] studied x-ray diffractograms of cellulose residues from 200-500°C at $100°C$ intervals. At 300°C a breakdown of the cellulose structure occurred, at $400°C$ a carbon repatterning was in evidence, and at $500°C$ the formation of a hexagonal network of carbon atoms on parallel sheets, similar to that of a graphite structure, was found.

Davidson and Losty [86] studied the carbonization of cellulose by programmed heating up to $1500°C$ and investigated the changes in cellulose by weight loss, infrared spectroscopy, x-ray diffraction, electron-spin resonance, and electrical, mechanical, and other physical properties. The structural changes were found to take place after 40% weight loss which corresponded to $250°C$ temperature. The crystalline structure was found to be destroyed rapidly coincident with the disappearance of an infrared band assigned to C-1 of the D-glucose residues and the appearance of the carboxyl group. This latter change was attributed to the cleavage of the oxygen ring and the formation of the carboxyl groups. The subsequent change involved the removal of the oxygen at C-1, resulting in the formation of aromatic units with the liberation of carbon monoxide, carbon dioxide, water, etc. This change corresponds to 40-60% weight loss in the temperature range of 240-370°C. During the weight loss between 60 to 68% (370-850°C) hydrogen was driven off resulting in a series of changes in the structure and arrangements of the aromatic residues. According to this theory, the residue obtained on complete carbonization of cellulose at 1500°C is a result of direct transformation of the cellulose chain molecule and not from cracking of the transitory volatilization products [1]. The carbonization of cellulose is shown in Fig. 6.

The effect of various salts and organic compounds on the relative amount of char formation is available in the voluminous literature concerning "flameproofing" or "flame retardancy." Several theories have been proposed

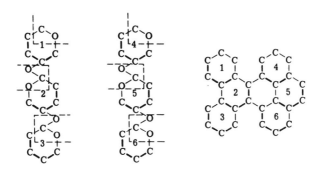

Fig. 6. Carbonization of cellulose (Shafizadeh [1]).

to explain the flame proofing of cellulose involving only speculation as to the mechanism of char formation.

II. GAS CHROMATOGRAPHY

A. Principles

Gas or vapor phase chromatography has been extensively used in identifying the thermal decomposition products of cellulose. The term "gas chromatography" includes all chromatographic techniques in which a mobile gas phase carries substances to be separated through a stationary phase packed into a suitable column. The classification of gas chromatography is based upon the nature of the stationary phase used: gas-solid chromatography (GSC) utilizes a solid with an active surface whereas gas-liquid chromatography (GLC), a light coating on a solid support. Gas-liquid chromatography was suggested by Martin and Synge [87] in 1941 and demonstrated through the experiments of James and Martin in 1952 [88] .

The term chromatography has been defined as follows: chromatography is a physical method of separation in which the components to be separated are distributed between two phases, one of the phases constituting a stationary bed of large surface area, the other being a fluid that percolates through the stationary bed. The principle of gas chromatography is based upon this general chromatographic principle. The evolution of gas-solid chromatography can be obtained from the literature, a bibliography of which has been published by Claesson [89] . An excellent review article has been published by Hardy and Pollard [90] on gas liquid chromatography. Several texts are available which review the technique both in theory and in practice [91-98] . The specific application of gas chromatography in the textile fiber field has recently been described [98] . GSC, the less flexible technique of the two, is primarily used in the case of "fixed" gases, whereas GLC has been widely used for a great variety of mixtures and systems.

Gas chromatography is almost always carried out by elution. However, there are two other methods available for transporting the sample through the column [99]. These have been termed frontal analysis [89] and displacement development [95]. In the frontal analysis, the vaporized sample is carried onto the chromatographic column continuously and at constant concentration in the mobile phase. Unlike the elution technique where peaks are recorded, in frontal analysis steps are recorded in the thermogram. The first step represents a pure compound and all subsequent steps will indicate mixtures. In displacement development, the sample mixture is introduced into the column, and the carrier gas is subsequently admitted. The chromatograph will consist of bands of relatively pure substance, separated by small intermediate zones of mixtures of two adjacent substances, and followed by displacing carrier gas. Since the majority of applications has been associated with the elution technique, this section will be devoted to a more detailed discussion of that particular technique.

B. Experimental Features of the Gas Chromatographic Elution Technique

A gas chromatographic measurement, in general, proceeds as follows: An inert gas, serving as an eluent carrier gas, is allowed to flow through a pressure regulating device, a buffer tank, a sample injector, one or more chromatographic columns, a detector, and finally a freeze-out trap as shown in Fig. 7. The apparatus basically consists of a handling system for the carrier gas, a sample injector device, one or multiple narrow columns constructed of either glass, stainless steel, or copper, and differential detector circuits. In temperature-programmed gas chromatography there is an additional heating system for increasing the column temperature at a required linear rate. The time of separation of the mixture is considerably reduced by running at constantly increasing column temperature.

A small sample of the volatile mixture to be separated is introduced into the column. The column temperature is either maintained isothermally or increased at a linear rate as mentioned above. A constant current of a selectively chosen inert gas is passed through the system and transports the components of the mixture in vapor state. Separation is achieved on the column by the partition function principle. The most widely recognized theories for the separation are (a) the plate theory and (b) the rate theory, which have been critically reviewed [99].

The components of the mixture emerge from the column as individual "bands" separated by zones of the carrier gas. The components leaving the column in the gas stream and their concentrations are recorded as a function of time, or of the volume of the carrier gas by a differential detecting circuit which registers some momentary property of the effluent gas. Figure 8 shows a typical elution diagram or elution curve from a gas-liquid chroma

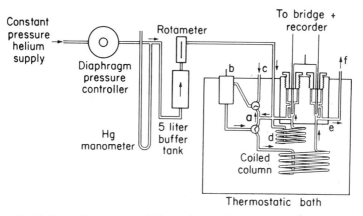

Fig. 7. Schematic representation of gas chromatography apparatus:
(a) by-pass sample injection valves, (b) by-pass sample injection chamber,
(c) vacuum, (d) helium pre-heater, (e) thermal conductivity detector,
(f) to flow measuring device or cold traps (Stross and Johnson [99]).

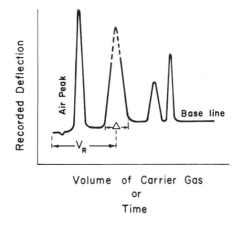

Fig. 8. A typical elution diagram from gas-liquid chromatogram
showing construction of peak width.

chromatograph with differential detector. For further analysis, the eluate
may be collected in suitable cold traps, a procedure used in preparative
gas chromatography.

C. Definitions

The experimental quantities obtained from the chromatographic record are peak width, peak area, and retention volume [99-101].

Peak width: This is defined as the segment of the peak base intercepted by tangents to the inflection points in both sides of a chromatographic peak, as shown in Fig. 8.

Peak area: The total area under the peak.

Retention volume: The volume of gas required to elute the compound under study and given by $V_R = t_R F_c$, where t_R is the retention time, the time for the emergence of the peak maximum after the injection of the sample and F_c is the volumetric flow rate of the carrier gas. This retention volume must be corrected for pressure drop and is known as the corrected retention volume V_R^o.

Specific retention volume: The term is defined as the corrected retention volume at $0\,°C/g$ of liquid phase. Its relationship with the partition coefficient is given by the following equation:

$$V_g = \frac{273k}{T\rho_L} = \frac{273\ (V_R^o - V')}{TW_L},$$

where V_g is the specific retention volume, T is the temperature of the column in degrees Kelvin, ρ_L is the density of the column liquid (grams/milliliter) at column temperature, and W_L is the weight (grams) of the liquid phase of the column. V' is the interstitial gas volume of the column.

Efficiency: The capability of a column to produce relatively narrow peaks.

Resolution: The capability of a column to separate a specific pair of materials under specific conditions.

D. Factors Affecting Gas Chromatography

The success in the separation and identification of compounds depends upon various instrumental factors, experimental procedures, and operator technique. Suggestions and guidelines given in the texts cited above are helpful for developing skill in producing improved chromatograms of difficult and complex mixtures. Reviews and specific examples are available [99, 102].

There are five main factors which are more important for the analysis of compounds. These are briefly discussed below:

(1) The carrier gas. The carrier gas must be inert and the following gases are used: argon, helium, hydrogen, nitrogen, and carbon dioxide. No one gas can represent the choice for all conditions. The selection depends upon the type of detector and the column efficiency requirement.

(2) Column efficiency. The efficiency of column depends upon its dimensions, the amount of stationary liquid, i.e., the liquid phase, per gram of solid support, the thermal stability and physical properties of the stationary

liquid, and the nature of the solid support in general, i. e., the pore size distribution, particle size distribution, port configuration, surface and overall physical properties of the particles, etc.

(3) Introduction of sample. The sample may be introduced into the apparatus as a gas, liquid, or a solid. Accessories are available with commercial equipment for introducing sample in any of the forms. Efficiency and resolution of the chromatogram are improved by reducing sample size and injection time, respectively. In the case of solid samples, the material is thermally degraded (pyrolyzed) by various techniques [103-106].

With commercial equipment two general types of pyrolyzers are available: hot wire loop or coil and hot chamber. In the hot wire pyrolyzer, the sample, if soluble in any solvents, is coated on the wire or, if not readily soluble, is placed in or on the wire. The coil is heated by means of an electric current. In the hot chamber pyrolyzer, the sample is placed in a ceramic boat and moved into and out of the chamber with the aid of a mechanical or magnetic device. The sample size is preferably on the order of a few milligrams, but microgram quantities can also be used.

(4) Detectors. The ideal requirements for a detector are high sensitivity to the presence of a component in the carrier gas, rapid response, linear respones, independence of operating variables and good baseline stability. There are two general types of detectors; integral and differential. The differential detectors are most widely used and are based on thermal conductivity measurements. However, for highest sensitivity the flame ionization detector [107] or the capillary tube detector [108] may be used.

E. Analytical Procedures: Qualitative Analysis

The constituents of a mixture can be identified by a simple comparison of their retention volumes or times with those of known compounds as reference samples. The identification is aided by the addition of a compound which had been expected to be present in the mixture, thereby increasing the peak height, and finally, by isolation and separate identification. If the compounds belong to a homologous series, the semilog plots of retention volume versus the number of carbon atoms are almost linear. This type of plot is also useful in identifying the compound.

There are at least three standard approaches for the quantitative analysis of the constituents of a mixture: (i) Measurements of peak height and area are compared with a calibration curve. (ii) Internal standard calibration, which consists of introducing a known reference sample and finding an empirical correlation between the peak height and/or the peak area and the amount introduced. (iii) Internal normalization which is an alternative of procedure (i) consisting of measuring the areas of all the peaks present and adding to determine a total area. The ratio of any individual areas to this total gives the estimation of weight in percentage amounts.

F. Applications

When cellulose is pyrolyzed it yields several fixed gases, volatile products having a wide range of boiling points from about -78 °C to 100° C, and char. Gas-solid chromatography is ideally suited for separating and analyzing the fixed gases whereas for the liquid volatile organic compounds, gas-liquid chromatography is more advantageous. Since the products of pyrolytic degradation of cellulose consist of more than 70% volatile products, gas-liquid chromatography has been more extensively used.

Martin and Ramstad [109, 110] and Berkowitz-Mattuck and Noguchi [111] have made use of both GSC and GLC. Martin and Ramstad [109] constructed a two-stage system which permitted the direct measurement of reaction products from the flash pyrolysis of cellulose using the technique developed by Nelson and Lundberg [112].

The gas-liquid column was propylene-glycol and the gas solid column was activated charcoal. The sensitivity of the system was such that in general it could provide quantitative values for sample sizes down to about 10^{-8} mole. The resolution of the column was claimed to be equivalent to about 2400 theoretical plates (refer to plate theory in Ref. [99]). Berkowitz-Mattuck and Noguchi [111] also used both types of gas chromatography but did not couple them. Instead, the volatile material was condensed and subsequently injected into a gas-liquid chromatograph. They pyrolysed some treated and untreated cotton samples in a helium stream by 1-sec exposure to thermal flux levels of 5 to 25 cal/cm^2 - sec. The reaction products were separated into four fractions by using cold traps and analyzed separately. Fractions comprising volatile products that failed to condense at -80°C were swept onto a gas solid (charcoal) column and analyzed immediately. Those fractions consisted of a mixture of CO, CO_2, CH_4, and ethylene. Greenwood et al. [113] pyrolyzed cellulose in nitrogen at 300°C and examined the most volatile fraction (0.2%) of the product by GLC.

Schwenker and co-workers have extensively studied cellulose thermal degradation products by various analytical methods in the past [32] and more recently by gas-liquid chromatography [16]. The GLC apparatus used was an F & M Model 500 linear-programmed high temperature gas chromatograph with a hot wire filament detector. Three different columns having different liquid phases but the same solid support, Halport F (perfluorocarbon polymer), were employed for the analysis. Cotton samples were pyrolyzed in conventional pyrolysis apparatus in air and in nitrogen at 370°C. The products were trapped at various temperatures and injected into the gas chromatograph. In some instances, samples were pyrolyzed for 10-12 sec within the injection port of the gas chromatograph on a hot nichrome wire loop to afford a more direct examination of the degradation products. The identification of compounds was made by comparison of retention volume, by inclusion of suspected knowns, and by several other confirmatory tests on the eluents. Numerous aldehydes, ketones, and acids were identified, and it was found that the decomposition at different rates of heating gave very similar products. The products are listed in Table 3 and chromatograms obtained are shown in Fig. 9.

TABLE 3

Pyrolytic Degradation Products of Cellulose Indicated by Gas-Liquid Chromatography [a]

Compound	Columns indicating [b]	Identification [c]
Volatile gases: CO, CO_2	9, 13, 14	P
Formaldehyde	9, 13, 14	P
Acetaldehyde	9, 13, 14	P
Acrolein	9, 13, 14	P
Propionaldehyde	9, 13, 14	P
n-Butyraldehyde	9, 13, 14	T
Glyoxal	9, 13, 14	P
Furfural	9, 13, 14	P
5-Hydroxymethyl furfural	13, 14	P
Acetone	9, 13, 14	P
Methyl ethyl keytone	9, 13, 14	P
Methanol	9, 14	T
Formic acid	9, 14	T
Acetic acid	9, 13, 14	P
Lactic acid	13, 14	T
Water	9, 13, 14	P
Levoglucosan	3% Carbowax-20M	T

[a] R. F. Schwenker and L. R. Beck, J. Polymer Sci., Part C, #2, 331 (1963).

[b] O = Octoil S, 13 = Silicone oil-550. 14 = Carbowax-20M.

[c] P = Positive identification, T = tentative identification.

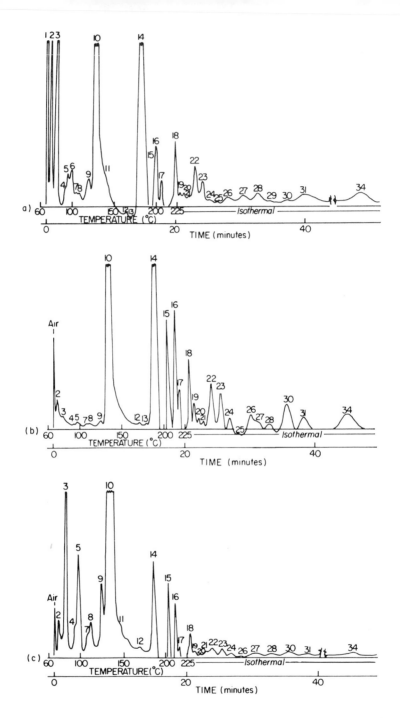

Fig. 9. Gas chromatograms of cellulose pyrolysis products on column 14 (Carbowax 20 M): (a) products of hot wire pyrolysis in the carrier gas stream, (b) products condensed at 75° C from pyrolysis of cellulose in nitrogen, (c) products condensed at - 78°C from pyrolysis of cellulose in nitrogen (Schwenker and Beck [16]).

Glassner and Pierce [114] compared the pyrolysis of cellulose and levo-glucosan by employing the chromatography techniques developed by Schwenker and Beck [16] . They found that the degradation products of cellulose and levoglucosan were essentially identical. Byrne et al. [115] further extended the investigation by studying the vacuum pyrolysis of cotton cellulose, alone and with flame retardants, at temperatures near 420°C. In the tar, levo-glucosan was confirmed by gas-liquid chromatography and other techniques. They employed a column which had a coating of a polyester stationary phase on small glass spheres and maintained at 124°C. Peaks in the chromatogram were characterized by specific retention volume [92] and area measurements.

In a recent article, a reinvestigation of cellulose (Avicel II) pyrolytic products was reported [61] , using a linear programming gas chromatograph equipped with a pyrolyzer. The amounts of the volatile compounds from cellulose were compared with cellobiose, glucose, levoglucosan, pyrocellulose, and about a dozen other polysaccharides. The samples, 15-30 mg, were pyrolyzed under helium. The volatiles from the pyrolysates were then introduced into the column by diverting the carrier gas through the pyrolysis chamber. In a preliminary investigation several column packings were tested, and finally, a column consisting of 15% diethylene glycol suc-cinate (Shimalite B) was employed. The relative amounts of volatile com-pounds from various samples were determined by peak height measurements.

The "hot wire" pyrolysis of cellulose, cellulose derivatives, and other polysachharides has also been further studied [116, 117] . The use of GLC for the analysis of phenylcarbamylated cotton has been reported [118] as well as for benzhydrylated cotton. Gas chromatography has also been applied to characterize the pyrolytic products of cellulose nitrate [73, 119] .

III. PAPER AND THIN LAYER CHROMATOGRAPHY

A. Principles

Paper chromatography is a well-established technique for the analysis of organic mixtures. The method is useful in isolating and identifying cellulose degradation products. The technique was developed by Martin and Synge [87] and Consden et al. [120] . In paper chromatography, a drop of the liquid mixture to be separated is applied at a point near one end of a strip of chromatography paper. The solvent is allowed to flow along the paper, over the spot of the mixture and toward the far end of the paper. In actual practice the paper is hung vertically and the solvent is applied either from the top of the paper or from the bottom of the paper. In the former case the solvent flows under gravity, "descending technique," and in the latter case, "ascending technique," via capillary and wicking action.

The components of the mixture migrate at different rates depending upon the paper-compound interactions, partition coefficients, and diffusion char-acteristics, in the direction of the solvent flow and thus they are separated from each other. At a suitable time, the solvent supply is stopped and the paper is dried. On drying, the separated components are fixed at individual places in the chromatogram. For identification pruposes, the paper is sprayed with a developing agent and the compounds are identified by comparing their R_f values of the substance and that of a known compound. The term R_f is defined as

$$R_f = \frac{\text{distance substance has traveled from the origin}}{\text{distance solvent front has traveled from the origin}}.$$

For further identification, the individual substance may be solvent extracted from the paper and analyzed.

Thin-layer chromatography is a relatively recent technique where, instead of a cellulosic paper support, a thin adsorbent such as silica gel or alumina layers are used to achieve the separation. The adsorbent plate is prepared by spreading a thin and uniform layer of the adsorbent together with a suitable binding agent on a glass plate. The experimental procedures used for applying the samples and developing the chromatogram are similar to those employed in paper chromatography. The advantages of this technique over the conventional paper chromatographic technique are (i) only minute quantities of samples are required for analysis and (ii) those compounds which cannot be separated by paper chromatography, but could be separated by silica gel or alumina column chromatography when large quantities are available, can be separated and analyzed in small quantities by thin-layer chromatography.

The detailed theory and experimental techniques are available in standard texts [102, 121]. An excellent review of the terms and the general applicability in research has been published by Stewart [122]. Specific applications in the fiber field to include cellulose have been reviewed [98].

B. Applications

Schwenker and Pacsu [32] utilized paper partition chromatography to determine the pyrolytic degradation products of cotton fabrics which were heated in air up to 375°C. Products condensable at 0°C were trapped as an aqueous distillate. The more volatile gases were allowed to escape and the water-insoluble material was rejected. Chromatography was carried out by the capillary descent technique using Whatman No. 1 paper. The chromatographic identification of volatile and nonvolatile compounds along with a series of confirmatory tests led to a conclusion that the cellulose pyrolyzate was a complex mixture of organic acids, aldehydes, ketones, water, and levoglucosan representing 14 or more different compounds as shown in Table 4.

Thin-layer chromatography was used by Byrne [123] for examining the carbonyl compounds resulting from cellulose pyrolysis. The actual experimental procedure and setting up of the chromatographic plate was that

TABLE 4

Cellulose Pyrolyzate Analysis by Paper Chromatographic [a] Technique

Compound	Percent
Levoglucosan	12.5
Carbonyls	About 15.0
Formaldehyde	
Gloyxal	
Unknown I	
Unknown II	
Unknown III	
Acid	About 7.5
Glycolic	
Lactic	
Dilactic	
Formic and/or acetic	
Unknown I	
Unknown II	
Unknown III	
Water	About 55.0

[a] R. F. Schwenker and E. Pacsu, Chem. Eng. Data Series, 2, 83 (1957).

described in a Shirley Institute report [124] and in a subsequent journal article [115]. Cellulose was pyrolyzed under vacuum and volatile products collected in traps at different temperatures. Tar was reacted with 2, 4 ditreitrophenylhydrazine and resulting dinitrophenylhydrazones were separated by TLC and identified by several additional instrumental techniques such as ultraviolet, visible, and infrared spectroscopy. The pyrolyzate was found to contain as many as 19 carbonyl compounds. The most abundant were glyoxal, glyco aldehyde, pyroaldehyde, and 5- (hydroxymethyl) furfural. They have also examined the pyrolyzates from cotton cloth treated with some conventional fire retardant agents. These results confirmed the original work of Schwenker and Pacsu [32].

IV. MASS SPECTROSCOPY

A. Principles

The mass spectrometer is most suitable for qualitative analysis of the small amounts of material separated by gas chromatography or by thermo-analytical techniques. It is an excellent tool for the identification of volatile pyrolytic products. In general, any sample with sufficient volatility for gas chromatography is volatile enough for introduction into the mass spectrometer. The minimum sample requirement is as small as for gas chromatography. In gas chromatography, analytical samples may vary from 10^{-2} to 10^{-13}g, depending upon the methods and equipment; in modern mass spectrometry the sample may vary from 10^{-2} to 10^{-11} g.

The principle of mass spectroscopy is based upon the ionization and fragmentation of the required molecules. Accelerated electrons are allowed to interact with gaseous molecules. This results in an energy transfer from the electrons to the molecules. This activation, in turn, results in the loss of an electron (to produce a molecular ion) and/or fragmentation of the molecule. The positive ion thus produced is separated according to mass (or mass-to-charge ratio) by a combination of electrical and magnetic fields and the intensity of the ion beams is measured electrometrically.

The five basic components of a mass spectrometer are the inlet system, ionization chamber, mass separating and mass analyzing system, amplifiers, and recording systems. Descriptive illustrations of various instruments are available [125-130].

There are three types of mass analyzers generally used in organic analysis. These are known as magnetic deflection, time-of-flight [131], and radio frequency. The principles and operation have been comprehensively presented by Duckworth [132].

B. Analysis of Mass Spectrograms

The use of mass spectrometry for qualitative analysis depends upon the unique mass pattern generated by individual compounds via electron bombardment. The pattern of fragment ions and relative intensities of these constitutes a fingerprint which can be used for positive identification of an unknown compound by comparison with mass patterns of known standard materials. A set of standard mass patterns is an invaluable aid to qualitative identification. There are several sources from which mass spectral data can be obtained [9, 133-135].

Even in the absence of a standard mass pattern, a great deal of information may be obtained in conjunction with infrared, ultraviolet, and NMR spectroscopy measurements to elucidate the structure of unknowns [125, 126, 136, 137]. Biemann [126] tabulated a list of papers up to 1962 containing a plethoria of mass spectral data.

Like many other analytical techniques, mass spectrometry also has certain limitations [138]. The position of double bonds and, particularly, cis or trans configurations are often not revealed. In some cases, long chain mono-substituted esters give different mass spectra [139]. In general, aromatic systems are fairly easy to elucidate, but often it is not possible to determine the ortho, meta, or para positions of side chains. The branched isomers of some ketones may not always be distinguished from each other. A particular branched system in, say, aldehydes, may give a spectrum that is similar to that expected from certain ketones. These are only a few examples of the limitations.

C. Use of the Mass Spectrometer with the Gas Chromatograph

Two basic systems of sample introduction are used for obtaining mass spectral analysis from chromatographic fractions. In one, the fraction is collected, or the chromatographic system temporarily stopped so that sample can be more conveniently analyzed. In the second system, the effluent is introduced into the mass spectrometer as it emerges from the chromatograph, known as tandem GC-MS. The time-of-flight mass analyzer (TOF) is usually used in tandem GC-MS applications. The tandem GC-MS analysis has several advantages. The application is ideally suited in the following cases: (i) if the chromatographic systems are not readily amenable to sample collection, (ii) if low collection efficiencies are attained, (iii) if the samples are readily oxidized or polymerized when condensed in the pure state. A detailed description of the use of the mass spectrometer with gas chromatography for organic materials is given by McFadden [138].

D. Use of the Mass Spectrometer with Differential Thermal Analysis and Thermogravimetric Analysis

The application of mass spectrometry with Differential thermal analysis and Thermogravimetric analysis is increasing although the literature available is very limited. The potential, however, is as good as for gas chromatography and the analytical principle is also similar. The time-of-flight analytical procedure would be most desirable for the study of pyrolytic degradation products. The principle of the technique for the application of the mass spectrometer with DTA and TGA can be obtained from a series of articles by Gohlke and co-workers [140-143].

E. Mass Spectroscopy With Other Instrumental Methods

Mass spectrometry is used routinely for the identification of products separated by the paper chromatography or thin-layer chromatography of various samples. However, there are a great many instances where mass spectral data are inadequate, so that further investigation of the sample is necessary by such methods as ultraviolet and infrared spectrophotometry or nuclear magnetic resonance techniques. In general, if the identification is of great significance, confirmatory evidence must be obtained by other analytical methods.

F. Applications

Madorsky and Straus [144] and Wall [145] and their collaborators at the National Bureau of Standards made early use of mass spectrometry to characterize polymer degradation products. In the case of cellulose, the decomposition was carried out with samples weighing 5 to 85 mg by heating for various periods of time at constant temperature under high vacuum in a specially designed apparatus [13, 144, 146, 147] . The volatile products were collected, separated into fractions, and analyzed by mass spectroscopy, and also in some instances by infrared spectroscopy.

The pyrolysis of purified cotton, modified cotton, and also cotton impregnated with sodium carbonate and sodium chloride were studied. From the mass spectra of the pyrolyzed products of cotton, cotton hydrocellulose, and viscose rayon, the identification of H_2O, CO_2, CO, and levoglucosan was made. The analysis of the tar fraction was performed by a technique developed by Bradt and Mohler [129] . The sample was heated in a vacuum tube furnace that opened directly into the ionization chamber, and the temperature was increased stepwise until there was no further evaporation. In some instances, the cellulosic samples were also pyrolyzed directly into the mass spectrometer.

Madorsky and co-workers also studied the pyrolysis of cellulose triacetate and NO2 -oxidized cellulose in the same way. A comparative study indicated that the tar yields were in the decreasing order: cotton, fortisan, cellulose triacetate, and oxycellulose. The volatiles detected consisted mainly of acetic acid, carbon dioxide, and carbon monoxide from the triacetate, and water, carbon dioxide, and carbon monoxide from the oxycellulose.

Broido and Martin [15] have studied the flash pyrolysis of cellulose (wood pulp) exposed in air or helium to radiation of intensity up to 30 cal/ cm^2/sec from a carbon arc at 5500°K. The volatile products, identified by gas-liquid chromatography and mass spectrometry, consisted largely of carbon monoxide, carbon dioxide, water, aldehydes, and ketones.

V. DIFFERENTIAL THERMAL ANALYSIS

A. Principles

Differential thermal analysis, DTA, is a method whereby the enthalpy or heat changes that accompany both physical and chemical transformations of matter as samples are heated or cooled, may be detected and measured as a function of temperature and/or time. Experimentally this is accomplished by measuring the temperature difference ΔT between the sample and a thermostable, inert reference material as both are heated (or cooled) simultaneously at a constant rate of heating in the same environment. The plot of ΔT as the ordinate versus temperature as the abscissa yields a curve or thermogram featuring peaks or bands representing the various transformations, as shown by the idealized thermogram in Fig. 10.

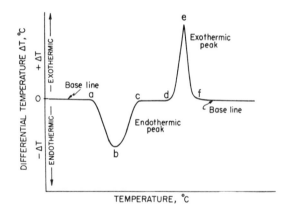

Fig. 10. Idealized thermogram for DTA (Schwenker [148]).

In usual practice, the sample and the reference are contained in a single metal block with a thermocouple junction in the sample and one in the reference. These thermocouples are connected in series. The metal block is heated at a programmed rate, and the voltage developed between the thermocouple junctions is continuously recorded as a function of either sample temperature, furnace temperature, or the reference temperature. For the correct interpretation of the curve it must be noted whether ΔT is plotted against the sample temperature, furnace, or the reference temperature. The voltage (emf) is proportional to ΔT, the temperature difference between the thermocouple junctions. As long as there is no reaction or transition of any kind, the temperature difference ΔT remains constant and consequently the emf is also constant. Therefore, in this case, the plot is a straight line, parallel to the base line. As soon as the reaction or a physical change starts, the temperature difference causes a deflection from the thermally steady state resulting in a base line shift, and finally a peak reflecting the absorption (endothermic) or the evolution (exothermic) of heat.

The temperature location, slope, shape, and area of a peak are characteristic for a particular physical or chemical transformation. The DTA curve then is effectively a thermal spectrum which may be used for identification as well as for the qualitative and quantitative analyses of polymeric materials such as fibers, films, and resins [148-151].

B. Theory

It has been shown that the endothermal and the exothermal changes observed under a given set of experimental conditions are both reproducible and unique. It is important, however, to locate the correct transition temperature manifested on the thermogram. Is it at the first deviation from base line, at the inflection point, or at the peak temperature? In the literature, suggestions for locating the correct transition temperature on the thermogram vary widely. For example, Frederickson [152] reported that melting temperature is the initial deviation from the base line, Partridge et al. [153] concluded that it is the inflection point, whereas Morita and Rice [154] said that it was the endothermic peak temperature. This confusion can be greatly clarified by (i) critically analyzing the temperature as measured [148] and (ii) a better understanding of the principle and heat theory involved in the technique.

The heat flow equation in the case of unidirectional flow is as follows:

$$\left(\frac{\partial^2 T}{\partial x^2}\right)_t = \frac{\rho c}{k} \left(\frac{\partial T}{\partial t}\right)_x$$

where x is the distance from the reference point, ρ is the density, c is the heat capacity, k is the thermal conductivity, and $\partial T/\partial t$ is the rate of heating.

The density, heat capacity, and thermal conductivity are related to give the thermal diffusivity $\underline{a}$ as follows:

$$a = \frac{k}{c}.$$

The DTA theory based on this fundamental heat equation, developed by Speil et al. [155] and modified by Kerr and Kulp [156] indicates that the peak area enclosed by the differential curve is related to the heat of reaction as follows:

$$\frac{M(\Delta H)}{gk} = \int_a^c \Delta T \, dt \, ,$$

where M is the mass of reactive sample, ΔH is the heat of reaction, g is the geometrical shape constant of the apparatus, a and c are the integration limits, k is the thermal conductivity and ΔT is the temperature difference. The preceding treatment neglects the differential terms and the temperature gradients in the sample and also considers the peak area to be independent of the specific heat of the sample. However, it allows ΔH to be expressed in terms of calories per gram as follows:

$$\Delta H = k' \int_{t_1}^{t_2} \Delta T \, dt = k' A,$$

where k' is the proportionality constant, experimentally determined for the particular instrument under a given set of experimental conditions, and A is the peak area.

In the case of differential scanning calorimetry (DSC) the peak area is directly proportional to ΔH, whereas in the case of DTA, consideration must be given to the changes in heat capacity and thermal conductivity of sample during heating as well as to unaccounted heat losses. Special DTA cells have been developed which are capable of much greater precision [157], although the indirect measurement and thermal lag involved in DTA still constitute problems.

Vold [158] derived an expression which required transformation of data obtained from the DTA curve as follows:

$$\frac{\Delta H}{C_S} \left(\frac{df}{dt} \right) = \left(\frac{d(\Delta T)}{dt} \right) + A \left[(\Delta T) - (\Delta T)_S \right]$$

where C_S is the heat capacity, f the fraction of sample transformed at any time t, (ΔT) the differential temperature, $(\Delta T)_S$ the steady-state value of the differential temperature after transformation has been completed, and A is a constant.

The constant A can be evaluated from a plot of log $(y-y_S)$ vs time, beginning at the top of a peak. The points yield a curve which becomes linear at the end of the transformation and thus gives a value of the time at which the transformation is over. There are certain limitations of the theory which are worth considering; a critical discussion on those limitations is available [159].

According to Boersma's findings, [160] the peak area is related to the heat of reaction by

$$\int_{t_1}^{t_2} \Delta T \ dt \ = \ \frac{2}{4} \, qak \ ,$$

where q is the heat of transformation per unit volume, a the radius of the sample cavity, and k the thermal conductivity.

More basic treatments, based on the heat equation for an infinite cylinder, were derived by Strella [161] and Tsang [162] .

It is implied in the foregoing theories that the equations developed are based upon the assumption that the sample is in equilibrium with its surroundings, and therefore the entire sample undergoes a transition instantaneously without any thermal lag. In actual practice, this situation is difficult, if not impossible, to achieve. However, the proper placement of sample and reference within the same block or enclosure can facilitate more rapid heat flow with consequent minimization of thermal lag. But, at any boundary there is a discontinuity so that appropriate new values of material density, heat capacity, and thermal conductivity will exist. In DTA, such boundaries are between the furnace and the specimen block, between block and specimen, and even within the specimen [148, 163] . Therefore, in actual practice there is a difficult situation to achieve, but with a knowledge of the factors and conditions that affect equilibrium conditions, accurate measurement can be made after calibration.

Kinetics by DTA

Borchardt and Daniels [164] derived a kinetic expression for analyzing the differential thermal analysis curve. The expression was derived specifically for a system in which a liquid sample and reference were employed and is as follows:

$$k = \left[\frac{KAV}{n_o} \right]^{x-1} \frac{C_p \ \frac{d(\Delta T)}{dt} \ + K \ \Delta T}{\left[K(A-a) - C_p \ \Delta T \right]^x} \ ,$$

where C_p is the heat capacity of the sample, K the heat transfer coefficient of reaction cells, A the total area under the curve, a the area under the curve that has been swept out at time t, k the reaction constant, x the order of reaction, n_o the initial number of moles reacted, and V the volume of sample.

Also, Kissinger [165] and more recently Reich [166] and Reed et al. [167] have derived new expressions which they applied to solids. The analysis of the kinetics of solid samples by DTA must be approached with caution since many of the basic assumptions may not be applicable to the transformation of solids.

C. Techniques of Measuring Energy Changes

The energy changes in the sample during heating or cooling can be determined by the measurements of (i) direct heating – cooling of sample, (ii) inverse rate of change of temperature, or (iii) differential temperature. The types of curves obtained by the above three methods for an endothermic reaction are shown in Fig. 11. The sample temperature increases linearly as shown by curve A, until the transformation starts at A where a deviation from linearity occurs as a result of temperature lag due to absorption of heat by the sample. At B, the reaction is essentially complete with a concomitant increase in sample temperature back to the furnace temperature. In the case of rate measurement (curve B) or differential temperature measurement (curve C) this phenomenon is reflected by a constant base line followed by a peak and return to a constant base line. However, in the case of rate measurement, when the transformation is complete, the return of the sample to the temperature of its surroundings produces a deflection in the opposite direction to that of the transformation. Of these three methods, the differential temperature measurement has been found most suitable.

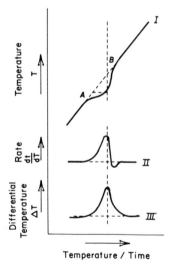

Fig. 11. Techniques of measuring energy changes (Wendlandt [159]).

D. Instrumentation

The basic apparatus consists of (1) temperature programmer, (2) furnace, (3) sample and reference block, (4) differential temperature circuit,

(5) sample, reference, or block temperature circuit, (6) signal amplifier circuit, (7) chart recorder, time base or X-Y, and (8) atmosphere control systems. A schematic diagram of typical DTA instrumentation is shown in Fig. 12 [159]. Detailed descriptions of most current instruments and discussions of instrument components including apparatus construction are provided by Smothers and Chiang [162], Slade and Jenkins [168], Garn [163], and Wendlandt [159].

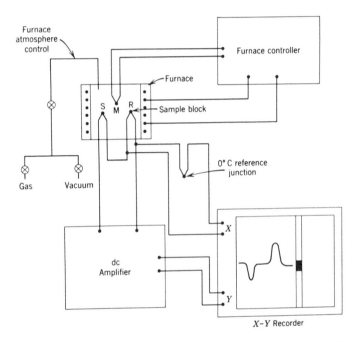

Fig. 12. Schematic diagram of typical differential thermal analysis instrumentation. S-sample thermocouple, R-reference thermocouple, M-heating rate monitor thermocouple (Wendlandt [159]).

E. Factors Affecting Differential Thermal Methods

The major factors affecting base line deviation and peak asymmetry are mismatched heat capacities, improper heat transfer, sample packing, particle size, characteristic properties of diluent and reference materials, etc. Garn [163] and Wendlandt [159] have gone into a detailed description of many of those effects. In this article only the major causes of base line deviation are discussed.

1. Sample Size, Holder and Reference Material

The sensitivity and resolution of peaks are greatly influenced by the sample size. As a simple rule, the larger the sample the greater the sensitivity and the poorer the resolution. Large samples are satisfactory if the transformation is sudden and the heat of transformation is large. In the case of small energies such as the glass transition, the smaller the sample the easier is the identification and reproducibility. In such a case the large sample size gives a broad peak. However, peaks can be sharpened by increasing the heating rate, but a practical problem arises due to poor thermal diffusivity in a large sample. The best way, therefore, to obtain maximum sensitivity and resolution is by maintaining a small sample size and increasing the efficiency of the heat transfer of the sample to the thermocouple. This can be accomplished by careful consideration in the design of the sample holder, thermocouple, and thermocouple placement. Various sample holder designs as discussed by Wendlandt [159] are shown in Fig. 13.

As has been pointed out elsewhere [148, 163] , in taking small samples care must be exercised since in the case of highly heterogeneous systems, a very small sample may not be representative, leading to poor reproducibility and other problems. The surface area and thickness of the sample can also have considerable effect. The sample preparation largely depends upon the nature and the purpose of the work. Fiber samples, cut into 2-3 mm lengths and fabrics cut into 2-3 mm squares can be used for DTA studies. As cautioned before [148] , if low temperature transitions are sought, it is not good practice to grind or mill samples since in the process sufficient heat can be generated to change the sample history.

The techniques for sample packing can be broadly classified as (a) unadulterated sample, (b) sample sandwiched between layers of reference material, and (c) admixing the sample with the reference material as a diluent. Usually, relatively large samples ($<$ 10 mg) are diluted with inert reference material whereby the undesirable factors such as the differences in heat capacity, thermal conductivity, and volume changes are greatly minimized. In both the sandwich method of sample packing and the diluent techniques, the inertness of the reference material, particle size, and sample packing will influence the thermogram. The reference materials alumina, silica, and silicon carbide are widely used [154, 169] .

The reference material should be inert and thermostable over the temperature range used. The commonly used reference materials are calcined alumina (aluminum oxide, Al_2O_3), silica, silicone carbide, glass beads, carborundum, glass wool, and quartz wool. However, if a diluent is mixed with the sample, the same material should be used as the reference.

In the case of fibers and other polymeric materials, because of their gross difference with inorganic materials in thermal characteristics, drifting base lines, artifacts, and spurious peaks can occur. In most instances where

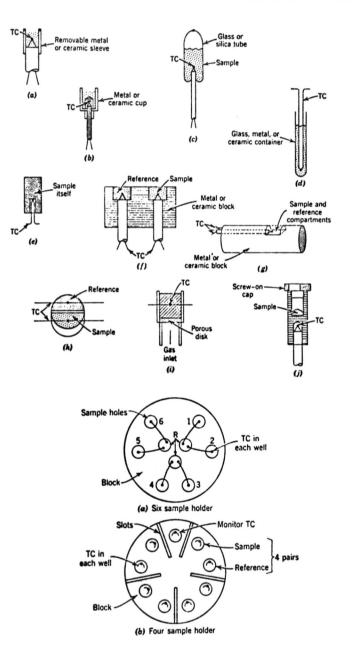

Fig. 13. Various sample holder designs for DTA instruments including multiple sample holders (Wendlandt 159).

sample size is 0.1 to 10 mg, no reference material is needed, other than the empty pan on the reference side.

2. Thermocouple Placement

In DTA, ΔT is plotted against the system temperature T. There are basically three methods for plotting the system temperature. By locating the thermocouple (a) in the center of the block, (b) in the reference cell, or (c) in the sample cell. Smyth [170] and Barrall and Rogers [171] have studied the effect of thermocouple placement. They found that the transition temperature corresponds to the peak temperature when ΔT is plotted against the temperature at the center of the sample and to the point of initial departure from the baseline when ΔT is plotted against the block or furnace temperature.

David [172] reviewed the above work and concluded that the use of the reference material temperature as the system temperature would be unsound. According to Garn [163], if the reference material is used as the system temperature, the transition or decomposition temperature may be obtained by adding or substracting the difference temperature. Recently, Schwenker [148] reviewed the situation, contradicted David's comment [172], and supported Garn's [163] analysis.

It is concluded that either the sample, reference, or block temperature may be used as the system temperature. It is also apparent that sample temperature is the best choice, and most commercial instrumentation uses it as system temperature. However, reference temperature is in fact a close second choice, whereas block or furnace temperature is a poor third. In any event it is imperative that the investigator be cognizant of the system temperature employed not only in his own apparatus but in any literature of interest.

3. Heating Rate

As reported by many investigators [173, 155], many systems exhibit peak shifting as a function of heating rate. In general, fast rates yield very sharp major peaks but lose details involving minor reactions. On the other hand, very slow rates of less than 5°C/min facilitate resolution of the peaks. The heating rate should be selected depending upon the type of information required in the particular experiment. In polymers, the phase-transition temperature is influenced by the heating rate. Therefore, to estimate the true transition temperature the DTA is run at several heating rates and the values are extrapolated for the hypothetical zero heating rate. As a general rule, reversible transition temperatures such as fusion temperatures are essentially independent of the heating rate, whereas irreversible transformation temperatures are dependent on heating rate.

The importance of linear temperature programming cannot be underestimated. Any nonlinearity in the heating rate may affect the curve and consequently be misinterpreted. If nonlinearity occurs the same instant that a small sample transformation occurs, it can obscure the true effect as well as to bias quantitative analysis of peak areas. Currently available commercial instruments do provide fairly linear temperature programming.

4. Atmosphere

The thermal decomposition of polymeric materials is very much dependent upon the atmosphere. The mechnaism or degradation of decomposition of a material changes with the change of the atmosphere. Therefore, atmosphere is an important variable. Further, the diffusivity and thermal conductivity of the atmospheric gas plays a role in the DTA heat conduction mechanism. Because of this, the thermogram obtained in nitrogren may significantly differ from that obtained in helium. Again, the selection of atmosphere depends upon the type of information required. The term "dynamic gas flow" refers to the continuous flow of gas at low flow rates through the sample.

F. Differential Scanning Calorimetry

Differential scanning calorimetry (DSC) is a technique whereby the energy transformation during a reaction can be directly determined from the recorded measurements. The technique was developed by Clarebrough et al. [174] in 1952. Although differential thermal analysis does proivde information regarding the energy transformation or enthalpy of a reaction, DSC provides the same information more directly. In DTA, the temperature differences between the sample and reference are measured as functions of time and/or temperature, whereas in DSC, the differences in energy required to maintain the same temperature in both the sample and the reference are measured in millicalories per second. The recorded tracing of DSC has the same appearance as that of the DTA except the ordinate, which reads the heat flow in millicalories or calories per second instead of ΔT. Thus, the area of a peak represents the enthalpy of transformation. The same technique has been designated variously as differential power analysis (DPA) and differential enthalpic analysis (DEA).

The general application and principle of the technique are available in articles by Eyraud [175], O'Neill [176], and Watson et al. [177]. Recently Schwenker [148] discussed the application of DSC for the analysis of textile fibers.

The instrument [177] consists of sample and reference holders enclosed in a chamber, each with a platinum resistance thermometer and a heater therein embedded. The temperature responses from sample and reference are amplified and passed through an electronic circuit whereby the power is adjusted as required. The differential power is then measured as a function of temperature and recorded.

A recent study by Schwenker and Whitwell [149] indicates that the precision of the technique for determining heat of fusion of pure metals (indium, lead, and tin) is closely comparable with the precision of the adiabatic calorimetric technique.

G. Applications

The application of DTA in the textile field has been recently reviewed by Schwenker [148] where many references on pioneering work pertinent to the present topic have been incorporated. The application, specific to cellulose degradation has also been reviewed by Shafizadeh [1] .

In the literature it will be found that the degradation of cellulose has been studied in several atmoshperes such as air, oxygen, nitrogen, and helium. In comparing the work done by one investigator with another it is important to consider all experimental conditions. The endotherms and exotherms of the DTA curve corresponding to the degradation of cellulose are very much influenced by the experimental variations previously discussed.

A careful analysis of the cellulose thermograms obtained by Schwenker et al. [178] , Berkowitz [84] , Morita and Rice [154, 179] , and Broido and Martin [15] reveal that the position and appearances of the endo- and exotherms are also dependent on the heating rate and the impurities in the material. Further, it should also be noted that the earlier investigators used homemade apparatus requiring large sample size which obviously somewhat insensitized the response. Currently, commercial instrumentation has been developed to such a point that only milligram samples are needed and can be analyzed with extremely high precision. Because of these facts the comparison of earlier data from the literature is difficult. It is believed that the current efforts of ICTA (International Confederation of Thermal Analysis) and the NATAS (North American Thermal Analysis Society) will help to establish needed standardization in methods and instrumentation, so that the data obtained by one investigator can be precisely compared with the data obtained by others.

The DTA of cellulose pyrolysis in general suggests that in nitrogen atmosphere the pure cellulose gives an endotherm at about $100°C$, an exotherm at about $330°C$, and a large endotherm at about $372°C$ at a heating rate of $10°C/min$. The DTA curve of α-cellulose in nitrogen by Tang and Neill [68] showed only a single endotherm at $335°C$.

The degradation and decomposition thermograms of substituted celluloses obtained by DTA lead to a method of identification and characterization of the cellulose compounds [154, 180] . Attmepts have also been made to characterize various reactions with cellulose by DTA [1, 118] .

The differential thermal analysis curves recorded by Conrad and Stanonis [40] on benzhydrated cotton cellulose indicate the utility of combining both TGA and DTA techniques. Whereas the TGA curves showed apparently simple weight loss, the DTA curves are far more revealing.

It was noted in the mass spectroscopy section that the application of mass spectrometry with DTA and TGA is gaining popularity.

The application of DTA for calculating heat of reaction of cellulose decomposition can be obtained from the work of Tang and Neill [68] . The heat of pyrolysis (in an inert atmosphere) was calculated by integrating the area under the curve and comparing it with the value obtained for a reference substance according to the following equation:

$$(\Delta H)_s = \frac{A_s \, M_k}{A_k \, M_s} (\Delta H)_k$$

where $(\Delta H)_s$ is the heat of reaction of the sample, $(\Delta H)_k$ the heat of transformation of the reference substance, M_s and M_k the weight (in grams) of the sample and a reference substance, and A_s and A_k are the integrated areas under the differential thermal analysis curve.

The heat of pyrolysis was found to be 88 ± 3.6 cal/g. The heat of combustion from the differential thermal analysis data was -3540 ± 140 cal/g, which is about 12% less than the value -4030 cal/g obtained by using an oxygen bomb calorimeter.

VI. THERMOGRAVIMETRIC ANALYSIS

A. Principles

Thermogravimetric analysis (TGA) or thermogravimetry (TG) is a method for the continuous, recorded measurement of the weight of a material as a function of time or temperature as it is heated, preferably at a linear rate. The technique was originally developed by Honda [181] . Significant advancement in the application of the techniques for the characterization of materials were due to Guichard [182] and Duval [183] .

To illustrate the principle of the technique, consider a hypothetical compound which undergoes a series of reactions accompanied by weight loss on heating. If the individual reactions initiate at different temperatures not too close to each other, a typical thermogravimetric stepped curve, as shown in Fig. 14, will be obtained.

In the curve the plateau regions indicate no reaction and therefore no weight change. As soon as the reaction starts and some volatile compound escapes from the system, the curve moves downward and becomes horizontal to the x axis, again when the reaction is completed. The various steps in the TGA curve provide information on thermal stability, original sample composition, thermal stability of intermediates formed, and kinetic data. However, the interpretation of the curve becomes more complicated when simultaneous reaction occurs or when two or more reactions initiate in the same range of temperature [75] .

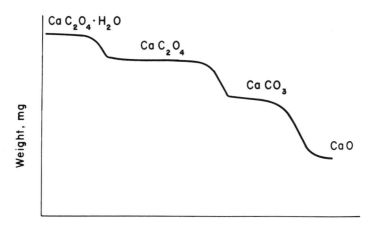

Fig. 14. A typical thermogravimetric curve showing the decomposition of $CaC_2O_4 \cdot H_2O$.

The technique is dynamic in nature, as is DTA, with the advantages of simplicity, speed, and a variety of quantitative information obtainable from a single experimental thermogram. However, at times complementary methods must be used in order to interpret the curve. Whereas TGA itself is a valuable complementary technique to DTA and/or DSC since TGA distinguishes between reactions or changes involving weight change as opposed to those that do not. However, when such a comparison is made, considerations must be given to providing the identical experimental conditions and thermocouple location as closely as possible.

The term DTG refers to differential thermogravimetry. Just as differential thermal analysis makes use of transient differences in temperature as a means of detection of a reaction, so one may use transient differences in weight. The DTG plotter in an instrument records the first derivative of the conventional TGA curve against temperature. The weight loss is then represented by peaks instead of steps. Garn [163] discussed the general principle and the control system for differential thermogravimetry.

B. Kinetics by TGA

The determination of kinetic parameters for solid state reactions by the analysis of TGA curves has been extensively reviewed [175-188].

Further, new kinetic techniques have been reported in the literature [77, 78, 188], including the expression for characterizing the chain reaction mechanism of cellulose pyrolysis [77]. The kinetic expressions derived so far are based on the assumption of the applicability to a solid-state reaction of three fundamental equations: (i) the rate expression, $dc/dt = f(c)$, derived from the law of mass action, (ii) the linear rate of heating, $dT/dt = \beta$, and (iii) the Arrehenius energy of activation expression, $k = Ae^{-E/RT}$. The only exception can be found where a nonlinear rate of heating has been suggested [87]. In addition to the above, a few other assumptions (such as heat equilibrium and homogeneous reactions in solid particles) are also involved, as discussed by Garn [163].

Therefore, the application of TGA in determining kinetic parameters should be made with sufficient caution. Various methods suggested for determining the kinetic parameters have been critically reviewed by Flynn and Wall [186]. They pointed out that all the general methods can be classified in the following groups: (a) integral methods utilizing weight loss versus temperature directly, (b) differential methods utilizing the rate of weight loss, (c) difference-differential methods involving differences in rate, (d) methods specially applicable to initial rates, and (e) nonlinear or cyclic heating rate methods.

The major conclusions reported were as follows: The integral methods involving a single thermogram are applicable only in special cases in which the isothermal order is known and meaningful. Of the differential methods, only those involving several thermograms are generally applicable in most cases. Methods involving a single thermogravimetric trace should be restricted to systems where a simple kinetic process is occurring. Difference-differential methods appeared to have limited applicability to polymeric systems where the kinetics often are complex. Determination of initial rates by methods (d) needs accurate data at low conversions, which are not always experimentally realizable. Nonlinear or cyclic heating rate methods greatly simplify integral methods, rendering approximate equations exact.

C. Instrumentation

The basic instrument consists of (1) a recording balance and its control system, (2) temperature programmer, (3) furnace, (4) sample holder, (5) temperature-measuring circuit, (6) signal amplifier circuit, (7) chart recorder, time base or X-Y, and (8) atmosphere control system.

The most important thing in the TGA system is the balance. Although various kinds of balances have been used in the current commercial equipment, a beam balance system is ideally suited for work in the fiber and polymer areas [148]. Both high accuracy and high sensitivity are possible with a beam balance system. In order to illustrate the general principle of the system a schematic description of the DuPont 950 thermogravimetric analyzer is shown in Fig. 15. Garn [163] has discussed the construction details of the miscellaneous balance systems.

Differential thermal balances used in DTG which provide a plot of dw/dt vs temperature or time instead of the conventional plot of W vs temperature or time, have been described and the general principles discussed, [159-193].

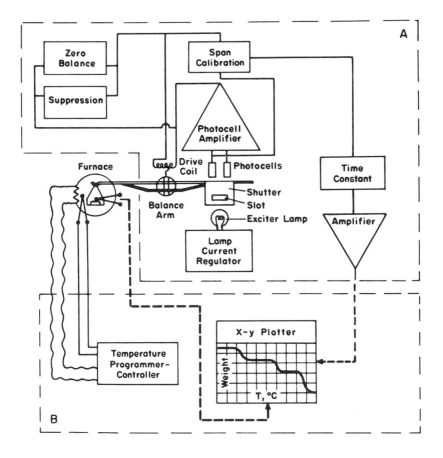

Fig. 15. A schematic representation of Du Pont 950 thermogravimetric analyzer (from Du Pont 950 TGA Instruction Manual).

The technique for one commerical instrument is as follows [159] : There are two identical samples in two identical furnaces which are suspended from a balance. The rate of heating of the two furnaces is kept constant but the temperature difference maintained at 5°C. As the reaction involving weight changes occurs in the two samples, the magnitude of weight differential would be proportional to the heating rate of the furnace. The resulting differential curve is continuously recorded via strip chart.

D. Factors Affecting Thermogravimetry

The TGA curve is affected by many factors which can be broadly divided into two categories: (i) instrumental factors and (ii) sample characteristics and experimental conditions. Of the instrumental factors, geometry of sample holder, sensitivity of recording mechanism, composition of sample container, etc., are the more important. The factors belonging to category (ii) are sample size, sample geometry, sample characteristics, heating rate, atmosphere and its control, thermocouple placement, solubility of evolved gas in the sample, thermal conductivity, etc. A brief discussion of the major factors is given here. More detailed discussions are available in several reviews cited.

1. Heating Rate

As heating rate is increased, the apparent decomposition temperature increases. This apparent higher decomposition temperature is due to the fact that the sample temperature cannot catch up to the environmental temperature, and therefore the recorded temperature always indicates a higher value than the sample temperature. However, with a material that undergoes a fast irreversible reaction, the heating rate will have little effect upon the weight loss curve.

The heating rate is very important in kinetic studies, since kinetics are based upon the assumptions of certain classical theories which are applicable to thermal equilibrium conditions only. Therefore, the lower the heating rate, the better the approximation of the equation will be. However, if heating rate is too low, the very advantage of dynamic thermogravimetry over static thermogravimetry is lost, hence, a compromise needs to be made between the two. In general, heating rates of 2 to 5°C/min are satisfactory and up to 10°C or higher should only be used for scanning purposes.

2. Sample Considerations

As in DTA, the smaller the sample size, the better the expected results. Further, to reduce the heat of reaction effect, the smaller sample size is desirable. However, for a heterogeneous material, consideration must be given as to the minimum size requirement for a representative sample.

3. Atmosphere

It is evident that the characteristic properties of atmosphere influence the TG curve. Consideration must be given to the fact that the sample under investigation evolves gaseous or volatile substances which might react with the atmosphere and influence further decomposition. The sample itself on the other hand may adsorb or react with the atmospheric gas. Newkirk [194] has shown that small changes in the composition of the atmosphere have a significant effect on the weight loss curve of $CaC_2O_4 \cdot H_2O$. The effects of different gaseous atmospheres and of pressure on the TGA curves has been discussed [163].

Usually, TGA runs are made in an inert atmosphere, such as nitrogen or helium under dynamic flow conditions. The flow of gas is maintained at approximately 0.5 liters p min. As has been pointed out elsewhere [148], spurious weight changes can result from too high rates of flow. A high rate of flow would also influence the balance system by increasing the buoyancy force on the pan resulting in apparent weight gains.

4. Temperature Measurement

Unlike DTA, the TGA system does not permit the placement of the thermocouple in direct contact with the sample or the sample pan. Generally, the temperature of the sample is measured by a thermocouple located in close proximity. Therefore, the true temperature of the sample either lags or leads the thermocouple temperature. As a consequence of this instrumental limitation, careful temperature calibration is essential for best results in thermogravimetry. Usually, the temperature is calibrated by running samples with known decomposition temperatures.

5. Other Effects

The TGA curve may also be affected by random fluctuation of the recording mechanism, furnace induction effects, nonlinearity of heating, and changes in thermobalance enivronment. Since the most important part of the instrument is the balance component, all the precautions that are required for operating a precision balance in the semimicro range are absolutely essential in this case also. Another source of error is the condensation of volatile decomposition products on the sample holder support rod, since this can lead to errors in weight change. Soulen and Mockrin [195] stated that this problem is intensified when a too-rapid inert gas flow is employed with consequent cooling.

E. Applications

Isothermal thermogravimetry of cellulose has been extensively studied primarily via homemade thermobalances, but, at least in one instance, a commercial TGA instrument was used [74]. In that work, the sample was heated at a definite rate until the desired temperature was reached. Thereafter, the temperature was maintained constant throughout the remainder of the experiment. The thermograms so obtained were normalized by considering the total weight loss as the difference between the weight corresponding to the beginning of the constant temperature and the weight at the limiting equilibrium stage (or carbonization end point).

Tang and Neill [68] studied the decomposition kinetics of α-cellulose and fire-retardant treated α-cellulose by TGA using the analytical technique of Freeman and co-workers [196, 197]. The same technique was applied to

the thermal decomposition of benzhydrylated cellulose [40] but apparently
with little success. Recently, untreated and tritylated celluloses of various
degrees of substitution were studied by TGA [78] where each thermogram
was analyzed by methods suggested by Chatterjee [75] and Anderson and
Freeman [197] , and also by a trial and error curve-fitting technique. A
critical analysis of each kinetic method as regards its applicability to cellulose decomposition was made. A more extensive kinetic treatment of the
TGA curve of cellulose has been proposed by Chatterjee and Conrad [77].
This treatment facilitates the analysis of the individual major decomposition
such as initial glucosidic bond scission and the propagation reaction which
results in the major volatile products.

Akita and Kase [198] reported the differential thermal analysis and the
thermogravimetric analysis of untreated cellulose and cellulose samples
treated with ammonium phosphate. They proposed a new kinetic treatment
of the DTA thermogram, whereas the thermogram obtained from TGA was
analyzed according to the principle suggested by Kissinger [165] and others
[199, 200].

In conjunction with DTA and DSC, TGA has been applied to study the
effects of bases on the thermal degradation and pyrolysis of cotton cellulose
[201, 202]. Highly substituted cyanoethyl cellulose invariably gives a
colored by-product, and the structure of this by-product has been investigated
using infrared and TGA [203] . The characterization and identification of
cellulose precursors, cellulose, and cellulose derivatives are readily accomplished by TGA [151, 180].

VII. TORSIONAL BRAID ANALYSIS

Torsional braid analysis (TBA) is a dynamic thermomechanical technique
developed by Gillham [204] whereby the temperature-dependent mechanical
properties of polymers can be determined not only in the solid state, but
also through melting and decomposition. It is an ingenious extension of
the classical torsional pendulum method with the primary difference that the
sample specimen is a composite consisting of a multifilament inert braid
impregnated with the polymer to be examined. Changes in the mechanical
properties of the composite can be attributed to changes in the polymer
which has been deposited on and in the multifilament substrate from a melt
or from solvent. In the conventional torsion pendulum, the constraint of
dimensional stability has limited that technique to those transformations
occurring in the solid state.

The TBA apparatus developed by Gillham [204] is shown schematically
in Fig. 16. The inertial mass of the torsional pendulum is a linear transmission (with respect to light) glass disk which is supported by the braid-
polymer composite specimen. The braid used is 8 to 10 in. in length and
consists of thousands of glass fiber filaments made by twisting several strands
of glass cloth together. The amount of polymer deposited on a 10-in. braid

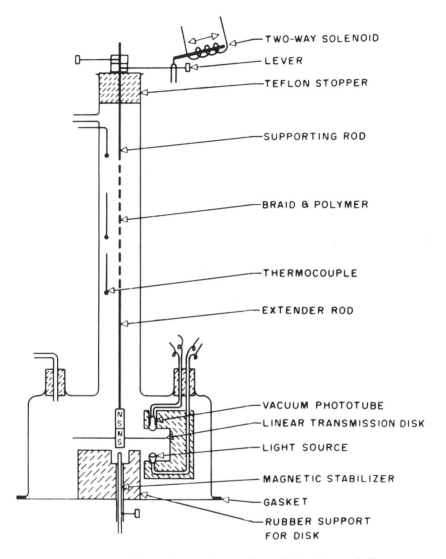

Fig. 16. Torsional braid apparatus: schematic instrumentation (Gillham [204]).

is about 100 mg. By means of a lever at the top of the apparatus, the pendulum is put into torsional oscillation. The specimen is heated or cooled at a linearly programmed rate of heating, usually 2°C/min. The glass column design permits control of the atomsphere so that experiments can be run in nitrogen, oxygen, air, vacuum, etc., depending upon the purpose. The temperature range available is -190° to +500°C.

The frequency (0.05 to 1 Hz) and the oscillation decay of the freely twisting specimen provide data on the modulus and the mechanical damping of the polymer. An electrical analog of the mechanical oscillations is obtained and recorded graphically with a strip-chart, time-base recorder. This involves the glass disk acting as an optical wedge featuring a linear relationship between light transmission and displacement angle. The graphical output appears as a decaying sine wave from which the period of oscillation, P, and the logarithmic decrement (decay of damped wave) 1/N, where N is the number of cycles between two boundaries, can be determined. From these data, the thermomechanical spectra are displayed as two curves as a function of temperature: (1) the relative rigidity $1/p^2$ (a measure of relative shear modulus) vs temperature and (2) mechanical damping index 1/N vs temperature.

Gillham and Schwenker [205] applied TBA in a study of the thermal behavior of fiber-forming polymers. The thermomechanical spectra as well as DTA and TGA curves obtained on cellulose triacetate are shown in Fig. 17. The TBA curves are the bottom set. For TBA the cellulose triacetate was deposited onto the glass braid substrate from a methylene dichloride solution. The TBA heating rate was 2°C/min, whereas 5°C/min was the rate for the DTA and TGA experiments. The runs were carried out in a nitrogen atmosphere.

In the TBA curves, the relative rigidity of modulus curve shows polymer modulus decreasing gradually until at 165°C a sharp drop begins which stops at 204°C. The corresponding mechanical damping curve shows a sharp maximum at 190 C. The DTA curve shows a baseline shift in this region. These results indicate a glass transition. Immediately above 200°C an increase in modulus is shown by the TBA curve reflecting stiffening of the polymer structure. An exothermic process is seen in the corresponding temperature region of the DTA curve. This behavior was interpreted as crystallization. In the 280° to 290°C range the TBA curves show a sharp decrease in polymer modulus and a maximum in mechanical damping, which phenomena correlate with a well-defined DTA endotherm. These changes are associated with melting. Above 320°C the polymer modulus increases in an exothermic process as shown by the DTA curve. This behavior is attributed to either crosslinking or intramolecular chain stiffening.

TBA has many potential applications for cellulosic materials, such as the characterization of cellulose derivatives where it provides unique data on the mechanical properties of such materials. It should also be noted that the apparatus can be used as a conventional torsional pendulum. Further, it is possible to use a cellulosic fiber braid for example whereby the interactions of finishing agents might be studied.

A commercial version of the TBA apparatus described is available from The Chemical Instruments Division, Metavac-Lenscote Corporation.

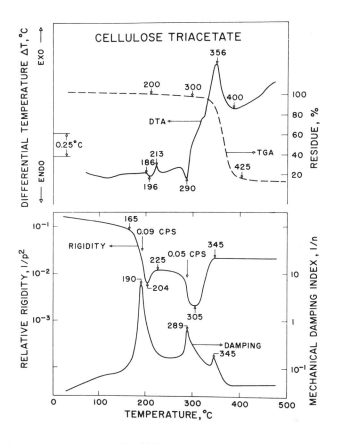

Fig. 17. Thermograms of cellulose triacetate in nitrogen (Gillham and Schwenker [205]).

VIII. INFRARED SPECTROSCOPY

The degradation or decomposition processes in cellulose are accompanied by the alteration of functional groups and physical structures (amorphous-crystalline ratio) of cellulose, creation of new functional groups (carboxylic, carbonyl, aldehyde, etc.), evolution of gases and volatile products and/or nonvolatile acids and other liquids. The transformations and products thereof can be investigated by infrared spectroscopy (IR). The experimental technique for handling all the three general states of the product (gas, solid, liquid) are available in any standard text on infrared absorption spectroscopy. Recently, O'Connor [206] reviewed IR in terms of its various applications in the fiber field.

The techniques for obtaining the infrared spectra of cellulosic materials as reported in the literature are: (i) by pressing the specimens as finely ground particles in an oil mull [207] , (ii) in pressed disk of KBr [43] or KC1 [82] , (iii) by sifting fibers uniformly over a polished die and pressing to form a coherent wafer which is then treated with a suitable liquid mounted in a sandwich cell [208, 209] , or (iv) by the technique of internal reflection spectroscopy (also known as attenuated total reflection or frustrated multiple internal reflection) [210-213] , in which case the samples are in yarn fabric, or sheet form.

For gaseous or volatile products and powdery solid material such as char, the techniques are those described in standard references [214-219] . The application of the KBr pellet technique in the study of cellulose oxidation has been investigated by many, as recently summarized by O'Connor in a review article [206] . This author and his co-workers have also extensively studied the oxidation of cellulose by the KBr pellet technique [220-222] .

Following are some typical examples of important results and/or conclusions on the mechanism of cellulose oxidation which were obtained through infrared spectroscopic studies [206] :

(a) The primarly alcohol group on the number six carbon atom can be oxidized to a carboxyl group by nitrogen dioxide treatment, and the secondary alcohol groups on carbons 2 and 3 can be converted to aldehyde groups by periodic acid treatment.

(b) Although chemical methods of oxidation result in a continuous increase of carbonyl groups, the γ irradiation of cotton does not show similar effects, at least in the initial stage of irradiation. With increased dosages, carbonyl bands of the -COOH group begin to appear.

(c) In periodate oxidized cellulose, -C=O stretching band of the aldehyde group at 1730 cm^{-1} becomes considerably weaker as the moisture content is increased, possibly due to the formation of hydrates or hemiacetatals or both, and/or hemialdol groups.

A current review on the infrared spectroscopy of carbohydrates by Tipson [223] contains much valuable information which is useful for the interpretation of the infrared spectra of cellulose and cellulose derivatives.

In a study on the mechanism of cellulose oxidation, Friedlander and co-workers [224] suggested an approach to determine the location and amounts of specific aldehyde, ketone, and carboxyl groups using a combination of chemical treatments and infrared analysis by the KBr pellet technique. During any degradation process the crystallinity of cellulose decreases. Therefore, the measurement of degree of crystallinity would be one way to follow degradation. A few quantitative infrared spectroscopic techniques have been reported to be useful for estimating the degree of crystallinity [205, 225-227] .

Composition of the tar fraction from pyrolysis of cellulose was examined using infrared spectroscopy by many investigators [13, 51, 111, 124] . Madorsky [13] reported a method where pyrolyzed fractions of cellulose were dissolved in methanol and deposited on rock salt as a thin film by evaporating the solvent, and the infrared spectra of the resulting films were obtained. The tar was also examined as a solution in NN-dimethylformamide by others [124]. The residue left from the pyrolysis of cotton after 16% weight loss was studied [146, 147] and it was found that the infrared spectrum remains virtually unchanged, except for the appearance of the carbonyl bond. Honeyman [124] and Gardiner [51] analyzed gaseous products by first collecting them in cold traps and then diluting with nitrogen to atmospheric pressures in gas cells having potassium bromide windows. An infrared spectrum characteristic of carbon monoxide was observed. The volatile organic components resulting from pyrolysis were examined either as vapor under similar conditions in a cell with barium fluoride windows or as a liquid between barium fluoride plates. For char analysis, the KBr pellet technique was used. Similar techniques to study the cellulose decomposition were followed by several other investigators [86, 228].

The appearance of carbonyl groups and the disappearance of characteristic cellulose bands were observed on heating cellulose pulp at 250°C in air [82]. Furthermore, a rapid decline in the intensity of the CH_2 deformation band suggested that the primary hydroxyl group is preferentially oxidized. On prolonging the heat treatment (12 h at 250°C), stable unsaturated rings containing carboxyl groups were observed in the infrared spectrum of the material. In the course of a thermal degradation study, Otani and co-workers [79, 80] examined the infrared spectra of cellulose heated below and above 275°C in nitrogen atmosphere. In the former case, C=C bands appeared in addition to carbonyl bands, whereas in the latter case there was loss of hydroxyl groups and occurrence of carbonization. In a similar study [83], it was concluded that cellulose structure was gradually replaced by an aromatic condensed ring system.

The kinetics of decomposition of cellulose nitrate was studied by following the changes in five infrared absorption bands during heating the nitrate in the range of 140 - 190°C [229]. Those bands were associated with the disappearance of the three nitrate groups and the glucositic linkage of the cellulose structure, as well as with the appearance of carbonyl groups.

IX. X-RAY DIFFRACTION

According to the most widely accepted concept, the molecular chains of cellulose lie parallel and in three-dimensional arrangements of high geometric order in the crystalline region of fibers. The other portions of the same molecular chains are assumed to lie in a less ordered state in the amorphous regions at the expense of the crystalline regions. This transformation should be evidenced by a change in the x-ray diffraction pattern.

Various methods are suggested for measuring the crystallinity of cellulose from x-ray diffractograms [230-232], and some of them have been critically reviewed [233]. A photographic recording of the diffracted bean

was employed by Hermans and Weidinger [230] for the measurement of crystallinity. To represent the proportion of amorphous cellulose in the sample,
they used a ball milled cellulose [234] as a reference sample, representing
completly amorphous material. More recently, Wakelin et al. [235] and
others [236, 237] have suggested alternative techniques and made use of the
diffractogram tracings instead of photographs.

The methods of measuring crystallinity from diffractogram tracings can
be broadly classified into two categories. By either measuring the areas
under the appropriate diffractogram curves or by measuring the relative
intensities of the appropriate portion of the curve, it is possible to arrive at
a value for the weight ratio between the amorphous and crystalline region
in the sample. The merits of the two different approaches have been discussed [238].

The degradation of cellulose by acid hydrolysis has been extensively
studied by x-ray diffraction [230-243]. The periodate and chromic acid
oxidation of cellulose was studied by Davidson using x-ray diffraction as
early as 1941 [17]. The study was extneded by Nevell [18] to nitrogen dioxide oxidation mechanism using a similar technique. Recently, the loss of
crystallinity during the periodate oxidation was measured [244].

Conrad and Creely [245] studied the thermal transformation and degradation of cellulose acetate by x-ray diffraction. Annealing of cellulose
acetate (above DS = 1.50) at temperatures of 170°C and above gives rise to
the cellulose acetate I crystal structure, which breaks down at temperatures
of 280 to 380°C. They related this destruction of crystal structure with the
combined process of melting and decomposition. Cynoethyl cellulose of DS
2.60 showed increasing crystallinity between 70 and 170°C, above which it
decomposed.

The glass transition temperatures of triethyl and trimethyl cellulose were
examined using x-ray diffraction and electron microscopy [246].

The x-ray photographic technique was used to study the mass evolution
and transfer within the samples by heating α-cellulose [247]. Evidence
was found of multiple processes such as dehydration, rearrangement of
structures, and volatalization of products. The carbonization process during
cellulose decomposition was studied with the aid of the x-ray technique by
Cobb [85]. He found that, at 300°C, the ring in the x-ray photograph of
unheated cellulose had given way to a general diffuse band which indicated
breakdown of the original structure. At 500°C the characteristic pattern
of graphite appeared.

Degradation during the milling of cellulose has been studied by several
investigators who found that some mechanical treatments impart extensive
structural damage to cellulose [248, 249]. The most significant effect was
observed by ball milling cellulose whereby the x-ray pattern of cellulose I
rapidly changes to a broad band characteristic of amorphous cellulose. The
mechanism of mechanical degradation has been assumed to be the combination
of several phenomena, such as oxidation, hydrolysis, and mechanical rupture.

X. OTHER TECHNIQUES

A. Sonic Pulse Propagation

The instrument, the pulse propagation meter, is available commercially; its major use has been largely in the synthetic fibers. Recently Back et al. [22] applied a relatively new technique, the sonic pulse propagation measurement, to characterize cellulose softening and autocrosslinking phenomena in cellulosic sheets on heating. Crosslinking results in an increase of Young's modulus of elasticity of the material, which can be determined using a pulse propagation meter. The modulus calculation procedure for a cellulosic sheet with this technique has been described by Craver and Taylor [250, 251].

The principle of the technique is as follows: The sample is placed on a flat foam platten in contact with two ceramic piezoelectric transducers. A peak voltage of very short time duration is emitted electronically and is transmitted to a transducer which is resonant at 5-10 kHz. An internal timing circuit is instantaneously initiated and the pulse is transmitted into the sample sheet and is propagated through it. A second transducer is placed at a measured distance away, which detects the pulse and converts it back to an electronic signal, thus closing the timing circuit. The elapsed time between the transmission and the reception of the signal is recorded on a strip chart. This elapsed time is obtained at various transducer separations. A linear plot can be prepared and used for the calculation of sonic velocity.

The calculation of Young's modulus in the case of a homogeneous, continuous filament or film is given by [250]:

$$E = \rho C^2, \tag{1}$$

Where E is Young's modulus, ρ the density of the material, and C the sonic velocity. In the case of sheets having paperlike structure, the relationship is more complicated. An approximate form of the general relationship, as derived by Chatterjee [252], is

$$E = kdc^2, \tag{2}$$

where k is dependent upon the density of the single fiber and the frequency of interfiber bonding, d is the apparent density of the sheet, E is the Young's modulus of the paper (specific to the interfiber bonded region), and c is the sonic velocity.

Back et al. [22] used a paperlike sheet (a heterogeneous structure) for the study of the crosslinking phenomenon but calculated Young's modulus (also called the dynamic or sonic modulus) by using an equation which is applicable only in the case of a homogeneous material. They used Eq. (1),

where ρ was arbitrarily changed to "d" for calculating the dynamic modulus
of the sheet. Thus, they employed an equation which differs from Eq. (2)
by a factor of k. It is likely that k is also a function of temperature. However, it is not known if this discrepancy would affect the essential observations of Back et al.

B. Mechanical and Electrical Properties

Aging, i. e. mild degradation, occurs to cellulose when it is exposed to
the atmosphere for a long period of time. As a result of this aging, mechanical properties deteriorate and electrical properties are often changed.
Therefore, many mechanical testing instruments such as the Instron, the
Mullen burst tester, tear tester, and folding tester have been employed to
study the aging mechanism, Murphy [66] , Stamm [67] , and others [253,
254] employed these various techniques to elucidate the mild degradation
mechanism of cellulosic materials and to follow the kinetics of the process.
Many investigators have attempted to correlate the electrical properties with
the mechanical properties.

C. Calorimetry

The classical technique of determining heat of combustion and heat of
reaction of many organic substances is described in the literature [255] .
A standard oxygen bomb calorimeter has been used in determining the heat
of combustion of wood and wood products [253] . Such an approach would
be useful in the study of the mechanism of cellulose pyrolysis.

REFERENCES

[1] F. Shafizadeh, in Advances in Carbohydrate Chemistry, Vol. 23 (M. L.
     Wolfrom and R. S. Tipson, eds.) Academic, New York, 1968, p. 419.
[2] G. A. Richter, Ind. Eng. Chem. , 26, 1154 (1934).
[3] R. L. W. Farquhar, D. Pesant, and B. A. McIaren, Can. Textile J. ,
     73, No. 3, 51 (1956).
[4] F. M. Clark, Am. Inst. Elec. Engrs. , 60, 778 (1941).
[5] F. M. Clark, Trans. Electro-chem. Soc. , 83, 143 (1943).
[6] C. M. Conrad, V. W. Tripp, and T. Mares, Textile Res. J. , 21, 726
     (1951), V. W. Tripp, T. Mares, and C. M. Conrad, ibid, 21, 840 (1951).
[7] E. Heuser, The Chemistry of Cellulose, Wiley, New York, 1944.
[8] L. F. McBurney, in Cellulose and Cellulose Derivatives, (E. Ott, H. M.
     Spurlin, and M. E. Grafflin, eds.), 2nd ed. P. I, Wiley-Interscience,
     New York, 1954, p. 99.
[9] Manufacturing Chemists' Assoc. Research Project, Mass Spectral Data,
     Chemical Thermodynamics Properties Center, Texas A & M University,
     College Station, Texas, (1959 to date).

[10] R. C. Waller, K. C. Bass, and W. E. Roseveare, Ind. Eng. Chem., 40, 138 (1948).

[11] J. G. Wiegerink, Textile Res. J., 10, 493 (1940); J. Res. Natl. Bur. Std., 25, 435 (1940).

[12] F. L. Browne, "Theories of the Combustion of Wood and its Control. A survey of Literature", U. S. Forest Prod. Lab Rept. No. 2136.

13 S. L. Madorsky, Theraml Degradation of Polymers, Wiley-Interscience, New York, 1964, p. 238.

[14] H. E. Rutherford, in Flameproofing Textile Fabrics (R. W. Little, ed.), Reinhold, New York, 1947, p. 27.

[15] A. Broido and S. B. Martin, U. S. Naval Radiological Defense Lab. Tech. Rept. USNRDL-TR-536, Oct. 1961, also available as AD 268, 729, U. S. Dept. Comm.

[16] R. F. Schwenker, Jr., and L. R. Beck, Jr., J. Polymer Sci., P. C, 2, 331 (1963).

[17] G. F. Davidson, J. Textile Inst., 32, T109 (1941).

[18] T. P. Nevell, J. Textile Inst., 42, T130 (1951).

[19] A. J. Stamm and L. A. Hansen, Ind. Eng. Chem., 29, 831 (1937).

[20] F. J. Kilzer and A. Broido, Pyrodynamics, 2, 151 (1965).

[21] E. L. Back, Trans. Inst. Consolidation Paper Web, Cambridge, 3rd. Sym., Sept. 1965, London, BP and BMA, 1965, p. 545 (pub. 1966).

[22] E. L. Back, M. T. Htun, M. Jackson, and F. Johanson, Tappi, 50, 542 (1967).

[23] E. L. Back and L. O. Klinga, Svensk Papperstid., 66, 745 (1963).

[24] E. L. Back and L. O. Klinga, Tappi, 46, 284 (1963).

[25] C. A. I. Goring, Pulp Paper Mag. Can., 64, T517 (1963).

[26] C. A. I. Goring, Trans. Inst. Consolidation of Paper Web, Cambridge, Sym. Sept. 1965, London BP and BMA, (1966), p. 555.

[27] P. R. Gupta, A. Rezanowich, and D. A. I. Goring, Pulp Paper Mag. Can., 63, T21 (1962).

[28] U. C. Chorley and W. Ramsey, J. Soc. Chem. Ind., 11, 872 (1892).

[29] P. Klason, G. von Heidenstan, and E. Norlin, Angew. Chem., 22, 1205 (1909).

[30] A. Pictet and J. Sarasin, J. Helv. Chim. Acta, 1, 87 (1918).

[31] S. Coppick, in Flameproofing Textile Fabrics (R.W. Little ed.), ACS Monograph 104, Reinhold, New York, 1947.

[32] R. F. Schwenker, Jr., and E. Pacsu, Chem. Eng. Data Ser., 2 (1), 83 (1957).

[33] J. A. S. Gilbert and A. J. Lindsey, Brit. J. Cancer, 11, 398 (1957).

[34] F. Shafizadeh and M. L. Wolfrom, J. Am. Chem. Soc., 80, 1675 (1958).

[35] F. Shafizadeh, M. L. Wolfrom, and P. McWain, J. Am. Chem. Soc., 81, 1221 (1959).

[36] M. L. Wolfrom et al, J. Am. Chem. Soc., 77, 6573 (1955); ibid.,
78, 4695 (1956); ibid. A. Chaney and P. McWain, 80, 946 (1958); ibid.
K. S. Ennor, 81, 3469 (1959).

[37] M. L. Wolfrom and G. P. Arsenault, J. Am. Chem. Soc., 82, 2819
(1960).

[38] G. Gaelernter, L. C. Browning, S. R. Harris, and C. M. Mason,
J. Phys. Chem., 60, 1260 (1956).

[39] C. H. DePuy and R. W. King, Chem. Rev., 60, 431 (1960).

[40] C. M. Conrad and D. J. Stanonis, Appl. Polymer, Symp. No. 2, 121
(1966).

[41] J. C. Irvine and J. W. H. Oldham, J. Chem. Soc. (London) Trans.,
119, 1744 (1921).

[42] W. G. Parks, R. M. Esteve, Jr., M. H. Gollis, R. Guercia, and
A. Petrarca, Abstr. of Papers 127 Meeting, ACS, Cincinnati, Ohio,
April 5, 1955, p. 6E.

[43] V. I. Ivanov, O. P. Golova, and A. M. Pakhomov, Izv. Akad. Nauk
SSSR, Otd. Khim. Nauk, No. 10, 1266 (1956), p. 6E.

[44] K. Tamaru, Bull Chem. Soc. Japan, 24, 164 (1951).

[45] O. P. Golova and R. G. Krylova, Dokl. Akad. Nauk SSSR, 116, 419
(1957).

[46] O. P. Golova, R. G. Krylova, and I. I. Nikolaeva, Vysokomolekul.
Soedin., 1, 1295 (1959), Eng. Trans.: Russian Trans. Program
(RTS1650) DSIR.

[47] O. P. Golova, A.M. Pakhomov, and E.A. Andrievskaya, Dokl. Akad.
Nauk SSSR, 112, 430 (1957).

[48] O. P. Golova, A. M. Pakhomov, E.A. Andrievskaya, and G.R. Krylova,
Dokl. Akad. Nauk SSSR, 115, 1122 (1957).

[49] T. V. Gatovskaya, O. P. Golova, R. G. Krylova, and V. A. Kargin,
Zh. Fiz. Khim., 33, 1418 (1959); Chem. Abs., 54, 8212 (1960).

[50] E. Pacsu and R. F. Schwenker, Jr., "Pyrolytic Degradation of
Cellulose and Chemical Modification of Cellulose for Flame and Glow
Resistance," Final Progress Rept., Contract No. DA19-129-QM-122,
Textile Research Institute, Princeton, New Jersey, 1957.

[51] D. Gardiner, J. Chem. Soc. Pt. C1, 1473 (1966).

[52] H. A. Schuyten, J. W. Weaver, and J. K. Reid, Advan. Chem. Ser.,
No. 9, p. 7 (1954).

[53] C. D. Hurd, The Pyrolysis of Carbon Compounds, Chemical Catalogue
Co., New York, 1929.

[54] A. M. Pakhomov. Izv. Akad. Nauk SSSR., Otd. Khim. Nauk, 1497
(1957); CA., 52, 581le (1958).

[55] S. Martin, Research and Development Tech. Rept. USNRDL-TR-102
"The Mechanism of Ignition of Cellulosic Materials by Intense Thermal
Radiation," Feb. (1956).

[56] A. Day, K. A. Friger, and J. R. E. Hoover, Progr. Rept. No. 1.
(Oct. 1954), QM Contract No. DA 19-129-QM-211, University of Penn.,
Philadelphia.

[57] H. B. Bull, Physical Biochemistry, Wiley, New York, 1943.
[58] I. N. Ermolenko, R. N. Sviridova, and A. K. Potapovich, Vestsi Akad. Navuk Belarusk, SSR, Ser. Khim. Navuk (1966), 111, Chem. Abs. 66-24303 (1967).
[59] J. C. Arthur and O. Hinojosa, Textile Res. J., 36, 385 (1966).
[60] R. M. Esteve and G. E. Wright, Abstr. of Papers, 138th Meeting ACS, No. 43, p. 17E (1960).
[61] K. Kato, Agr. Biol. Chem. (Tokyo), 31, No. 6, 657 (1967).
[62] R. Y. Pernikis, O. V. Kiselis, V. A. Sergeev, Y. A. Surna, and U. K. Stirna, Izv. Akad. Nauk Letv. SSR, Ser. Khim, No. 3, 373 (1967).
[63] M. M. Tang and R. Bacon, Carbon (Oxford), 2, 221 (1964).
[64] A. E. Lipska and W. J. Parker, J. Appl. Polymer Sci., 10, 1439 (1966).
[65] E. J. Murphy, Trans. Electrochem. Soc., 83; 161 (1943).
[66] E. J. Murphy, J. Polymer Sci., 58, 649 (1962).
[67] A. J. Stamm, Ind. Eng. Chem., 48, 413 (1956).
[68] W. K. Tang and W. K. Neill, J. Polymer Sci., Pt. C. 65, No. 6 (1964).
[69] J. F. v. Gizycki, Chem. -Zt., 74, 649 (1950).
[70] K. K. Andraev and B. S. Samsonov, Dokl. Akad. Nauk, SSSR, 114, 815 (1957).
[71] K. K. Andraev, Explosivstoffe, 8, 275 (1960); CA, 55, 10891 (1961).
[72] K. K. Andraev, Ind. Chim. Belge, Suppl., 2, 222 (1959); CA, 55, 22824 (1961).
[73] K. Ettre and P. F. Varadi, Anal. Chem., 35, 69 (1963).
[74] P. K. Chatterjee and C. M. Conrad, Textile Res. J., 36, 487 (1966).
[75] P. K. Chatterjee, J. Polymer. Sci., Pt. A., 3, 4253 (1965).
[76] P. K. Chatterjee, J. Appl. Polymer. Sci., 12, 1859 (1968).
[77] P. K. Chatterjee and C. M. Conrad, J. Polymer. Sci., Pt. A-1, 6, 3217 (1968).
[78] C. M. Conrad and P. Harbrink, Textile Res. J., 38, 366 (1968).
[79] S. Otani, Kogyo Kagaku Zasshi, 62, 871 (1959).
[80] S. Otani, A. Ogura, and J. Hurudatsu, Kogyo Kagaku Zasshi, 73, 1448 (1960).
[81] H. G. Higgins, Australian. J. Chem., 10, 496 (1957).
[82] H. G. Higgins, J. Polymer. Sci., 28, 645 (1958).
[83] W. Hofman, T. Ostrowski, T. Urbanski, and M. Witanowski, Chem. & Ind. (London), 1960, 95.
[84] N. Berkowitz, Fuel, 36, 355 (1957).
[85] J. W. Cobb, Fuel, 23, 121 (1944).
[86] H. W. Davidson and H. H. W. Losty, Conf. Ind. Carbon Graphite Papers, 2nd, London, p. 20 (1965), (pub. 1966).
[87] A. J. P. Martin and R. L. M. Synge, Biocem. J., 35, 1358 (1941).

[88] A. T. James and A. J. P. Martin, Analyst, 77, 915 (1952).

[89] S. Claesson, Arkiv Kemi, Mineral Geol., 23A, No. 1 (1946).

[90] C. J. Hardy and F. H. Pollard, J. Chromatography, 2, 1 (1959).

[91] S. Dal Nogare and L. W. Safranski, in Organic Analysis, Vol. 4
      (J. Mitchell, Jr., I. M. Kolthoff, E. S. Proskaren, and A. Weissberger,
      eds.), Wiley-Interscience, New York, 1960, p. 91.

[92] D. H. Desty, Vapor Phase Chromatography, Academic, New York,
      1957.

[93] A. I. M. Keulmans, Gas Chromatography, Rheinhold, New York, 1957.

[94] A. B. Littlewood, Gas Chromatography, Academic, New York, 1962.

[95] C. S. G. Phillips, Discussions Faraday Soc., No. 7, 241 (1949).

[96] S. Dal Nogare and R. S. Juvet, Gas Liquid Chromatography, Wiley-
      Interscience, New York, 1962.

[97] C. S. G. Phillips, Gas Chromatography, Academic, New York, 1956.

[98] D. M. Cates and T. H. Guion, in Analytical Methods for a Textile
      Laboratory, AATCC Monograph No. 3, 2nd ed. (J. W. Weaver, ed.),
      1968, p. 239.

[99] F. H. Stross and H. W. Johnson, Jr., "Gas Chromatography" in
      Encyclopedia of Chemical Technology (R. E. Kirk and D. E. Othmer,
      eds.), 2nd supplement Vo. A. A. Stenden, ed., Wiley-Interscience,
      New York, 1960, p. 377.

[100] D. Ambrose, A. I. M.R. Keulemans, and J.H. Purnell, Anal. Chem.,
      30, 1582 (1958).

[101] H. W. Johnson, Jr., and F. H. Stross, Anal. Chem., 30, 1586 (1958).

[102] J. M. Cassel, Analytical Chemistry of Polymers (G. M. Kline, ed.),
      Wiley-Interscience, New York, 1962, p. 359.

[103] R. S. Lehrle and J. C. Robb, Nature, 183, 1671 (1959).

[104] A. Barlow, R. S. Lehrle, and J. C. Robb, Polymer, 2, 27 (1961).

[105] F. A. Lehman and G. M. Brauer, Anal. Chem., 33, 673 (1961).

[106] J. Strassburger, G. M. Brauer, M. Tryon, and A. F. Forziati,
      Anal. Chem., 32, 454 (1960).

[107] I. G. McWilliam and R. A. Dewar, Nature, 181, 760 (1958).

[108] J. E. Lovelock, Nature, 182, 1663 (1958).

[109] S. B. Martin and R. W. Ramstad, Anal. Chem., 33, 982 (1961).

[110] S. B. Martin, Tenth Symposium (International) on Combustion, The
      Combustion Institute, 1965, p. 877.

[111] J. B. Berkowitz-Mattuck and T. Noguchi, J. Appl. Polymer. Sci.,
      7, 709 (1963).

[112] L. S. Nelson and J. L. Lundberg, J. Phys. Chem., 63, 433 (1959).

[113] C. T. Greenwood, J. H. Knox, and E. Milne, Chem. Ind., 1878
      (1961).

[114] S. Glassner and A. R. Pierce, III., Anal. Chem., 37, 525 (1965).

[115] G. A. Byrne, D. Gardiner, and F. H. Holmes, J. Appl. Chem., 16,
      81 (1966).

[116] W. Kast and L. Flaschner, Kolloid-Z., 111, 6 (1948).

[117] V. Gokcen and D. M. Cates, Appl. Polymer. Symp., No. 2, 15 (1966).

[118] C. H. Mack and S. R. Hobart, Appl. Polymer. Symp., No. 2, 133 (1966).

[119] H. Schubert and F. Volk, Explosivsivstoffe, 14, 1 (1966); CA, 64, 15667 (1966).

[120] R. Consden, A. H. Gordon, and A. J. P. Martin, Boichem. J., 38, 224 (1944).

[121] J. C. Giddings and R. A. Keller, ed., Advances in Chromatography, Vols. 1-4, Dekker, New York, 1965-67.

[122] G. H. Stewart, in Encylopedia of Chemical Technology, Vol. 5 (R. E. Kirk and D. W. Othmer, eds.), 2nd ed., Wiley-Interscience, New York, 1964, p. 413.

[123] G. A. Byrne, J. Chromatogr., 20, 528 (1965).

[124] J. Honeyman, "A Fundamental Study of the Pyrolysis of Cotton Cellulose to provide information needed for improvement of flame resistant treatments for cotton," Final report, Proj. UR-E29-(20)-9, Shirley Institute, Manchester, Eng. (1964), F. H. Holmes and C. J. G. Shaw, J. Appl. Chem., 11, 210 (1961).

[125] J. H. Beynon, Mass Spectrometry and Its Applications to Organic Chemistry, Van Nostrand, Princeton, New Jersey, 1960.

[126] K. Biemann, Mass Spectrometry: Organic Chemical Applications, McGraw-Hill, New York, 1962.

[127] F. W. McLafferty, ed., Mass Spectrometry of Organic Ions, Academic, New York, 1963.

[128] D. W. Stewart, in Techniques of Organic Chemistry, .Weissberger, ed.), 3rd ed., Pt. IV, Vol. I, Wiley-Interscience, New York, 1960, p. 3449.

[129] P. Bradt and F. L. Mohler, J. Res. Natl. Bur. Std., 55, 323 (1955).

[130] C. A. McDowell, ed., Mass Spectrometry, McGraw-Hill, New York, 1963.

[131] L. A. Wall, in Analytical Chemistry of Polymers (G. M. Kline, ed.), Pt. II, Wiley-Intersceince, New York, 1962, p. 251.

[132] H. E. Duckworth, Mass Spectroscopy, Cambridge Univ. Press, London, 1958.

[133] American Petroleum Institute, Res. Project 44, Catalogue of Mass Spectral Data, Chemical Thermodynamics Properties Center, Texas A & M University, College Station, Texas, 1948 to date.

[134] ASTM Committee E-14 File of Uncertified Mass Spectra, A. H. Struck, Chairman, Perkin Elmer Corp., Norwalk, Conn, 1958 to date.

[135] R. S. Gohlke, ed., Uncertified Mass Spectral Data, Dow Chemical Co., Framingham, Mass., 1963.

[136] A. L. Ryland, in Encyclopedia of Chemical Technology (R. E. Kirk and D. E. Othmer, eds.), 2nd ed., Vol. 13, Wiley-Interscience, New York, 1967, p. 87.

334   P. K. CHATTERJEE and R. F. SCHWENKER

[137] R. M. Silverstein and G. C. Bassler, Spectrometric Identification of
Organic Compounds, 2nd ed., Wiley, New York 1967.

[138] W. H. McFadden, Advan. Chromatogr., 4, 265, (1967).

[139] B. Hallgren, R. Ryhage, and E. Stenhagen, Acta Chem. Scand., 13,
845 (1959).

[140] R. S. Gohlke, Paper No. 23, Meeting of ASTM Committee E-14 on
Mass Spectroscopy, New Orleans, La.

[141] R. S. Gohlke, Chem. Ind. (London), 41, 946 (1963).

[142] H. G. Langer and R. S. Gohlke, Anal. Chem., 35, 1301 (1963).

[143] H. G. Langer, R. S. Gohlke, and D. H. Smith, Anal. Chem., 37,
433 (1965).

[144] S. L. Madorsky and S. Straus, J. Res. Natl. Bur. Std., 40, 417
(1948).

[145] L. A. Wall, J. Res. Natl. Bur. Std., 41, 315 (1948).

[146] S. L. Madorsky, V. E. Hart, and S. Straus, J. Res. Natl. Bur,
Std., 56, 343 (1956).

[147] S. L. Madorsky, V. E. Hart, and S. Straus, J. Res. Natl. Bur. Std.,
60, 343 (1958).

[148] R. F. Schwenker, Jr., in Analytical Methods for a Textile Laboratory
(J. Weaver, ed.), AATCC Monograph No. 3, 2nd ed., 1968, p. 385.

[149] R. F. Schwenker, Jr., and J. C. Whitwell, in Analytical Calorimetry,
Proc. Am. Chem. Soc. Symposium (R. S. Porter and J. F. Johnson,
eds.), Plenum, New York, 1968, p. 249.

[150] R. F. Schwenker, Jr., and L. R. Beck, Jr., Textile Res. J., 30,
624 (1960).

[151] R. F. Schwenker, Jr., L. R. Beck, Jr., and R. K. Zuccarello,
Am. Dyestuffs Reptr, 53 (No. 19, p. 30), 817 (1964).

[152] A. F. Frederickson, Am. Mineralogist, 39, 1023 (1954).

[153] E. P. Partridge, V. Hicks, and G. W. Smith, J. Am. Chem. Soc.,
63, 454 (1941).

[154] H. Morita and H. M. Rice, Anal. Chem., 27, 336 (1955).

[155] S. Speil, L. H. Berkelhamer, J. A. Pask, and B. Davis, U. S. Bur.
Mines, Tech. Paper, 664 (1945).

[156] P. F. Kerr and J. L. Kulp, Am. Mineralogist, 33, 387 (1948).

[157] B. Ke, ed., Thermal Analysis of High Polymers, J. Polymer Sci.,
Polymer Symp., No. 6 (1964).

[158] M. J. Vold, Anal. Chem., 21, 683 (1949).

[159] W. W. Wendlandt, Thermal Methods of Analysis, Wiley-Interscience,
New York, 1964.

[160] S. L. Boersma, J. Amr. Ceram. Soc., 38, 281 (1955).

[161] S. Strella, J. Appl. Polymer. Sci., 7, 569 (1963).

[162] W. F. Tsang, W. J. Smothers, and Y. Chiang, eds., Handbook
of Differential Thermal Analysis, Chap. 5, 2nd ed., Chemical
Publishing Co., New York, 1966.

[163] P. D. Garn, Thermoanalytical Methods of Investigation, Academic, New York, 1965.

[164] H. J. Borchardt and F. Daniels, J. Am. Chem. Soc., 79, 41 (1957).

165 H. E. Kissinger, Anal. Chem., 29, 1702 (1957); J. Res. Natl. Bur. Std., 57, 217 (1956).

[166] L. Reich, J. Appl. Polymer. Sci., 11, 161 (1967); ibid., 10, 1801 (1966).

[167] R. L. Reed, L. Weber, and B. S. Gotterfield, Ind. Eng. Chem., 4, 38 (1965).

[168] P. E. Slade, Jr., and L. T. Jenkins, eds., Techniques and Methods of Polymer Evaluation, Vol. I: Thermal Analysis, Dekker, New York, 1966.

[169] E. M. Barrall, II, J. F. Garivert, R. S. Porter, and J. F. Johnson, Anal. Chem., 35, 1837 (1963).

[170] H. T. Smyth, J. Am. Ceram. Soc., 34, 221 (1951).

[171] E. M. Barrall, II, and L. B. Rogers, Anal, Chem., 34, 1106 (1962).

[172] D. J. David in Techniques and Methods of Polymer Evaluation, (P. E. Slade, Jr., and L. T. Jenkins, eds.), Vol. 1, Dekker, New York, 1966, p. 43.

[173] L. P. Arens, Soil Sci., 72, 406 (1951).

[174] L. M. Clarebrough, M. E. Hargreaves, D. Mitchell, and G. E. West, Proc. Roy. Soc. (London), A215, 507 (1952).

[175] C. Eyraud, Compt. Rend., 238, 1511, 507 (1954).

[176] M. J. O'Neill, Anal, Chem., 36, 1238 (1964).

[177] E. S. Watson, M. J. O'Neill, J. Justin, and N. Brenner, Anal. Chem., 36, 1233 (1964).

[178] R. F. Schwenker, Jr., L. R. Beck, Jr., and W. J. Kauzmann, "Thermal Characteristics of Coated Textile Fabrics," Final Report Contract N140 (132) 57442B, Textile Res. Institute, Princeton, N. J., Feb. 15, 1960.

[179] H. Morita, Anal. Chem., 29, 1095 (1957).

[180] P. K. Chatterjee and R. F. Schwenker, Jr., Tappi, 55, III (1972).

[181] K. Honda, Sci. Rept. Tohoku Univ., 4, 97 (1915).

[182] M. Guichard, Bull. Soc Chim. France, 37, 62 (1925).

[183] C. Duval, Anal. Chim. Acta., 1, 341 (1947).

[184] C. B. Murphy, Anal. Chem., 36, 347R (1964).

[185] C. D. Doyle, in Ref. 161, p. 113.

[186] J. H. Flynn and L. A. Wall, J. Res. Natl. Bur. Std., A Phys. Chem., 70A, 487 (1966).

[187] J. Sestak, Silikaty, 11, 153 (1967).

[188] J. Zsako, J. Phys. Chem., 72, 2406 (1968).

[189] W. L. deKeyser, Bull. Soc. Ceram. France, 20, 1 (1953).

[190] W. L. deKeyser, Nature, 172, 364 (1953).

[191] F. Paulik, J. Paulik, and L. Erdey, Z. Anal. Chem., 160, 241 (1958).

[192] P. L. Waters, J. Sci. Instr., 35, 41 (1958).
[193] C. Campbell, S. Gordon, and C. L. Smith, Anal. Chem., 31, 1188
       (1959).
[194] A. E. Newkirk, Anal. Chem., 32, 1558 (1960).
[195] J. R. Soulen and I. Mockrin, Anal. Chem., 33, 1909 (1961).
[196] E. S. Freeman and B. Carroll, J. Phys. Chem., 62, 394 (1958).
[197] D. A. Anderson and E. S. Freeman, J. Polymer. Sci., 54, 253
       (19   ).
[198] K. Akita and M. Kase, J. Polymer. Sci., Pt. A-1, 5, 833 (1967).
[199] A. W. Coats and J. P. Redfern, Nature, 201, 68 (1964).
[200] H. H. Horowitz and G. Metzger, Anal. Chem., 35, 1464 (1963).
[201] C. H. Mack and D. J. Donaldson, Textile Res. J., 37, 1063 (1967).
[202] R. M. Perkins, G. L. Drake, Jr., and W. A. Reeves, J. Appl.
       Polymer. Sci., 10, 1041 (1966).
[203] P. K. Chatterjee and C. M. Conrad, J. Polymer. Sci., Pt. A-1, 4,
       233 (1966).
[204] J. K. Gillham, Appl. Polymer. Symp. No. 2, 45 (1966).
[205] J. K. Gillham and R. F. Schwenker, Jr., Appl. Polymer. Symp. No.
       2, 59 (1966).
[206] R. T. O'Connor in Analytical Methods for a Textile Laboratory,
       J. W. Weaver, ed., AATCC Monograph No. 3, 2nd ed., 1968, p. 295.
[207] F. H. Forziati and J. W. Rowen, J. Res. Natl. Bur. Std., 46, 38
       (1951).
[208] C. D. Smith and J. K. Wise, Anal. Chem., 39, 1968 (1967).
[209] H. T. Smyth, J. Am. Ceram. Soc., 34, 221 (1951).
[210] J. Fahrenfort, Spectrochim. Acta, 17, 698 (1961).
[211] N. J. Harrick, Ann. N.Y. Acad. Sci., 101, 928 (1963).
[212] W. W. Wendlandt and H. G. Hecht, Reflectance Spectroscopy, Wiley-
       Interscience, New York, 1966.
[213] P. A. Wilks, Jr., and M. R. Iszard, Develop. Appl. Spectry., 4,
       141 (1964) (Pub. 1965).
[214] H. A. Szymanski and N. L. Alpert, IR-Theory and Practice of
       Infrared Spectroscopy, Plenum, New York, 1964.
[215] R. G. J. Miller, Laboratory Methods in Infrared Spectroscopy,
       Heyden and Sons, London, and Sadtler, Philadelphia, 1965.
[216] K. Nakanishi, Infrared Absorption Spectroscopy - Practical, Holden-
       Day, San Francisco, 1962.
[217] W. J. Potts, Jr., Chemical Infrared Spectroscopy, Vol. I: Techniques,
       Wiley, New York, 1963.
[218] R. G. White, Handbook of Industrial Infrared Analysis, Plenum,
       New York, 1964.
[219] R. G. Zhbankov, Infrared Spectra of Cellulose and Its Derivatives,
       Consultant Bureau, New York 1966, p. 23.
[220] R. T. O'Connor, E. F. DuPre, and E. R. McCall, Anal. Chem.,
       29, 998 (1957).

[221] R. T. O'Connor, E. F. DuPre, and E. R. McCall, Textile, Res. J., 28, 221 (1958).

[222] R. T. O'Connor, E. F. DuPre, and D. Mitchem, Textile, Res. J., 28, 382 (1958).

[223] R. S. Tipson, Infrared Spectroscopy of Carbohydrates, A Review of the Literature, National Bureau of Standards Monograph 110 (June 1968).

[224] B. I. Friedlander, A. S. Dutt, and W. H. Rapson, Tappi, 49, 468 (1966).

[225] H. G. Cassidy., Fundamentals of Chromatography, Vol. 10 in Techniques of Organic Chemistry, (A. Weissberger, ed.), Wiley-Interscience, New York, 1957.

[226] M. L. Nelson and R. T. O'Connor, J. Appl. Polymer. Sci., 8, 1311 (1964).

[227] M. L. Nelson and R. T. O'Connor, J. Appl. Polymer. Sci., 8, 1325 (1964).

[228] T. Urbanski, W. Hofman, T. Ostrowski, and M. Witanowski, Bull. Acad. Polon. Sci., Ser. Sci., Chim., Geol. et Geograph, I, 851 (1959); CA, 55, 23049 (1961).

[229] R. W. Phillips, C. A. Orlick, and R. Steinberger, J. Phys. Chem., 59, 1034 (1955).

[230] P. H. Hermans and A. Weidinger, J. Appl. Phys., 19, 491 (1948).

[231] W. Kast and L. Flaschner, Kolloid-Z., 111, 6 (1948).

[232] J. B. Nichols, J. Appl. Phys., 25, 840 (1954).

[233] K. Ward, Jr., Textile Res. J., 20, 363 (1950).

[234] K. Hess, H. Kiessig, and J. Gundermann, Z. Phys. Chem., B49, 64 (1941).

[235] J. H. Wakelin, H. S. Virgin, and E. Crystal, J. Appl, Phys., 30, 1654 (1959).

[236] Ø. Ellefsen, E. W. Lund, B. A. Tønnesen, and K. Øien, Nor. Skogind, 11, 284, 286, 349, 355 (1957).

[237] L. Segal, J. J. Creely, A. E. Martin, Jr., and C. M. Conrad, Textile Res. J., 29, 786 (1959).

[238] W. Parrish, Philips Tech. Rev., 17, 206 (1956).

[239] G. F. Davidson, J. Textile Inst., 34, T87 (1943).

[240] H. G. Ingersoll, J. Appl. Phys., 17, 924 (1946).

[241] J. A. Howsmon, Textile Res. J., 19, 152 (1949).

[242] R. F. Nickerson, Ind. Eng. Chem., Anal. Ed., 13, 423 (1941); Ind. Eng. Chem., 33, 1022 (1941); ibid., 34, 85, 1480 (1942).

[243] R. F. Nickerson and J. A. Habrle, Ind. Eng. Chem., 39, 1507 (1947).

[244] S. P. Rowland and E. R. Cousins, J. Polymer Sci., Pt. A-1, 4, 793 (1966).

[245] C. M. Conrad and J. J. Creely, J. Polymer Sci., 58, 781 (1962).

[246] V. B. Ryzhov, N. D. Burkhanova, and P. V. Kozlov, Vysokomolekul. Soedin., 6, 1471 (1964).

[247] K. A. Murty and P. L. Blackshear, Pyrodynamics, 4, 285 (1966).

[248] K. Hess, E. Steurer, and H. Fromm, Kolloid-Z., 98, 148 (1942); ibid., 98, 290 (1942).

[249] M. L. Nelson and C. M. Conrad, Textile Res. J., 18, 155 (1948).

[250] J. K. Craver and D. L. Taylor, Tappi, 48, 142 (1965).

[251] J. K. Craver and D. L. Taylor, Pulp Paper Mag. Can., 67, T331 (1967).

[252] P. K. Chatterjee, Tappi, 52, 699 (1969).

[253] V. Bush, J. Am. Inst. Elec. Engrs., 44, 509 (1925); F. L. Browne and J. J. Brenden, U. S. Dept. Agr. Forest Service Res. Paper FPL 19 (Dec. 1964).

[254] V. M. Montsinger, Trans. Am. Inst. Elec. Engrs. 49, 294 (1930); see also Elec. Eng., 58, 512 (1939).

[255] Parr Instrument Co., Oxygen Bomb Calorimetry and Combustion Methods, Tech. Manual 130, Molini, Ill.

Chapter 6

X-RAY DIFFRACTION

Verne W. Tripp and Carl M. Conrad*

Southern Regional Research Laboratory
New Orleans, Louisiana

I. INTRODUCTION

Historically, cellulose was one of the earliest polymeric materials to
be examined by x rays. Diffraction patterns were reported by Nishikawa
and Ono [1] in 1913, at a time when the macromolecular structure of such

---

*Deceased.

materials was scarcely suspected. In this sense, cellulose has enjoyed
the benefit of a long, and in some respects, exhaustive investigation by
x rays in efforts to understand and define its structure. Such extensive
study reflects the economic importance of the polymer as a widespread
and renewable technological resource. The principal role of x-ray diffraction studies has been to establish the secondary or supramolecular
organization of the polymer, leaving the primary or molecular structure
to the more fruitful methods of chemical analysis.

In principle, the pattern of scattering of x rays by the cellulosic substrate contains all of the information necessary to describe the arrangement in space of the atomic groupings present, the extent to which these
groupings are ordered in crystalline or quasicrystalline form, and their
orientation with respect to some arbitrary or external axis. In a useful
paper on the mechanics of deformation of polymers, Hearle [2] has
suggested a number of parameters of the secondary structure which are
pertinent in characterizing the physical nature of such materials and in
relating the submicroscopical structural details to technologically important properties. These parameters include, among others, (a) the
type and spectrum of three-dimensional order present, (b) the local-
ization and dimensions of the ordered structures, and (c) the extent of
orientation of the polymer chains. Measurement of these characteristics
of cellulose and its derivatives is accessible, to a large extent, by means
of x-ray diffraction methods.

It is well to emphasize that the characterization of cellulosic materials
is best accomplished by the application of a variety of physical and chemical techniques, including their variations. It will be clear that each
approach to measuring a given property of the substrate is unique. In
the case of secondary polymer structure, each technique is capable of
"seeing" a particular aspect of the arrangement present, and is most
sensitive to substrate volume elements of a certain size. Thus, x-ray
diffraction requires that the volume elements be larger than a minimum
size to be characterized as "crystalline," while infrared absorption will
reflect aspects of the physical arrangement of the polymer in terms of
much smaller groupings. Taking into account such limitations, however,
x-ray diffraction remains a powerful tool for arriving at the secondary
structure of polymers.

The objective of nearly all x-ray diffraction procedures is the recording and evaluation of the scattering direction and intensity or radiation
diffracted by atom planes a fixed distance apart, according to the well-
known Bragg law [3], illustrated in Fig. 1,

$$n\lambda = 2d \sin \theta ,$$

where λ is the wavelength of the radiation, d the distance between parallel planes, θ the angles of incidence and reflection of the x rays with

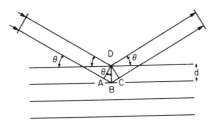

Fig. 1. The Bragg law, indicating conditions under which reinforce-
ment of diffraction occurs, i.e., when the path difference of the rays
diffracted by parallel planes is an integral number of wavelengths of the
radiation. After Ref. [12], by courtesy of John Wiley and Sons, Inc.

respect to the planes, and n an integer. The details of the basic theory
of x-ray diffraction, as well as those of most of the necessary apparatus
and equipment for the generation, collimation, and recording of the
radiation, together with means for presentation of the specimen, are described in number of excellent texts [3-5]. Of particular value also is
the review of Rudman [6], which lists available commercial equipment
and includes recent price ranges of most of the items described.

   X-ray diffraction procedures are attended by numerous hazards.
Aside from the high voltages associated with the generation and detection
of x-rays, there are many possibilities for the accidental exposure of
operators to both stray and directed radiation. Radiation of 10-100 kV
energies is damaging to mammalian tissue, and maximum permissible
dosages for workers have been recommended by a number of official
agencies. Earlier recommendations of the National Committee on Radiation Protection give the maximum permissible total weekly dose as 0.3R
[7]. The International Commission on Radiation Protection, however,
lowers this value to 0.1 R per week (5 R per year) [8]. Tube housings of
commercial x-ray diffraction instruments are well-designed to prevent
radiation from reaching the operator, but particular caution must be
taken to assure that port shutters, collimating slits, and pinholes are
properly installed to eliminate leakage. Scattering of radiation from
specimens, sample holders and associated apparatus, cameras, and detection devices can occur readily. It must be minimized by proper positioning of these components, together with addition of shielding in an
appropriate way. The use of radiation survey meters in any installation
or experimental arrangement is mandatory, and the employment of film
badge services or their equivalent is highly recommended.

II. CELLULOSE FINE STRUCTURE

Cellulose may be considered to be made up of linear chain molecules consisting very largely, if not exclusively, of anhydro-D-glucopyranose units linked through β -1-4 linkages. The length of the chain shows enormous variation, ranging from 40 monomer units (200 Å) or less to 10,000 units (50,000 Å) or more in typical specimens. Commercial regenerated celluloses average a few hundred andydroglucose units in length, while those of native fibrous materials may be more than 10 times as long. The chains are aggregated laterally by polar and van der Waals forces in such a way as to build up the elementary fibrils, microfibrils, and larger assemblies visualized by various forms of microscopy in virtually all cellulose preparations. Discrete microfibrillate character is not as evident in most regenerated products, a reflection of the lesser extent of regular intermolecular hydrogen bonding. Cellulose displays three-dimensional order of varying degrees of perfection, and at least four distinct polymorphic crystalline forms are well-recognized, with evidence for others. The great length of the molecules argues against the likelihood that a high degree of lateral order can be maintained continuously except over small volumes. More probable is a situation in which the lateral packing of the chains is characterized by distortions and defects to a greater or lesser degree. Minor deviations from regular spacing would present a paracrystalline arrangement or "sloppy lattice," while other areas would display a state of disorder or random array indistinguishable from that of a classically amorphous structure. This concept of a spectrum of order-disorder holds true, as a practical matter, for virtually all natural and synthetic fibers.

The distribution of ordered and disordered regions in cellulose has been the subject of many proposed schemes. They range from models in which the chains are considered to form a continuum consisting essentially of a single phase, which may be mainly noncrystalline or paracrystalline, to assemblies of one-phase units, and to two-phase structures, such as the fringed micelle model. The latter were among the first to be advanced and have been generally useful in serving as schemes for explaining the observed properties of celluloses. At the other extreme is the view of Kargin [9] that cellulose is amorphous, i.e., noncrystalline, with a high degree of correlation in direction between neighboring chain units. The paracrystalline model of Hosemann [10] , which portrays the chains arranged so as to have short-range order approaching the crystalline state, but without long-range order, is another plausible concept. More recently, Manley [11] has advanced a folded-chain (crystalline) elementary fibril model, based on the recognition that many linear polymers will naturally crystallize by a folding process. The occurrence of chemically homogeneous cellulose in an immense variety of physical forms makes it unlikely that a single structural model will serve to portray the secondary arrangement of the molecules. Native celluloses may be laid down in the form of macroscopic films, as in the product of Acetobacter xylinum, or in the cell wall of the marine alga Valonia ventricosa, or as fibers as the cell walls of seed hairs and wood tissues. The secondary molecular structure of these products of relatively slow biological processes tends to display higher lateral order than that of regenerated materials, which ordinarily are coagulated or precipitated from solution more rapidly under a variety of conditions, and are subjected to concomitant mechanical working. Such products will have a spectrum of secondary arrangements, many of them of low three-dimensional order. The protean technological forms of cellulose are strong evidence for its ability to take arrangments well-suited to serve as structural members of plants, textile fibers, self-bonded sheets, and packaging films.

A dominant feature of most celluloses is the molecular orientation both in elementary fibrils and along the cell wall axis of fibers or the machine direction of films and sheets. Strong orientation along these axes is associated with high tensile strength and has a profound effect on the mechanical properties of the bulk material. In native celluloses, high molecular orientation is associated with high lateral order; in regenerated materials, Fortisan, a saponified cellulose acetate fiber, displays both strong orientation and order. In many plant fibers the microfibrils are arranged in a spiral fashion along the length of the cell wall; in cotton, this arrangement is particularly complex, with the spiral alternating in sense from S to Z many times in a single fiber. Orientation can be increased by application of stress when the fiber elements or molecules are in a plastic condition, as during the spinning of viscose. In most cases, poorer orientation results from allowing swelling of the cellulose without imposing external stresses.

III. CRYSTAL FORM AND DIMENSIONS

A. Diffraction Patterns

The typical arrangement for obtaining a diffraction pattern from a bundle of parallelized cellulosic fibers or other sample is illustrated in Fig. 2 [12] . A diffraction photograph from ramie fibers is shown in Fig. 3, in which the fiber axis is vertical and normal to the direction of the x-ray beam. The characteristics of the pattern, including the number of interferences, their distances from the center, their relative intensity, width, and concentration or extension into arcs, and general background scattering permit considerable information on the nature of the specimen to be obtained. In such a fiber diagram, the pattern is symmetrical, with the equatorial (paratropic) and meridional (diatropic) interferences occurring twice and the other reflections four times. In effect, the pattern is repeated in each quadrant of the diagram. Although the relatively high

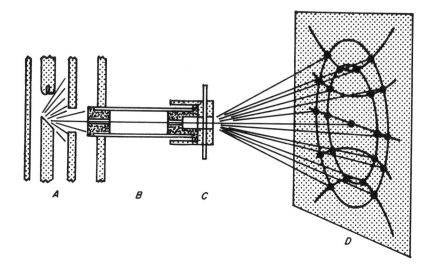

Fig. 2. Schematic representation of arrangement for obtaining fiber diagrams: A. x-ray source; B. pinhole collimator; C. fiber sample; D. photographic film. After Ref. [12] , by courtesy of John Wiley and Sons, Inc.

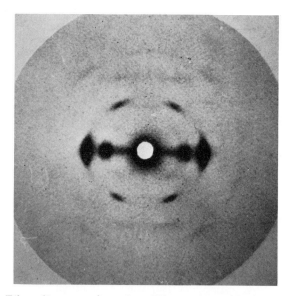

Fig. 3. Fiber diagram of ramie. Fiber axis vertical.

birefringence of many cellulose samples would lead to the conclusion that it is strongly crystalline, the size and perfection of the crystalline regions are such that diffraction is weak compared to many other substances, and ordinary diagrams show rather few reflections.

In a forward-reflection or "flat-film" pattern of the type illustrated, the lattice spacing corresponding to an interference may be arrived at from the Bragg law, determining the angle θ from the relation $\tan 2\theta = R/S$, where S is the specimen-to-film distance and R the distance from the reflection to the center of the film. The intensity of the reflections is a measure of the atomic population of the lattice planes, which for cellulose are those containing the chain molecules (equatorial interferences). The crystal class may be determined from the presence, absence, and general arrangement of the diffraction spots. For most celluloses, the crystalline form appears to be monoclinic. The width of the interferences is inversely proportional to the size and perfection of the crystalline areas, while the extension of the spots into arcs is a measure of the distribution of the alignment of the crystal with respect to a given axis in the fiber. Finally, the intensity of diffuseness of the background scattering of the x-rays is related to the amount of fiber substance which can produce nondiscrete reflection and thus possess noncrystalline characteristics.

While the flat-film technique is capable of producing the maximum amount of information obtainable in a single observation, the relatively long photographic exposure and development times required make electronic counting methods attractive. For the most part these are carried out with diffractometers built in the parafocusing arrangement, illustrated in Fig. 4 [3] . This method is most frequently used with pressed (flat) specimens, and the reflecting (diffracting) surface of the specimen is rotated at a rate one-half that of the detector slit. In this way, the sample surface is kept tangent to the focusing circle, as required by the parafocusing principle. Detection of the diffracted radiation after transmission can also be easily accomplished with this arrangement. In the reflection mode, only a relatively thin depth of the specimen is responsible for the recorded diffraction, and particular care to ensure the desired orientation of the crystalline bodies must be taken in presenting the specimen. In effect the diffractometer measures the intensity of diffraction along a single diameter of the flat-film photograph and is most commonly used with powdered or ground specimens.

B. Polymorphic Forms and Unit Cells

Chemically unmodified cellulose is generally recognized to occur in four polymorphic forms. There is some evidence for the existence of others [13, 14] , and by appropriate treatment, the discrete crystalline reflections usually observed can be eliminated and the cellulose appear to have no lateral order. The introduction of additional groupings through

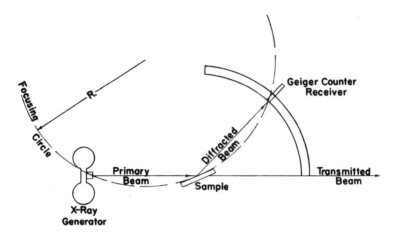

Fig. 4. Schematic diagram of diffractometer employing parafocusing optics.

chemical modification may, of course, also disrupt the original three-dimensional order of the polymer, and the formation of a new lattice will depend on the regularity and crystallographic compatability of the modified molecular chains. The crystal structures of the various cellulose polymorphs represent arrangements arrived at on the basis of often limited experimental data because of the weak diffraction of the polymer. One result of this paucity of data is that it has not been possible to put forward unique structures for the modifications observed. It is also probable that, as a result of the numerous possibilities for hydrogen bonding between adjacent cellulose chain segments, a single (crystalline) arrangement will not exist over large volume elements.

In discussing the polymorphs of cellulose, the crystallographic convention of Meyer and Misch [15] is used, in which the b axis is the fiber axis, and the β angles are the supplements of those assigned in standard practice. In mentioning this discrepancy, Woods [16] has concluded that attempts to rectify the situation would create serious confusion because of the widespread use of the Meyer and Misch unit cell parameters.

1. Cellulose I

The characteristic pattern obtained by the flat-film method from an oriented bundle of native cellulose fibers is shown in Fig. 3 (ramie) and in Fig. 5 (cotton). On the basis of such diagrams, the unit cell of cellulose I is generally, but not unanimously, accepted to be monoclinic, and of space group P2, as originally proposed by Meyer and Misch [15] . Wellard [17] carried out a careful investigation of the variation in lattice

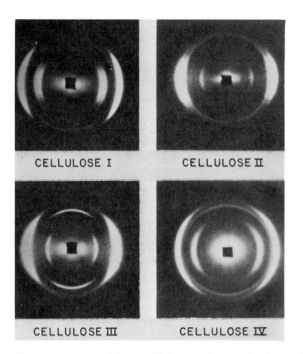

Fig. 5. Fiber diagrams of four cellulose polymorphs derived from cotton. Fiber axis vertical.

spacings and found that there were slightly different interplanar distances in the equatorial reflections, depending on the source of the cellulose. Table 1 lists the pertinent data with respect to the unit cell dimensions and lattice spacings of the cellulose I modification. The relative intensities of a large number of interferences were measured by Mann et al. [18] for ramie; the equatorial reflections, 002, 10$\bar{1}$, and 101 gave intensities of 25,000, 3500, and 3500, respectively; the meridional reflections 040 and 020 measured 4500 and 250, respectively: the off-meridional spot 021 had an intensity of 980. The chief alternative to the Meyer and Misch unit cell is that proposed by Ellis and Warwicker [19], based on the assumption of a P1 space group and having the following parameters for a monoclinic cell: a, 10.85 Å, b, 10.3 Å, c, 120.8 Å, and β, 93.2°. Examination of the cellulose of the marine alga Valonia ventricosa by electron diffraction procedures [20, 21] has led to the conclusion that the unit cell of this material (and probably that of bacterial cellulose) is slightly different from that of the native fibers. It has been assigned a triclinic crystal from, for reasons of symmetry, and the a and c parameters of the Meyer and Misch cell doubled; the β angle is 82°. The diffraction patterns of a large number of natural cellulose fibers are presented in publications of Heyn [12] and Howsmon [22].

The diffractometer trace of a cellulose I specimen (cotton) over the range 3 - 10 Å is shown in Fig. 6. Such a pattern is obtained from the reflection diffraction of a pressed pellet of fibers ground to pass a 20-mesh sieve, and only the 101, 10$\bar{1}$, and 002 interferences are distinguished; the 021 reflection appears as a weak shoulder on the strong 002 peak. No indication of the extension of the interferences into arcs can be gained from such a recording, but the simple and rapid location of the major lattice spacings makes this procedure the method of choice for much diffraction work.

2. Cellulose II

The treatment of native cellulose with strong alkali solutions, e.g., NaOH, KOH, LiOH, trimethylbenzyl ammonium hydroxide, and others, causes a transformation of the cellulose I lattice to a greater or lesser extent. A similar change is noted in the regeneration of the cellulose in the viscose process, by the saponification of cellulose acetate or other esters, or by the precipitation of the polymer from nearly any solution or dispersion. The swelling accompanying exposure of native cellulose to concentrated acids (HC1, HNO_3, H_2SO_4) and strong solutions of certain lyotropic salts can also effect the same change [23]. A diffraction pattern of the cellulose II polymorph is presented in Fig. 5, and a diffractometer scan in Fig. 6, both obtained from mercerized cotton. Similar patterns may be found in Ref. [22]. The unit call advanced by Andress [24] has stood the test of criticism well, and is generally accepted. Its parameters are listed in Table 2. The observed integrated intensities of the reflections of interest are: 101, 2000; 10$\bar{1}$/002, 18000; 020, 400; 040, 1200; 021, 2350 [18]. These values again point to the high population of the planes containing the chain molecules. The b-axis repeat distance, 10.3 A, does not change from that observed for native cellulose. The cellulose II lattice may be regarded as a flattened version of the native lattice with respect to the ac plane. The relative ease with which the I-II transformation takes place implies the greater stability of the lattice II form, and the reverse change has not been demonstrated directly as yet.

The widespread practice of treating fibers and pulps with caustic soda solutions for technical purposes has led to the development of x-ray analytical methods to follow the effects of the processes involved. For example, mercerization, in which cotton or other native fibers are exposed to 20% NaOH solution as a means of improving strength, luster and dyeability, is characterized by the formation of a soda-cellulose complex and its breakdown to produce a variable amount of lattice II. The conversion is much more extensive when the cellulose is not under physical restraint during the mercerization step. Figure 7 compares the results of

TABLE 1

Dimensions of the Cellulose I Unit Cell (Monoclinic)

Parameter	Average Value	Range
a	8.20 Å	8.17 - 8.28 Å
b	10.3 Å	-
c	7.90 Å	7.85 - 7.96 Å
β	83.3°	81.7 - 83.6°
d(101)	6.01 Å	5.98 - 6.08 Å
d(10$\bar{1}$)	5.35 Å	5.28 - 5.42 Å
d(002)	3.94 Å	3.90 - 3.96 Å
d(040)	2.58 Å	-

TABLE 2

Dimensions of Cellulose II Unit Cell (Monoclinic)

Parameter	Typical Value	Observed Range
a	7.9 Å	7.88 - 8.05 Å
b	10.3 Å	-
c	9.1 Å	9.09 - 9.38 Å
β	63°	61.8 - 63.2°
d(101)	7.19 Å	7.17 - 7.38 Å
d(10$\bar{1}$)	4.42 Å	4.40 - 4.44 Å
d(002)	4.06 Å	4.02 - 4.13 Å
d(040)	2.58 Å	-

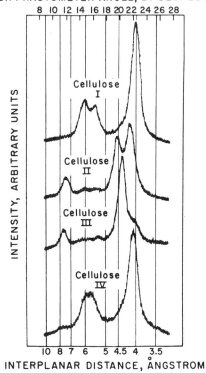

Fig. 6.   Diffractometer traces of four cellulose polymorphs.

mercerizing cotton in the form of loose yarn and as fabric [25]. The limitation of swelling by the fabric weave does not permit the expansion necessary for complete formation of the soda-cellulose complex, and the conversion to lattice II is far less than in the yarn specimen. A number of methods for quantifying the extent of lattice conversion have been proposed. The simplest is that described by Ellefsen and co-workers [26], in which the intensity of diffraction at 5.8 Å of the specimen is compared with those of known samples which consist entirely of cellulose I and II. The 5.8-Å position in the diffractogram is the location of marked intensity change as a result of lattice conversion, as it is a maximum for cellulose I and a minimum for cellulose II. The fraction of native lattice present in the specimen may be arrived at from the relation:

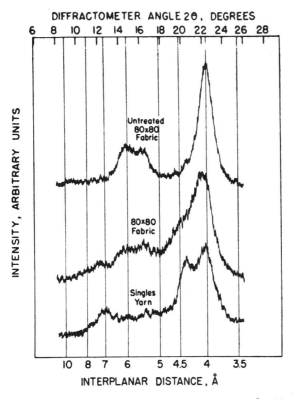

Fig. 7. Diffractometer traces showing response of cotton yarn and fabric to a mercerizing treatment.

$$I_s = I_1X + (1 - X)I_2$$

where I_s is the intensity of the specimen, I_1 and I_2 of the specimens free of the other lattice, and X the fraction of cellulose I in the sample. The error of the measurement in a typical diffractometer arrangement is approximately 3%. The values obtained in such a procedure are, of course, not absolute, being simply reflections of the response of the sample with respect to the standard materials chosen to represent the 100% lattice form. The method of Patil et al. [27] for crystallinity estimation (see Section IV) is more complicated, but is capable of yielding considerably more information on the crystalline phase fractions of samples containing mixed I and II lattices. Gjonnes and Norman [28] have suggested a method involving the ratio of intensities of the 101 and 002 interferences, with correction for the line width (crystalline order). The relative intensity of the 101 interference at 7.2 Å in cellulose II is another parameter by which the extent of lattice conversion may be estimated. Any of these methods, while not absolute, are useful in characterizing the nature of the specimens particularly in a limited series of sample types. It is important to remember that lattice conversion, in nearly all cases, is accompanied by a loss of crystalline order. An exception, brought about as a result of the strong alignment of the chains during its manufacture is the regenerated (saponified) acetate filament, Fortisan.

Demonstration of complete lattice conversion becomes difficult when only small amounts of residual lattice I are present. In most cases, lowering the temperature of the mercerizing alkali is effective in promoting conversion, as is repetitive exposure to the caustic solution. Commercial mercerization techniques are often carried out with warm solutions as a result of dilution of the concentrated alkali, and this diminishes the extent of lattice conversion. Manjunath and Peacock [29] have shown that lattice I may be regenerated in both mercerized cotton and viscose rayon, presumably as a result of the presence of residual cellulose I nuclei which grow upon stretching the fibers. Cuprammonium rayon and Fortisan do not exhibit any cellulose I interferences.

3. Cellulose III

When cellulose is treated with liquid ammonia or short-chain aliphatic amines, and the complex formed decomposed in the absence of water or other polar liquids, the crystalline modification called cellulose III is formed [30-32] . The diffraction pattern and diffractometer scans of cotton cellulose treated with ethylamine are shown in Fig. 5 and 6, and the unit cell parameters listed in Table 3 for both the monoclinic and hexagonal cells. The loss of lateral order that accompanies the swelling, complex formation, and decomposition makes the precision of the determination of lattice constants poorer for cellulose III specimens than for those of native and regenerated samples. As a result, the flattening of the ac plane of cellulose I gives a pattern for lattice III specimens not unlike that of lattice II, and the distinction between monoclinic and hexagonal symmetry becomes difficult. Legrand [33] proposed the monoclinic arrangment, while Hess and Gundermann [31] prefer the hexagonal, a conclusion with which Wellard [17] agrees on crystal density considerations. Wellard points out that the unit cell parameters for cellulose III show a slight but real difference depending on whether the lattice transformation was accomplished starting with native cellulose or cellulose II. It has not been established whether the presence of lattice III as such has a significant effect on the technological behavior of cellulose. As in the case of other modifications, it is most probable that the overall lateral order and chain orientation are the controlling factors.

4. Cellulose IV

Cellulose I or II heated in water, glycerol, or formamide to high temperatures forms a crystalline polymorph whose diffraction pattern is similar to, but distinctive from, the native lattice [34, 35]. This form, variously called cellulose T, cellulose IV, and high temperature cellulose, displays the pattern and diffractogram presented in Fig. 5 and 6, obtained from cotton treated with ethylenediamine to form the complex which was subsequently decomposed by boiling in N, N-dimethyl formamide [36]. The pattern analysis has led to a rhombic unit cell, the parameters of which are listed in Table 4. The similarity of the diffraction patterns of celluloses I and IV, as in the case of lattices II and III, have persuaded some authors to regard III and IV as disordered forms of II and I, respectively [35]. Sharper patterns obtained from hydrocelluloses of appropriate specimens, however, strongly support the existence of III and IV as true polymorphs. The extent of conversion of native and regenerated celluloses to lattice IV is strongly dependent on the method used, the ethylenediamine-dimethyl formamide route appearing to give the most complete effect. The technological significance of the presence of lattice IV does not appear to be great; its presence of mixtures of other lattice forms, however, interferes with evaluation of degree of crystallinity or other measurements which require good resolution of specific interferences.

5. "Amorphous" Cellulose

While not a polymorph in the accepted sense, the limiting state of disorder (as observed by x rays) to which cellulose can be reduced by certain mechanical or chemical treatments constitutes an important form of the polymer. The diffraction pattern of such material (Fig. 8) shows no crystalline reflections, and with chemically unmodified cellulose, shows a broad scattering curve which is symmetrical about a poorly defined maximum at 4.4 Å. Hess et al. [37] showed that grinding cellulose in a vibratory ball mill progressively destroyed the crystalline interferences until a pattern resembling that of a truly amorphous substance was obtained. When agate or ceramic balls are used, the ground cellulose will be contaminated with these materials, and the use of hard steel (Cr-plated) balls is advisable. The rapid precipitation of dissolved cellulose from viscose or copper-amine base solutions, or by the saponification of dissolved cellulose esters in the absence of water, leads to products of extremely low order, indistinguishable by x rays from ball-milled cellulose or amorphous (lyophilized) cellobiose [26, 38, 39]. If kept dry, these products show little tendency to recrystallize to any of the polymorphic forms. Wahdera and co-workers [40] suggest that treatment

TABLE 3

Dimensions of Cellulose II Unit Cell

Parameter	Typical value	
	Monoclinic	Hexagonal
a	7.8 Å	8.6 Å
b	10.3 Å	10.3 Å
c	10.0 Å	-
β	58°	-
d(101)	7.6 Å	4.3 Å
d(10$\bar{1}$)	4.3 Å	-
d(002)	4.3 Å	-
d(100)	-	7.5 Å

TABLE 4

Dimensions of Cellulose IV Unit Cell

Parameter	Typical value
a	8.1 Å
b	10.3 Å
c	7.9 Å
β	90°
d(101)	5.7 Å
d(10$\bar{1}$)	5.7 Å
d(002)	4.0 Å

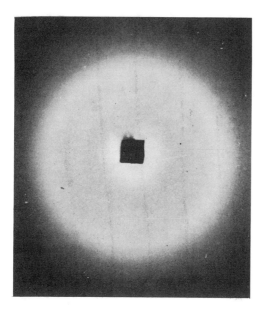

Fig. 8. Diffraction pattern of ball-milled cellulose.

of the disordered cellulose with formaldehyde to stabilize its structure with intermolecular crosslinks gives a product which will withstand exposure to boiling water without significant lattice formation. The milling of native fibers produces fragments which still exhibit strong birefringence, indicating a high degree of molecular orientation, and treatment of these materials with water tends to restore the original lattice to a considerable extent. Sometimes the formation of lattice II may occur.

The apparent loss of crystalline reflections can have its origins in the creation of random order in the polymer, the reduction of crystallite size to the extent that discrete interferences are extremely weak, or to the introduction of lattice defects and distortions which result in line broadening of a serious extent. The presence of good orientation of the molecular chains argues in favor of the latter two causes, for the case of ball-milled native fibers, and Hermans [41] is of the opinion that some 10% of the ball-milled materials are in crystalline order.

The presence of disordered cellulose in all cellulosic materials has a profound technological significance, as these regions are those which are most susceptible to sorption and chemical modification. A "100% amorphous" cellulose would not appear to have useful properties exceeding those of similar products, e.g., regenerated films, but is an excellent substrate for studies of the properties of the polymer free from the strong forces of hydrogen bonding, which are usually present.

6. Cellulose Derivatives

The formation of new chemical species in a cellulose substrate occurs by complex formation with or substitution at the hydroxyl groups in nearly all cases. When only a low level of reaction occurs, there is no change in the x-ray pattern of the product, for in this case, the ordered regions responsible for the observed reflections are not disturbed. Attack on the crystalline areas, however, produces profound changes in the observed diffraction, beginning with the destruction of the original lattice and the creation of disorder. It will be clear that the formation of a new lattice based on the modified molecule is often statistically improbable unless satisfactory homogeneity of the modified species obtains. As a rule, the larger the modifying group, the less likely lattice formation will occur. Nevertheless, a large number of modified cellulose lattices, many unstable, have been observed.

Complex formation with alkalis, amines, acids, and other substances readily occurs [42]. The excellent review of Warwicker [23] lists the effects of these complexing agents on the diffraction patterns obtained. Particularly in the case of alkali hydroxides, the reported lattice parameters are in poor agreement, presumably because of the sensitivity of the product to the tension existing in the fiber bundles being examined. The observation of such systems is complicated by the uptake of water or carbon dioxide from the air. A sample holder giving good results in diffractometer experiments with wet specimens has been described by Segal [43], in which the specimen is covered with thin films of polyethylene. Unit cell parameters for typical alkali and amine complexes are listed by Howsmon and Sisson [35]. The composition of the complexes may be stoichiometric. In the case of amines and diamines, the 101 interplanar spacing of the complex tends to show an increase proportional to the chain length of the complexing agent, implying the existence of lamination along this plane. In the n-butylamine complex, the 101 spacing reaches the value of 28.7 Å, with a β angle of 28°. Complex formation with amines is usually more easily accomplished with a cellulose II substrate. Decomposition of the complexes with water regenerates the original lattice. When extracted with nonpolar solvents or caused to evaporate, the amines may cause a marked lowering of the lateral order of the cellulose.

The effect of progressive esterification on the diffraction of cellulose is illustrated in Fig. 9, in which diffractometer scans of acetylated cotton demonstrate loss of lateral order. Up to a D.S. of 1 (20% acetyl), little change occurs in the native pattern. Near the triacetate level, virtually no crystalline diffraction is present. This type of behavior is fairly general, although in the etherification of cellulose with benzyl chloride, there is a persistence of new lattice up to D.S. 1 [44, 45]. In the case of the triacetate, formation of a strong lattice occurs when the ester is heat-stressed (Fig. 10). Sprague et al. [46] have examined the crystal structure of

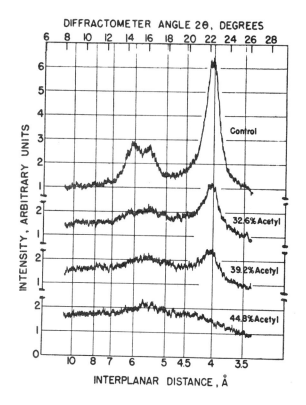

Fig. 9. Diffractometer tracings of cotton cellulose acetylated to in-creasing extents.

cellulose triacetate, distinguishing the origin and relationship of the two polymorphic forms, I and II, and clarifying the earlier work of Hess and Trogus [47, 48], Baker and co-workers [49], Happey [50], and others. Table 5 lists the parameters of the unit cells, both of which contain 4 cellobiose residues. Triacetate I is formed when native cellulose is esterified without loss of fibrous form; when the ester dissolves in the reagent and is regenerated, triacetate II is obtained, as is the case when the starting material is cellulose II. The changes in mechanical properties that occur when triacetate fibers are heat-stressed or annealed are of importance in the manufacture of commercial textile materials. Methods for following the increase and subsequent decrease in lateral order of the thermoplastic triacetate by diffraction techniques have been described by Creely and Conrad [51, 52]. The method essentially consists of record-ing the intensity of the 11.6-Å reflection by oscillating the diffractometer

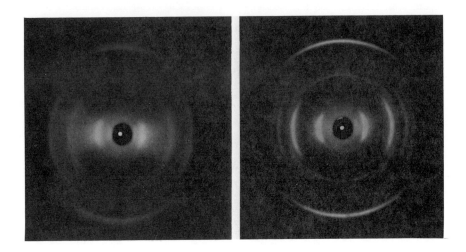

Fig. 10. Fiber diagrams of cotton cellulose triacetate before (left) and after (right) heat-stressing.

TABLE 5

Unit Cell Dimensions of Cellulose Triacetates

Parameter	Triacetate I	Triacetate II
a	22.6 Å	25.8 Å
b	10.5 Å	10.5 Å
c	11.8 Å	11.45 Å
β	79°	66.4°

over a small angular range in this region while the specimen is heated. The heating arrangement is that described by Rowland and co-workers [53]. Similar results were observed on the cyanoethyl and benzyl ethers of cellulose [54, 44].

As in the case of complex formation with amines, the 101 lattice spacing of cellulose shows the greatest susceptibility to change and reformation in the new lattice form of the derivative. The appearance and growth in intensity of this new interference often follows closely the increase in extent of derivative formation. In the case of small substituent groups, the relatively large lattice spacings noted must be ascribed to the exist- ence of unit cells containing more than the two cellobiose residues assigned to the common polymorphic forms of cellulose itself [46].

7. Lattice Transformations

The chief lattice change in chemically unmodified celluloses is that of the native lattice to cellulose II. To the extent that the presence of a mixture of lattice types influences the technological behavior of cellulose, it is of interest to consider the lattice transformations that have been observed. X-ray diffraction is the principal means for eliciting such information, although infrared absorption spectra also hold promise (see chapter 2). A major difficulty in obtaining satisfactory x-ray evidence for transformations lies in the often poorly resolved reflections arising from specimens with low lateral order. Ellefsen and co-workers [55] have examined the transitions reported, and their conclusions may be summarized as follows: Direct reversible transitions are observed between I and III, possibly I and IV, II and III, II and IV, and III and IV. The transition from I to II does not seem to be reversible, but may be accomplished by converting II to III or IV and then to I. Because of the difficulties mentioned, these observations need confirmation in some instances and represent a fertile field for studies of polymorphism in cellulose. With the exception of studies on the polymorphism of the triacetates, little work has been done in the area of cellulose derivatives, and as Warwicker et al. [23] have pointed out, in the case of cellulose complexes.

IV. DEGREE OF ORDER

A. The Problem

The three-dimensional order of aggregates of chain molecules in cellulose, whether considered as the extent to which crystallographic perfection is approached or in terms of response to reagent systems, is an important aspect of cellulose characterization. As a rule, high order is associated with high tensile strength and modulus, for example, while poorer order tends to increase extensibility, moisture sorption, swellability, and accessibility to dyes and reagents [35, 56]. The usefulness of these correlations in predicting end-use behavior of cellulose products has generated a large body of experimentation. Warwicker and co-workers [23] have gathered the results of some twenty-odd experimental approaches to order-disorder measurement, including physical, chemical, and sorption techniques. With few exceptions, the methods show general agreement, ranking native fibers, wood pulps and mercerized fibers, and regenerated celluloses in decreasing levels or order. The reported ordered fraction of a single cellulose type, however, may range from 40% to 90% or more, emphasizing a continuing problem in polymer-order measurement, and one which is necessary to keep in perspective when considering the application of any technique in this area. As pointed out, the volume

element of the substrate probed by various techniques will differ substantially. X-ray diffraction will be sensitive to well-ordered regions in
cellulose larger than 10-15 Å, while chemical and sorption methods, most
of which are dependent on the hydrogen bond-breaking capability of the
reagent used, will be operative over a wide range of volume elements. If
a cellulose specimen should consist entirely of extremely small crystal-
lographically perfect regions of order, x-ray diffraction would indicate a
rather low degree of order. Density measurements, on the other hand,
would lead to a relatively high estimate [57, 58] . Thus, the value of
the parameters variously called degree of crystallinity or order, accessibility, etc., each of which reflects in some way the extent of three-
dimensional order in the polymer, make an unequivocal picture of this
aspect of cellulose an elusive one. Statton [59] correctly emphasizes
the distinction between a method and the model used for making its interpretation.

The assessment of order in cellulose traditionally has been associated
with the evaluation of x-ray diffraction data [60] . Whether or not the
classical idea of a simple two-phase (crystalline-amorphous) system is
abandoned in favor of that of a wide spectrum of degrees of order, or that
of an ordered continuum with defects, evaluation of crystallinity requires
means for separating the observed intensity distribution into two complementary parts. The broadening of reflections by small crystallite size
and the generation of diffuse scatter by lattice defects render this problem
difficult. Means for assigning a proper fraction of the scattered intensity
to regions of specified three-dimensional order remains the subject of
continuing study. As a practical matter, it is assumed, in the simplest
approach, that the total scattering observed is the sum of that arising
from the "crystalline" and "amorphous" phases (after correction for air
scattering, etc.). It is convenient to discuss the division of the scattering
in terms of external and internal reference techniques.

B. External Standard Methods

In this method and its variations, the scattering intensity at some point
or region in the x-ray diagram is compared to that of one or more
"standard" materials. In the case of cellulose, the standards are arbitrarily assigned a high or low degree of crystallinity, as judged by the
sharpness of their reflections. Typically, a hydrocellulose prepared
from ramie [27] may be used as a "100% crystalline" standard where
determinations involving specimens containing lattice I are made, while
hydrocelluloses from completely mercerized ramie or Fortisan are suitable for cellulose II measurements. The high degree of disorder achieved
by ball-milling almost any form of cellulose makes this material acceptable as a standard to which "0% crystallinity" may be assigned.

A typical procedure employing an external standard is that of Ellefsen and co-workers [26] , who make the assumption that the scattering from the disordered fraction of the randomly oriented fiber sample is that observed at the minimum between the 10$\overline{1}$ and 002 interferences in cellulose I. In the case of cellulose II samples, the minimum between the 101 and 101 reflections is given the same significance. The intensity at this point in the scattering curve is compared to that at the corresponding point in the diagram of ball-milled cellulose and the ratio defined as the "degree of amorphity." The intensities are corrected for absorption, air scattering, polarization, etc. Taking the ball-milled cellulose as 1.00, the reported degrees of amorphity of other specimen types were: viscose staple, 0.65; acetate-grade wood pulp, 0.48; mercerized acetate-grade wood pulp, 0.51; cotton linters, 0.43; hydrolyzed surgical cotton, 0.38. It would appear that the values derived from this method tend to provide an upper limit for the degree of amorphity, as the intensities of the minima measured in the samples contain some contribution from adjacent reflections, and thus would give a higher value for the assumed disordered scattering. Because of the lack of a well-defined minimum in the scattering of specimens containing mixed lattices, the method is not suitable for such materials.

An effective and widely used procedure is that due to Hermans and Weidinger [61, 41] , which does away with the requirement for having a standard of known or arbitrarily assumed crystalline fraction by comparing the scattering from two samples of sufficiently different crystallinity. The scattering curves of the two samples, after correction, are separated into a "crystalline" area above a smooth line drawn in under the major diffraction peaks, and the remaining area under the line assigned to the "amorphous" scattering. If the ratio of the crystalline areas of the two specimens is denoted by x/y, and the ratio of the corresponding amorphous parameters by (1-x)/(1-y), then the values of x and y - the crystallinity of the specimens - are easily found. A regression nomograph of the crystalline and amorphous parameters may be constructed from which the degree of crystallinity as defined by the parameters may be obtained readily. The percentages of crystalline matter found in various celluloses as arrived at from this procedure include the following: native fibers, 70; wood pulp, 65; mercerized ramie, 45; regenerated celluloses, 40; ball-milled cellulose, 8. The Hermans method has much to recommend it, as it is relatively rapid once the nomogram has been prepared on the basis of the experimental arrangement used. More recently, Hermans and Weidinger [62, 63] have reexamined the results of their earlier work, using improved instrumentation and techniques. The conclusions of their previous work were largely substantiated. In the case of certain regenerated fibers, the complication of the background measurement by the presence of cellulose IV interferences was noted. The background radiation was found to have a complicated anisotropic distribution, but, in

disagreement with the assumption of Ellefsen et al. [26] , it was concluded that at no point in the scattering range examined was the background equal to the observed intensity of the cellulose sample.

The method of Wakelin and co-workers [64] compares the test sample
intensities, over the significant scattering angular range, with those of a
native fiber hydrocellulose and a ball-milled cellulose. The latter is assigned 100 and 0% crystallinity values. The differences between the intensities at corresponding points on the scattering curves of the test sample
and the amorphous standard, (U - A), are found and the corresponding intensity differences between the crystalline and amorphous curves, (C–A),
arrived at in a like manner. The slope of (U–A) on (C–A) is then calculated
and the slope of the line taken as the crystalline fraction of the sample. An
example of this treatment is illustrated in Fig. 11. The correlation
coefficient of the regression reflects the suitability of the standards chosen,
insofar as the test sample can be regarded as a mixture of the two. The
crystallinity of six native cottons examined by this procedure was approximately 70%. An adaptation of the method to evaluate cellulose II specimens
by using Fortisan hydrocellulose as a standard indicated that Fortisan
filaments had a crystallinity index of 74% and mercerized cotton 51% [65].

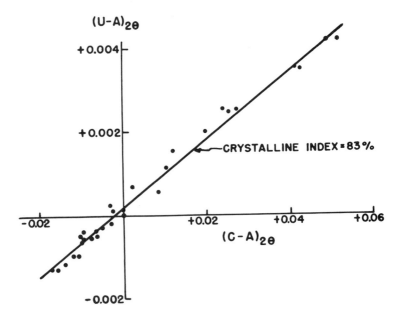

Fig. 11. Application of Wakelin technique for obtaining crystalline index.
Reprinted from Ref. [64] by courtesy of American Institute of Physics.

Patil and co-workers [27, 66, 67] extended the Wakelin approach to include lattice I and II standards simultaneously, using a bivariate regression analysis to obtain the two polymorphic crystalline fractions in samples containing both lattices. In principle, the method could be expanded to include the analysis of more complex lattice mixtures, although it would be far less effective because of overlapping interferences. It will be clear that the frequency of the data points chosen for obtaining the regression parameters should be greater in those regions where the intensity is changing rapidly with respect to interplanar spacing values. Typical crystalline fractions reported by the method of Patil and co-workers include: native cotton, 80%(I); cotton exposed to 24% NaOH solution, 6%(I), 50%(II); cotton swelled in 78% ethylenediamine, 46%(I).

The method of Wakelin and its elaborations are useful in relating the characteristic scattering of the specimen to that of the chosen standards. Parafocusing geometry may be used if adequate care is taken to avoid great differences in sample presentation. The study of Nelson and Schultz [68] contains useful information on this point. As might be expected, the results of such methods have been found to be dependent on the peak position and line breadth of the pertinent reflections, whether arising from crystallite dimensions or distortions. From a theoretical synthesis of the diffraction intensity curves for the cellulose I and II standards of crystallite dimensions approximating those observed, the degree of crystallinity of the standards was found to be well below 90% [67]. Unfortunately, it is scarcely feasible to calculate standard intensity distributions which take into account a range of crystallite sizes. A useful empirical standard intensity curve may be obtained by correcting the intensities of a test sample displaying moderate line broadening to values corresponding to 100% crystallinity in the Wakelin correlation. The error associated with the regression methods discussed here is about 2-3%.

C. Internal Standard Methods

The procedures stem from the demonstrable principle that the total radiation scattered by the cellulosic sample is independent of the crystalline-amorphous ratio. They are chiefly concerned with separating the sharp "crystalline" diffraction from the diffuse "amorphous" scatter. In arriving at means for achieving this separation, reference to the scattering characteristics is usually involved. In this sense, the result does not approach absolute values any more satisfactorily than does an external reference index.

Some internal reference methods are attractively simple. They are based on functions relating the intensity of the principal interferences in cellulose to that of a simultaneously varying minimum scattering point. The limited data used in these procedures, considered in terms of the overall scattering diagram of cellulose, produce indices rather than values which represent quantitative estimates of the ordered fraction present.

However, they permit rapid evaluation of order, particularly when used with electronic counting diffractometers. Clark and Terford [69], who examined a variety of pulps, used the ratio of the intensity of the 002 reflection to the sum of the intensities of this reflection and the minimum between the $10\bar{1}$ and 002 peaks of cellulose I. Ratios for the pulps ranged from 56% to 68%, with a small but consistent increase being evident in bleached specimens. The index of order proposed by Ant-Wuorinen and Visipaa [70, 71] takes the form $1 - (I_a/I_c)$, where I_a is the intensity of the principal minimum between the equatorial interferences and I_c the intensity of the 002 maximum (or the 101/002 maximum in cellulose II-containing specimens) above this minimum. Typical values obtained included: cotton linters, 0.83; sulfite pulp, 0.70; viscose staple, 0.41. In this work, the pressure applied in preparing the pellet of cellulose was found to be the most serious cause of variation in index values for a given specimen. Segal and co-workers [72] suggest a crystallinity index given by the ratio of the intensity of the 002 reflection of (ground) cellulose I above the scattering at 4.9 Å (approximate minimum) to the total intensity of the 002 peak. The function gave a value of 79% for untreated cotton, which decreased to 63% after swelling the cotton in ethylamine; subsequent boiling of the amine-treated cotton in water increased the index to 76%. Centola and Borruso [73] have employed the crystallinity index of Segal et al. to evaluate regenerated celluloses, employing the 101 reflection intensity and that of the minimum at 5.9 Å. Venkateswaran [74] applied both the Ant-Wuorinen and Segal functions to studies of the crystallinity of cellophane, but found that neither was satisfactory for arriving at a reliable value for the order in this material. However, the methods mentioned, when used with carefully standarized experimental procedures, are excellent for routine estimates of the crystalline character of many typical cellulose products.

In a technique proposed by Bonart and co-workers [75], a background line is drawn in the scattering curve approximately tangent to the minima at 10, 2, and 1.5 Å. The area above this line is regarded as arising from the crystalline fraction of the specimen. The reported results of this procedure indicate small differences in the crystalline fractions of native and regenerated samples of widely varying nature and history. The entire range of crystalline fractions was 36-44%. Similar results, encompassing an even smaller range of values, were obtained by Viswanathan and Venkatakrishnan [76], who examined a representative series of raw and mercerized cottons, polynosic and high tenacity rayons, and ramie fiber. In this study, the range of crystalline fractions for all cottons and rayons was 30-32%, and the value for ramie, 36%.

An important criterion of order is the line shape of the reflections observed. An interesting approach to order estimation bases on this consideration has been put forward by Gjonnes et al. [77]. They note that the corrected intensity distributions of the principal interferences in cellulose, e.g., 002, are similar to the long-tailed Cauchy distributions observed in patterns of cold-worked metals. On the basis of such comparisons, native cellulose specimens "consist of aggregates of distorted crystallites and only small quantities of highly disordered material are present" (about 5%). For mercerized celluloses, the line shape indicated that significantly higher fractions of disordered polymer were present. Such data emphasize the view that, for many celluloses, a paracrystalline molecular array is present in the crystallites, implying a spectrum of order in these regions as well as in the specimen as a whole. Gjonnes and Norman [78, 79] extended these observations to use of the peak position and line width of both cellulose I and II samples as a means of characterizing the lateral order. In a study by Parks [80] , the angular width at half-maximum intensity of the 002 interferences of a series of pulp and linters specimens ranged from 1.3 to 2.0°, but was not necessarily an accurate reflection of the state of their lateral order distribution.

V. SIZE OF THE ORDERED REGIONS

The dimensions of the volume elements over which three-dimensional order exists affect the response of a polymer to its environment. Large, regularly bonded molecular arrays will display higher modulus and lower accessibility to reagents. The microfibrillate texture of cellulose in most of its forms implies that the regions of high lateral order will be extremely small. Electron microscopy (Chapter 4) places an upper limit on the lateral dimensions of the crystallites in terms of their physical size, the diameter of the elementary fibrils being in the range 35-150 Å. X-ray diffraction offers estimates of crystallite dimensions which are complementary to microscopical observations, and which can be applied to specimens which are difficult to fibrillate so as to make direct measurement of the structural elements.

The breadth of the diffraction spot arising from a large (perfect) crystal depends primarily on the effective slit width of the optics employed and the monochromaticity of the radiation. As the crystal diminishes in size, broadening of the reflection will occur. These quantities are related by the Scherrer equation [3]

$$D_{hkl} = K \lambda / \beta \cos \theta ,$$

where D_{hkl} is the dimension of the crystal perpendicular to the hkl plane from which the reflection arises, λ the wavelength of the radiation, θ the Bragg angle of the reflection, β the pure broadening of the reflection in radians, and K a constant (approximately unity). The value of D is dependent on the accuracy with which β can be estimated, as the observed line width must be corrected for any increase due to the experimental procedure. The range over which the line broadening method can be applied is roughly 10 - 1000 Å. The measurement of extremely small line widths becomes difficult and thus the dimensions of large crystals questionable.

As strains in the crystal lattice also contribute to broadening of the reflections, it will be clear that the dimensions arrived at by the Scherrer
relation will represent minimum crystallite sizes.

The pure line broadening of a reflection is obtained by correcting the
observed line width (usually defined as the width at half-maximum intensity) for instrumental effects. The line width of a reflection from a
large crystal, at a Bragg angle close to that if the cellulose interference
is measured. Quartz is a satisfactory material. The pure broadening,
β , is found by the relation $\beta^2 = B^2 - b^2$, where B is the observed line
width and b the line width of the quartz reflection. At low Bragg angles,
the corrections may be negligible, but when considering the 040 interference, a substantial correction is involved. In a study of factors affecting line broadening determinations on typical cellulose preparations, Patil
and Radhakrishnan [67] found monochromaticity of the radiation and
correction for polarization and the Lorentz effect important points to be
observed. The elegant Fourier analysis method of Stokes [81] has been
applied by Nieduszynski and Preston [82] ; the computations involved in
this means for obtaining the pure line broadening make a digital computer
desirable. While the diffraction curves from a pressed powder specimen
are suitable for evaluation of the paratropic interferences, the use of reflection from the cut ends of a bundle of parallelized fibers gives good results for the meridional reflections [27].

The meridional reflections of cellulose are narrower than the equatorial
lines, indicating that the crystallites are longer in the direction of the b
(fiber) axis. The early observations of Hengstenberg and Mark [83] on
ramie gave a crystallite diameter of 50-100 Å and a length greater than 500
Å; the dimensions for rayon crystallites were 40 Å diameter and 300 Å or
more in length. Later work tends to yield shorter crystallite lengths in
native cellulose, with a rather constant value of 200-250 Å occurring in
fibers subjected to a variety of treatments, as observed in the usual
manner [27]. Elaborate background corrections were applied by
Viswanathan and Venkatakrishnan [84] , however, which yielded somewhat shorter lengths for native fibers (about 100-160 Å), and which decreased to 80-90 Å on mercerization. Electron microscope measurements
on the residual crystallites of native hydrocelluloses suggest lengths of
the order of several hundred Angstrom units, and Kulshreshtha and co-
workers [85], who observed the intensity of several orders of meridional
reflections in ramie, propose that this is a suitable value. They suggest
that the shorter lengths found by conventional line broadening techniques
are too low because of paracrystalline distortions [86, 87], and conclude
that 460 Å is a representative crystallite length.

The cross-sectional shape of the crystallite can be arrived at by
evaluation of the broadening of reflections of planes which extend in
approximately orthogonal directions, or at a known angle to each other.
In the cellulose I configuration, resolution of the 101 and 10$\bar{1}$ interferences

is required, and in cellulose II, separation of 002 and 10$\bar{1}$ reflections. In native cotton the thickness of the crystallite normal to the 002 plane is about 60 Å, and the corresponding dimensions for the 101 and 101 planes somewhat smaller. Considering the uncertainties involved, the cross section of the crystallite appears to be fairly symmetrical. Mercerization or other swelling treatments tends to lower the native dimensions, and hydrolysis to increase them. The 35-40 Å diameter reported by Heyn [88] from observations of cross sections of fibers must be regarded as showing rather good agreement with the values derived from line broadening measurement. Increase in diameter of the crystallite appears to be correlated with improvement in lateral order, as observed in Valonia cellulose, bast and seed hairs, wood pulps, and regenerated filaments.

Another approach to obtaining information on the spacing and dimensions of diffracting bodies is the evaluation of x-ray scattering in the region of the incident beam, i.e., at low Bragg angles [3] . The observation of the scattering pattern so close to the main beam requires improvement in the collimation of the radiation striking the specimen, as the scattering of interest lies in region corresponding to lattice spacings greater than 50 Å. To expand this region, longer specimen-to-film/detector distances are usually employed. Statton [89] has described the results obtained with a camera with a 32-cm specimen-film distance, and a vacuum path to eliminate the air scattering which may obscure the sample scattering is desirable. As the scattering is relatively weak in most instances, the use of more intense primary beams is of value.

The pattern surrounding the beam stop in low-angle scattering shows shapes and charcteristics varying with types of cellulose examined. There may be discrete diffraction of the type observed at wide angles, which in effect arises from a large lattice spacing; diffuse scattering which displays no maxima is also frequently observed. Heyn [90, 91] studied the scattering figures for a variety of celluloses and sought to interpret them in terms of a distribution of particle sizes in the range of 40-1000 Å. To provide sufficient separation of the particles and to avoid strong interparticle interference, it was proposed that the fibers be swollen in water. The method of converting the scattering angle-intensity data was that put forward by Guinier [92] . Statton [89] emphasized that particle size distributions in cellulose fibers could be interpreted equally as well as pore-size distributions, since it is only the differences in electron densities which are recorded in the low-angle scattering experiment. Hermans and co-workers [93] demonstrated that pore volume per unit sample volume was inversely correlated with low-angle scattering intensity. The technique was placed on an absolute basis by Heikens [94] , using the scattering power per unit volume introduced by Porod [95] .

As is evident from the outline of the studies mentioned here, the translation of low-angle scattering data from cellulose to crystallite demensions remains tenuous. Heyn [90] arrived at values derived from the equatorial scattering intensity distribution which are in agreement to some extent with those of line broadening measurements; the bast fibers give dimensions of 44 - 68 Å. However, the value for viscose rayon was 73 Å, and those for raw and mercerized cotton 95 and 146 Å, respectively; these dimensions indicate that a simple treatment of low-angle scattering data may be unacceptable in many cases. Statton [89] found discrete meridional reflections in viscose tire yarns at scattering angles corresponding to lengths of about 200 Å, a result not inconsistent with other types of observations. Roy and Das [96, 97] examined the patterns of a number of natural celluloses and Fortisan. The diameter of the scattering elements was in the range 50 - 60 Å, again in general agreement with line broadening data. Treatment of the scattering data by another procedure, however, led to lower values, e.g., 18 - 43 Å.

VI. ORIENTATION

A. Procedures

The physical properties of cellulosic materials are influenced strongly by the overall degree of alignment of the chain molecules with respect to each other and, more specifically, to the fiber, sheet, or film axis. Some of the properties which are correlated positively with such orientation include tensile strength, rigidity, and birefringence; negative correlations exist for extensibility, dye sorption, elastic recovery, and ability to withstand repeated creasing without failure. Since these encompass a broad range of technological behavior, the determination of molecular orientation forms an important part of polymer characterization [98] .

Refractive index measurements (q.v.) and the evaluation of x-ray diffraction patterns are convenient procedures to study this aspect of cellulose structure. The optical method yields a single value for the average orientation of all of the molecular chains of the specimen, while the x-ray method is capable of providing information on only those groupings of the molecules which are ordered well enough to give crystalline diffraction. The advantages of the x-ray method lie in the fact that it yields the statistical distribution of orientation about an axis and that it samples a relatively large volume of specimen, containing the entire bulk of numerous fiber bodies. Refractive index measurements are usually confined to observation of the surface condition of a small number of fibers. In specimens having nearly ideal orientation, both types of measurements should be in close agreement.

Sisson and Clark [99] , recognizing the application of the fiber diagram analysis of Polanyi [100] to cellulosics, proposed that the distribution of diffracted intensity along the azimuthal extension of the interferences would be a measure of crystallite orientation in the specimen. Thus, the fiber pattern of ramie (Fig. 3) indicates nearly ideal orientation, as the deviations of the reflections from the equator and meridian (fiber axis) are small.

Conversely, random orientation is characterized by extension of the reflections into complete circles, as seen in a powder diagram. In general,
the intensity of the diffraction arc at any angle to the fiber axis (or other
arbitrary direction) is proportional to the number of crystallites aligned
at that angle.

The procedure for measuring the intensity along the arc of a selected
interference of the fiber diagram is illustrated in Fig. 12. Typically, a
parallelized bundle of fibers, wrapped at each end to prevent movement
and held under sufficient tension to remove crimp, is placed normal to the
beam. In electronic counter diffractometer arrangements, the receiving
slit is set at the appropriate angle to detect the interference, and the intensity of diffraction recorded as the sample is rotated in a plane perpendicular to the incident beam (Fig. 13 and 14). A device for rotating the
bundle while held under tension is shown in Fig. 15 [101] .

The meridional interferences, e.g., 040, while providing a more direct
measurement of the dispersion of the crystallite b axis, are inherently less
intense as observed with the axis of the fiber normal to the beam, and in
all but the most highly oriented specimens are likely to be contaminated
with the reflections that appear at neighboring Bragg angles. For these
reasons, the stronger equatorial reflections are commonly used. The
azimuthal intensity distributions of the equatorial arcs will agree with those
of the meridional arcs only if the normals to the planes from which they
arise do not show selective orientation with respect to the b axis. Corrections based on reciprocal lattice considerations have been advanced
[102, 103] , but for purposes of characterizing a series of related materials, it is often sufficient to use the included azimuthal angle at which
the equatorial arc intensity has fallen to 50%(or similar fraction) of its
maximum as a measure of crystallite orientation. Thus, the "50% x-ray
angle" of a fiber specimen is the angle included between points on the arc

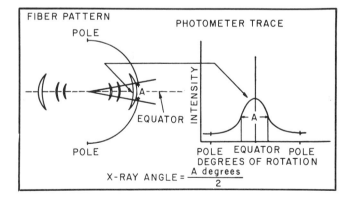

Fig. 12. Procedure for measuring (50%) x-ray angle.

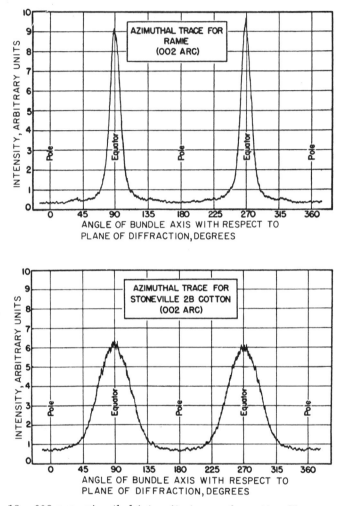

Fig. 13. 002 are azimuthal intensity traces for native fiber specimens.

where the diffracted intensity has fallen to 50% of its highest value (usually on the equator), as illustrated in Fig. 12. An orientation coefficient, defined as the reciprocal of the angle defined above, is also used [102]. Hermans [41] employs the orientation factor f, given by

$$f = 1 - \frac{3}{2} \sin^2 \beta ,$$

where β is the average angle of orientation obtained from a meridional

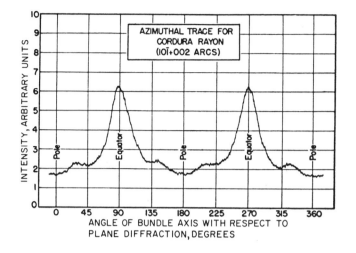

Fig. 14. 10$\bar{1}$-102 arc azimuthal intensity trace for regenerated fiber specimen.

Fig. 15. Device for rotating fiber bundles while held under tension.

arc. The intensity of the meridional arcs can be enhanced by tilting the fiber bundle axis to the x-ray beam at an angle equal to the Bragg angle of the reflection [102, 103] . The characteristic patterns of specimens displaying selective orientation have been analyzed by Howsmon and Sisson [35].

B. Native Fibers

The observed intensity distribution along the arcs arises from a combination of factors which include both gross and fine fiber structure. The well-oriented bast fibers (ramie, flax, hemp) show little despersion of the reflections, the 50% angle being less than 10°, as illustrated in the azimuthal scan in Fig. 13 [101] . These fibers appear to have little change in the angle at which the fibrils of the concentric cell wall layers are laid down, and a major part of the observed disperson may arise from the lack of perfect crystallite orientation in the microfibrils themselves. In the important case of cotton, the substantial spiral angle at which the fibrils are arranged reverses itself at short intervals; this complexity is augmented by the presence of superimposed fiber convolutions. There is considerable evidence that the spiral angle of the fibrils in the layers changes from outside to inside of the cell wall. As Howsmon and Sisson [35] point out, the observed intensity distribution in fibers possessing a spiral structure is a function of the spiral angle and the distribution about that angle. It follows that the spiral angle must be large and the dispersion about it small if the arc is to show separation into discrete maxima. Cotton, with an apparent spiral angle of 25°, shows no indication of separate maxima in 002 azimuthal scans of fiber bundles, although fiber patterns of single fibers [104] may show this effect. On the other hand, compression wood, in which the spiral angle is also about 25°, does show equatorial arc separation [105] , and this is also apparent in the diagrams of coir, yucca, and other fibers [12, 22] .

The extensive data of Berkley and co-workers [106] on the relationship of x-ray angle and tensile strength of cotton have firmly established this ·orientation parameter as an index of cotton quality. A range of 20-42° in 40% x-ray angle was accompanied by a breaking strength change over a similar range, with a good degree of correlation. The correlation could be improved by subdividing the cottons examined into species. Meredith [107, 108] reexamined and improved the earlier technique, obtaining essentially similar results. The more rapid diffractometer arrangement was applied by Segal, Creely, and co-workers [101, 109], and the 50% angle recommended over the 40% angle as being easier to evaluate and providing equivalent data. Over the range of cottons examined, the 50% angle was found to be about 3.3° less than the 40% angle. The contribution of fiber convolutions in increasing the measured x-ray angle of cotton has been examined [108, 110], with the suggested conclusion that the spiral angle for most cottons varies over a relatively narrow range. The analysis of the shape of the azimuthal intensity distribution curve was undertaken by DeLuca and Orr [111] who considered the intensity distribution of the 002 arc to be the sum of two Gaussian curves. Since it is improbable that the fibrils have the same spiral angle throughout the cell wall thickness, the constancy of the value of the derived spiral angle is doubtful. The change in angle in woody fibers is particularly notable [112].

The apparently paradoxical effect of increased crystallite orientation when fibers are subjected to the mercerizing process, with its attendant creation of disorder, arises in part from the removal of convolutions as the fibers untwist during swelling. The constricting effect of the primary wall may also lead to higher orientation. The mercerizing process is often accompanied by internal or external tensile stress, and when this is maintained at a high level throughout the drying step, the resultant x-ray angle may approach those of the strongly oriented bast fibers.

C. Regenerated Fibers

The orientation of the chain molecules in regenerated cellulose is dependent primarily on the degree of stretching undergone while in the gel state during the spinning process. A secondary orienting effect becomes operative during the lateral shrinking of the filament or sheet as it loses water. Howsmon and Sisson [35] have discussed the complexities of viscose regeneration in terms of the resultant orientation, and Hermans [41] presents a comprehensive review of the spinning process and its effect on the supramolecular state of the product. In Hermans' measurements [41, 113], the orientation factor, f_x (see p. 370) was found to increase regularly with filament stretch in similar regeneration processes. A highly orientated mercerized ramie fiber gave a value of 0.98; viscose filaments with stretches of 10 - 120% had f_x values of 0.78 - 0.91; for cuprammonium (Bemberg) rayon, f_x was 0.86. A notable result of Hermans' observations was the difference in azimuthal intensity distribution of the 101 and 10$\bar{1}$/002 arcs, this difference varying in viscose and Lilienfeld rayons, although negligible in reoriented mercerized ramie. Ellis and Warwicker [114] examined the orientation of saponified cellulose acetate filaments, observing the 101 (paratropic) and 020 (meridional) arcs, expressing the orientation as the reciprocal of the "50% x-ray angle," and making suitable corrections of the observed intensity distribution for lattice geometry and experimental method. Higher degrees of polymerization of the cellulose promoted increased orientation for equivalent amounts of stretching during preparation of the filaments. Typical orientation coefficients for a cellulose of degree of polymerization 729 were: 0.045 (no stretch), 0.123 (100% stretch). For a shorter chain length material (430), the comparable data were 0.037 and 0.097, respectively.

REFERENCES

[1] S. Nishikawa and S. Ono, Proc. Math. Phys. Soc. (Tokyo), 7, 131 (1913).

[2] J.W.S. Hearle, J. Polymer Sci., Pt. C, No. 20, 215 (1967).

[3] H.P. Klug and L.E. Alexander, X-ray Diffraction Procedures, Wiley, New York, 1954.

[4] H.S. Peiser, H.P. Rooksby, and A.J.C. Wilson, eds., X-ray Diffraction by Polycrystalline Materials, The Institute of Physics, London, 1955.

[5] E.W. Nuffield, X-ray Diffraction Methods, Wiley, New York, 1966.

[6] R. Rudman, X-Ray Diffraction Analysis, Division of Chemical Education, American Chemical Society, Easton, Pa., 1967.

[7] National Bureau of Standards, X-ray Protection, Handbook 60, U.S. Department of Commerce, Washington, 1955.

[8] International Commission on Radiation Protection, ICRP Publication 9, Pergamon, Oxford, 1966.

[9] V.A. Kargin, J. Polymer Sci., 30, 247 (1958).

[10] R. Hosemann, Polymer, 3, 349 (1962).

[11] R. St. J. Manley, Nature, 204, 1155 (1964).

[12] A.N.J. Heyn, in Mathews' Textile Fibers, 6th ed. (H.R. Mauersberger, ed.), Wiley, New York, 1954, pp. 1149 - 1213.

[13] R.H. Marchessault and A. Sarko, Advan. Carbohydrate Chem. 22, 421 (1967).

[14] O. Ellefsen and N. Norman, J. Polymer Sci., 58, 769 (1962).

[15] K.H. Meyer and L. Misch, Helv. Chim. Acta, 20, 232 (1937).

[16] H.J. Woods, in Recent Advances in the Chemistry of Cellulose and Starch (J. Honeyman, ed.), Wiley-Interscience, New York, 1959, pp. 134-147.

[17] H.J. Wellard, J. Polymer Sci., 13, 471 (1954).

[18] J. Mann, L. Roldan-Gonzalez, and H.J. Wellard, J. Polymer Sci., 42, 165 (1960).

[19] K.C. Ellis and J.O. Warwicker, J. Polymer Sci., 56, 339 (1962).

[20] G. Honjo and M. Watanabe, Nature, 181, 326 (1958).

[21] D.G. Fisher and J. Mann, J. Polymer Sci., 42, 189 (1960).

[22] J.A. Howsmon, in Handbook of Textile Fibers (M. Harris, ed.), Harris Research Laboratories, Washington, D.C., 1954, pp. 88-102.

[23] J.O. Warwicker, R. Jeffries, R.L. Colbran, and R.N. Robinson, A Review of the Literature on the Effect of Caustic Soda and Other Swelling Agents on the Fine Structure of Cotton, Shirley Institute Pamphlet No. 93, The Cotton Silk and Man-made Fibres Research Association, Manchester, 1966.

[24] K.R. Andress, Z. Physik. Chem., B4, 190 (1929).

[25] L. Segal and C.M. Conrad, Am. Dyestuff Reptr., 46, 637 (1957).

[26] O. Ellefsen, E. Wang Lund, B. A. Tonnesen, and K. Oien, Norsk. Skogind., 11, 284 (1957).

[27] N.B. Patil, N.E. Dweltz, and T. Radhakrishnan, Textile Res. J., 32, 460 (1962).

[28] J. Gjonnes and N. Norman, Acta Chem. Scand., 14, 683 (1960).

[29] B.R. Manjunath and N. Peacock, Textile Res. J., 39, 70 (1969).

[30] A.J. Barry, F.C. Peterson, and A.J. King, J. Am. Chem. Soc., 58, 333 (1936).

[31] K. Hess and J. Gundermann, Ber., 70B, 1788 (1938).

[32] L. Segal and M.L. Nelson, J. Am. Chem. Soc., 76, 4626 (1954).

[33] C. Legrand, J. Polymer Sci., 7, 333 (1951).

[34] T. Kubo, Z. Physik. Chem., A187, 297 (1940).

[35] J.A. Howsmon and W.A. Sisson, in Cellulose and Cellulose Derivatives, (E. Ott, H.M. Spurlin, and M.W. Graflin, eds.), 2nd ed., Pt. I, Wiley-Interscience, New York, 1954, pp. 231 - 346.

[36] L. Segal, J. Polymer Sci., 55, 395 (1961).

[37] K. Hess, H. Kiessig, and J.Gundermann, Z. Physik. Chem., B49, 64 (1941).

[38] W.A. Sisson, Ind. Eng. Chem., 30, 530 (1938).

[39] R. Jeffries, J. Appl. Polymer Sci., 12, 425 (1968).

[40] I.L. Wahdera, R. St. J. Manley, and D.A.I. Goring, J. Appl. Polymer Sci., 9, 2634 (1965).

[41] P.H. Hermans, Physics and Chemistry of Cellulose Fibers, Elsevier, New York, 1949.

[42] J.O. Warwicker, J. Appl. Polymer Sci., 13, 41 (1969).

[43] L. Segal, Textile Res. J., 32, 702 (1962).

[44] J.J. Creely, D.J. Stanonis, and E. Klein, J. Polymer Sci., 37, 43 (1959).

[45] B.R. Manjunath and C. Nanjundayya, Textile Res. J., 35, 377 (1965).

[46] B.S. Sprague, J.L. Riley, and H.D. Noether, Textile Res. J., 28, 275 (1958).

[47] K. Hess and C. Trogus, Z. Physik. Chem., B5, 161 (1929).

[48] K. Hess and C. Trogus, Z. Physik. Chem., B9, 160 (1930).

[49] W.O. Baker, C.S. Fuller, and N.R. Pape, J. Am. Chem. Soc., 64, 776 (1942).

[50] F. Happey, J. Textile Inst., 41, T381 (1950).

[51] J.J. Creely and C.M. Conrad, Textile Res. J., 32, 184 (1962).

[52] C.M. Conrad and J.J. Creely, J. Polymer Sci., 58, 781 (1962).

[53] R.A. Rowland, E.J. Weiss, and W.E. Bradley, Natl. Acad. Sci., Natl. Res. Coun. Publ., No. 456, 85 (1956).

[54] C.M. Conrad, D.J. Stanonis, J.J. Creely, and P. Harbrink, J. Appl. Polymer Sci., 5, 163 (1961).

[55] O. Ellefsen, K. Kringstad, and B.A. Tonnesen, Norsk Skogind., 18, 419 (1964).

[56] K. Ward, Textile Res. J., 20, 363 (1950).

[57] R.L. Miller, in Encyclopedia of Polymer Science and Technology, Vol. 4, Wiley-Interscience, New York, 1966, pp. 429-528.

[58] R.H. Marchessault and J.A. Howsmon, Textile Res. J., 27, 30 (1957).

[59] W.O. Statton, J. Polymer Sci., Pt. C, No. 18, 33 (1967).

[60] P.H. Hermans, Experientia, 19, 553 (1963).

[61] P.H. Hermans and A. Weidinger, J. Polymer Sci., 4, 709 (1949).

[62] P.H. Hermans and A. Weidinger, Textile Res. J. 31, 558 (1961).

[63] P.H. Hermans and A. Weidinger, Textile Res. J., 31, 571 (1961).

[64] J.H. Wakelin, H.S. Virgin, and E. Crystal, J. Appl. Phys., 30, 1654 (1959).

[65] M.L. Nelson and R.T. O'Connor, J. Appl. Polymer Sci., 8, 1325 (1964).

[66] N.B. Patil and T. Radhakrishnan, Textile Res. J., 36, 746 (1966).

[67] N.B. Patil and T. Radhakrishnan, Textile Res. J., 36, 1043 (1966).

[68] M.L. Nelson and E.F. Schultz, Jr., Textile Res. J., 33, 515 (1963).

[69] G.L. Clark and H.C. Terford, Anal. Chem., 27, 888 (1955).

[70] O. Ant-Wuorinen and A. Visapaa, Paperi Puu, 47, 311 (1965).

[71] O. Ant-Wuorinen and A. Visapaa, Norelco Reptr., 9, 47 (1962).

[72] L. Segal, J.J. Creely, A.E. Martin, Jr., and C.M. Conrad, Textile Res. J., 29, 786 (1959).

[73] G. Centola and D. Borruso, Ind. Carta (Milan), 16, 87 (1962).

[74] A. Venkateswaran, J. Appl. Polymer Sci., 13, 2459 (1969).

[75] R. Bonart, R. Hosemann, F. Motzkus, and H. Ruck, Norelco Reptr., 7, 81 (1960).

[76] A. Viswanathan and V. Venkatakrishnan, Proceedings 7th Tech. Conf., Section B, Ahmadebad Textile Industry Research Association, December 1965.

[77] J. Gjonnes, N. Norman, and H. Viervoll, Acta Chem. Scand., 12, 489 (1958).

[78] J. Gjonnes and N. Norman, Acta Chem. Scand., 12, 2028 (1958).

[79] J. Gjonnes and N. Norman, Acta Chem. Scand., 14, 689 (1960).

[80] L.R. Parks, Tappi, 42, 317 (1959).

[81] A.R. Stokes, Proc. Phys. Soc. (London), 61, 282 (1948).

[82] I. Nieduszynski and R.D. Preston, Nature, 225, 273 (1969).

[83] J. Hengstenberg and H. Mark, Z. Krist., 69, 271 (1928).

[84] A. Viswanathan and V. Venkatakrishnan, J. Appl. Polymer Sci., 13, 785 (1969).

[85] A.K. Kulshreshtha, N.B. Patil, N.E. Dweltz, and T. Radhakrishnan, Textile Res. J., 39, 1158 (1969).

[86] R. Hosemann, J. Polymer Sci., Pt. C., No. 20, 1 (1967).

[87] D.R. Buchanan and R.L. Miller, J. Appl. Phys., 37, 4003 (1966).

[88] A.N.J. Heyn, J. Cell Biol., 29, 181 (1966).

[89] W.O. Statton, J. Polymer Sci., 22, 385 (1956).

[90] A.N.J. Heyn, Textile Res. J., 19, 163 (1949).

[91] A.N.J. Heyn, J. Appl. Phys., 26, 519 (1955).

[92] A.J. Guinier, Proc. Phys. Soc. (London), 57, 310 (1945).

[93] P.H. Hermans, D. Heikens, and A. Weidinger, J. Polymer Sci., 35, 145 (1959).

[94] D. Heikens, J. Polymer Sci., 35, 139 (1959).

[95] G. Porod, Kolloid-Z., 124, 83 (1951).

[96] S.C. Roy and S. Das, J. Appl. Polymer Sci., 9, 3427 (1965).

[97] S.C. Roy and S. Das, J. Appl. Polymer Sci., 9, 3439 (1965).

[98] W.O. Statton, J. Polymer Sci., C, No. 20, 117 (1967).

[99] W.A. Sisson and G.L. Clark, Ind. Eng. Chem., Anal Ed., 5, 296 (1933).

[100] M. Polanyi, Z. Physik, 7, 149 (1921).

[101] L. Segal, J.J. Creely, and C.M. Conrad, Rev. Sci. Instr., 21, 431 (1950).

[102] K.C. Ellis and J.O. Warwicker, J. Polymer Sci., Pt. A, 1, 1185 (1963).

[103] T. Radhakrishnan, N.B. Patil, and N.E. Dweltz, Textile Res. J., 39, 1003 (1969).

[104] K.E. Duckett and V.W. Tripp, Textile Res. J., 37, 517 (1967).

[105] W.A. Sisson, Ind. Eng. Chem., 27, 51 (1935).

[106] E.E. Berkley, O.C. Woodyard, H.D. Barker, T. Kerr, and C.J. King, U.S. Dept. Agr., Tech. Bull. No. 949, 1948.

[107] R. Meredith, Shirley Inst. Mem., 25, 25 (1951).

[108] R. Meredith, Shirley Inst. Mem., 25, 41 (1951).

[109] J.J. Creely, L. Segal, and H.M. Ziifle, Textile Res. J., 26, 789 (1956).

[110] S.M. Betrabet, Textile Res. J., 33, 720 (1963).

[111] L.B. DeLuca and R.S. Orr, J. Polymer Sci., 54, 457 (1961).

[112] R.D. Preston, in Fibre Structure (J.W.S. Hearle and R.H. Peters, eds.), The Textile Institute-Butterworths, Manchester and London, 1963, pp. 235 - 310.

[113] P.H. Hermans, Contributions to the Physics of Cellulose Fibres, Elsevier, New York, 1946.

[114] K.C. Ellis and J.O. Warwicker, J. Appl. Polymer Sci., 8, 1583 (1964).

Chapter 7

WIDE-LINE NUCLEAR MAGNETIC RESONANCE SPECTROSCOPY

Robert A. Pittman and Verne W. Tripp

Southern Regional Research Laboratory
New Orleans, Louisiana

I. INTRODUCTION

Perhaps no other technique of chemical analysis has enjoyed the rapid
rise to prominence as has nuclear magnetic resonance (NMR) spectroscopy.
The development of NMR spectrometers capable of observing the spectra of
samples in the liquid and solid states and the recognition that the resonant

condition for nuclei of the same type was dependent on their chemical environment spurred the improvement of spectrometers to obtain ever higher resolution. As a result of these advances high resolution NMR spectrometers became one of the most important tools at the disposal of the analytical chemist. At the present time, however, the practical observation of these chemical shifts is limited to liquid materials or those capable of dissolution in suitable solvents at concentrations high enough to produce observable signals. It is these solubility requirements that have hindered and limited the use of high resolution NMR spectroscopy in cellulose chemistry.

Wide-line NMR spectroscopy does not suffer from the requirement that the sample be in liquid state, and consequently the study of cellulose in its ordinary state is feasible. In addition, the signal exhibited by liquids sorbed on cellulose is readily distinguishable from that of the cellulose substrate. For this reason, as well as the technological importance of such systems, much of the work on wide-line NMR spectra of cellulose has centered about the problems of water sorption. The parameters associated with wide-line spectra, such as the spin-spin relaxation time, spin-lattice relaxation time, and the second moment of the absorption curve, are reflections of the supramolecular morphology of a solid material, and information on the crystallinity, vibrational modes, internuclear distances and geometry, and related properties may be obtained from the study of these spectra.

A. Theory

The rigorous phenomenological interpretation of wide-line NMR spectra can be quite involved, and the considerably simplified and condensed version of the implications of these spectra given here should be regarded as purely introductory. For a more detailed and precise treatment of the interpretational aspects of wide-line NMR, the reader should consult a standard text or one of the classical articles on the subject [1 - 3].

Those nuclei which have a nuclear spin number I greater than zero will possess a magnetic moment. When such a nucleus is placed in a magnetic field, the nuclear angular momentum vector is oriented so that its component parallel to the magnetic field becomes quantized and the magnetic quantum number m may take any of $2I+1$ values from the series $I, I-1, I-2 \ldots, -I+1, -I$. If μ is defined as the maximum measurable component of the magnetic moment, then the possible components of the moment parallel to the magnetic field are the series $\mu, \mu(I-1)/I, \mu(I-2)/I, \ldots, \mu(-I+1)/I, -\mu$, which are the values of $m\mu/I$. The energy associated with these states in a magnetic field H_0 is given by $-m\mu H_0/I$. Thus, the allowable energy levels

are separated by an energy $\mu H_0/I$, and since the selection rules for a jump
from one state to another allow only a change of ± 1 for m it follows in this
simplistic view the NMR spectrum of a given nucleus in a given field consists of a single line. That this is not so is due to the fact that the nucleus
experiences not only the field superimposed externally but also those from
other nuclei in the environment as well as those produced by associated
electrons. The energy $\mu H_0/I$ is often given as $g\mu_0 H_0$ where μ_0 is the
nuclear magnetron derived from classical considerations and having the
value 3.1524 MeV/G and $g=\mu/\mu_0 I$.

For absorption to take place, transitions must be induced from a lower
to the next higher state by irradiating the sample with electromagnetic energy of frequency ν_0 such that $h\nu_0 = g\mu_0 H_0$. For a proton in a field of
10,000 G, the required frequency for resonance is 42.5759 MHz. Only ^{3}H
has a higher magnetic moment among the other nuclei, and thus the latter
are observed in easily obtainable magnetic fields at radio frequencies.

The quantum number I may be zero, an integer, or a half-integer. Thus,
for protons (I=1/2), the magnetic quantum number will be either +1/2 or
-1/2, while for deuterium (I=1), m may be +1, 0, or -1. A necessary condition for absorption of energy by the nucleus is that one or more of the a-
vailable energy levels are occupied by more nuclei than the next higher level.
The populations of the energy levels is determined by the Maxwell-Boltzmann
statistics, i.e., the ratio of the number of protons in the lower level, N_0,
to the number in the upper level, N_1, is

$$N_0/N_1 = \exp(2\mu H_0/kT), \tag{1}$$

where k is the Boltzmann constant and T is the temperature of the spin system. For a population of protons at 30°C and a field of 10,000 G, this ratio
is about 1.000007, corresponding to only 7 excess protons per million in the
lower level under these conditions. The absorption of rf energy at resonance
can raise the spin temperature rapidly to such a value that the populations
in the two levels become essentially equal unless some mechanism exists in
the system to transfer the spin energy to the lattice system. Such a mechanism is referred to as spin-lattice relaxation and is characterized by a
spin-lattice relaxation time T_1.

Since similar nuclei in a sample are precessing with a frequency which
is the same or very close to the resonant frequency of their neighbors,
these nearby nuclei may experience an electromagnetic signal at their resonant frequency, and under these conditions pairs of nuclei, one in the lower
state the other in the upper state, may exchange spins. The probability of
such a transition, that is, the lifetime of the nuclei in the upper state, is

related by the Heisenberg uncertainty principle to the line width of the emitted energy as

$$T_2 = 1/2 \pi \Delta \nu. \tag{2}$$

The time T_2 is called the spin-spin relaxation time and the quantity $\Delta \nu$ is a measure of the line width of the absorption curve in hertz. A more detailed discussion of the significance of these parameters will follow in appropriate sections.

B. Instrumentation

From the foregoing, it is evident that an NMR spectrometer must have (a) a magnetic field into which the sample is placed so as to remove the degeneracy among the various spin states that exists in zero magnetic field and (b) a source of radio frequency (rf) electromagnetic power of proper frequency. In principle, a rotating magnetic field rf vector is needed if one wishes to determine the sign of the magnetic moments of nuclei; however, most spectrometers employ a plane-polarized rf signal which is adequate for all but the most specialized work. Spectrometers have been devised which consist basically of a coil in which the sample is placed and which is supplied with a suitable rf signal. This assembly is placed in the field of a magnet which is equipped with an auxiliary set of coils so that the magnetic field may be swept back and forth through the resonant value. Appropriate electronic equipment is employed to measure the change in the quality factor Q of the rf coil as resonance is traversed. For various reasons this type of instrument has not found wide usage. The widely used double-coil nuclear induction arrangement of Bloch et al. [4,5] will be described in more detail.

The essential components of a NMR spectrometer of the nuclear induction type are shown in Fig. 1. The main magnetic field of the instrument is provided by a large magnet (M). Within the field of this magnet are a pair of coils (A), which are connected to a source of audiofrequency power, capable of adjustment over a range of frequencies and amplitudes, which serves to modulate the main magnetic field with an audiofrequency component. The coil T introduces the rf signal into the sample probe while the coil R receives the induced signal from the sample and feeds the signal to the receiver where it is amplified and detected. A pair of paddles (which are equivalent to inductors mutually coupled with the transmitter and receiver coils) permits a choice of either an absorption or dispersion mode signal from the sample by proper tuning. The paddles (not shown in Fig. 1) are tuned by observing the signal on an oscilloscope or meter, first tuning for minimum signal and then reintroducing the desired mode with the appropriate paddle. Since the receiver detects the audiomodulated signal from the receiver coil, the output from the receiver to the recorder will be proportional to the slope of the signal from the sample. Thus the signal displayed on the recorder will be the derivative of the signal introduced to the receiver coils by the paddles. Since in most instances this is the absorption mode, the derivative of the absorption curve is normally observed on the recorder.

In operation the sample is placed in the probe of the instrument (at point S in Fig. 1) and the rf generator set at an appropriate frequency for the nucleus to be observed and at magnetic field to be employed. The audio-frequency generator is then set to some arbitrary frequency (if the operator is familiar with the signal sought, the proper frequency can be chosen) and to maximum amplitude. The operator then attempts to observe a signal on the oscilloscope, 0, (Fig. 1) by varying either the radio frequency or the magnetic field. Once some indication of the signal is seen on the oscilloscope, the operator then proceeds to the detailed tuning of the instrumental parameters to provide the desired output to the x-y recorder. The main magnetic field is then set to some value just above or below the signal and made to scan through the absorption which is recorded on the x-y recorder in terms of signal strength vs magnetic field. The magnetic field is calibrated in gauss above or below the center of the resonance. A typical NMR spectrum of dry cellulose recorded under such conditions is shown in Fig. 2.

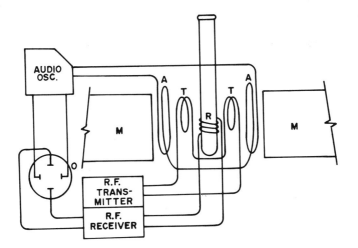

Fig. 1. Typical layout of wide-line NMR absorption spectrometer.

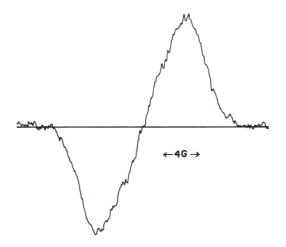

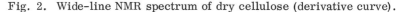

Fig. 2. Wide-line NMR spectrum of dry cellulose (derivative curve).

II. LINE SHAPE FACTORS

Although it is possible under favorable conditions to observe chemical shifts for certain nuclei in the solid state, this situation ordinarily does not obtain in cellulose systems. Under these circumstances most of the information to be gained by the observation of wide-line NMR spectra of cellulose must come from the parameters associated with the line shapes of the spectra. If the line shapes associated with the absorption curves of the samples can be described fairly well by Gaussian or Lorentzian functions, then most of the line shape factors normally used in describing the spectra can be related by simple expressions. Generally the spectra observed for cellulosic systems are not described adequately by such classic functions. The failure of these line shape factors to be described as demanded by these functions gives a simple test for the failure of the absorption curve to conform.

A. Line Width

The considerations in Section I,A which gave the resonance condition for a nucleus isolated in space as $H\nu_0 = g\mu_0 H_0$ would lead one to expect NMR spectra consisting of a single narrow line. When imbedded in matter however, the nucleus experiences interactions with the magnetic field of other nuclei and influences of the molecular orbitals in its vicinity. The latter influences are important in high resolution NMR spectroscopy but may be ignored for the cellulosic materials in the solid state to be considered here.

A nucleus at a distance r from an adjacent nucleus of moment μ can be shown by classical considerations to experience a magnetic field $H = (3 \cos^2\theta - 1)/r^3$ in the direction of H_0, where θ is the angle between the direction of r and the magnetic field H_0. For the methylene protons in cellulose, which are separated by about 1.7Å, the function has a value between maximum and minimum of about 11 G. This is about the line width observed for most celluloses.

The phenomenon called spin-spin relaxation is reflected primarily in the line width of the absorption curve. This process is characterized by the spin-spin relaxation time T_2 defined by

$$T_2 = g(\nu)^{\max}/2, \tag{3}$$

where $g(\nu)$ is the line shape function of the absorption curve normalized so that

$$\int_{-\infty}^{+\infty} g(\nu)\, d\nu = 1.$$

These line width effects may be modified by the motion of the nuclei; generally, greater motion leads to narrowing of the line width. It follows that most liquids have narrower line widths than solids. In order to obtain the spin-spin relaxation time as defined in Eq. (3) from the derivative of the absorption curve obtained from the spectrometer, it is necessary to integrate the experimental curve to arrive at the absorption curve and then to normalize this curve by integrating again to determine the area under the unnormalized curve.

The line width $\Delta\nu$, usually measured between the maximum and minimum points on the derivative curve, frequently is used to characterize the substance under study. If the absorption curve can be adequately described by the Gaussian line shape, T_2 may be obtained from

$$T_2 = 1/\sqrt{2\pi}\, \Delta\nu , \tag{4}$$

or if described by a Lorentzian line shape by

$$T_2 = 1/\sqrt{3}\, \pi\, \Delta\nu. \tag{5}$$

Cellulose spectra are usually neither Gaussian nor Lorentzian. The experimental curves are obtained by sweeping the magnetic field through resonance, and in this event the line width may be measured in Gauss. For protons, the conversion from Gauss, ΔH, to hertz, $\Delta\nu$, is given by

$$\Delta\nu = 4.2576 \times 10^3\, H. \tag{6}$$

B. Second Moment

Van Vleck [6] was able to show that, for a rigid structure, the second moment of the absorption curve about resonance can be calculated from the internuclear distances if they are known. In this expression, the second moment is related to a sum of terms involving the reciprocal of the sixth power of the internuclear distances. Thus, the experimentally determined second moment may be of value in selecting the correct structure from several alternatives when the theoretical second moments can be calculated. The experimental second moment is calculated from

$$<\Delta H>^{2} = \int_{-\infty}^{+\infty} h^{2}g(h) \; dh, \tag{7}$$

where $g(h)$ is the normalized absorption curve and h is the magnetic field deviation from resonance defined by $h = H_0 - H$, where H_0 is the resonant field and H the instantaneous field strength.

III. APPLICATIONS OF LINE SHAPE FACTORS

The spectrum shown in Fig. 3 is a typical NMR signal of a wet cotton. The instrumental parameters have been selected so as to optimize the wide component of the spectrum, hence the narrow component is distorted as to line width, line shape, and intensity. The dotted line indicates the approximate division of the spectrum into cotton and water components. It is quite possible, of course, that some of the water protons, which are most tightly bound to the cellulose and thereby largely immobilized, contribute to the wide component of the signal.

If the instrument is tuned to optimize observations of the narrow component of the spectrum, then the wide component of the spectrum will no longer be seen, and the narrow component will have a smaller line width and greater intensity than that represented in Fig. 3. Dry cotton will show only a trace of a narrow component arising from the last traces of water, which can be removed only with extreme difficulty.

A. Dry Cellulose

Glazkov [7] in early work on dry cellulose and related compounds, observed the line widths and second moments of their spectra. Since the location of the protons in the samples studied were not precisely known, the theoretical second moment of the samples could not be calculated, and instead the spectra of the monomeric analogs were obtained and compared to the polymers. Thus maltose, as the monomer of amylose, gave a second

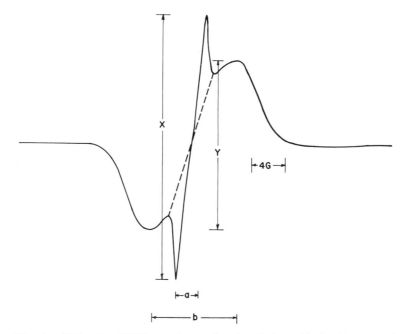

Fig. 3. Wide-line NMR spectrum of wet cellulose (derivative curve).
Dotted line indicates arbitrary separation into water and cellulose
components.

moment of 10.8 G^2, compared to 7.3 G^2 for amylose. Similarly, cellobiose
gave 10.7 G^2 compared to cellulose's 8.8 G^2. The author concluded that
the results were explainable in terms of reduced rotational freedom of the
pyranose rings in the monomers. The higher value for the second moment
of cellulose as compared to amylose was attributed to the larger radius of
gyration of the pyranose ring in cellulose which is reflected in an inhibition
of motion. Small increases in line width and second moment were noted as
the temperature was decreased from 25° to -100°C.

Statton [8] studied the NMR spectra of a variety of fibers over a range of
temperatures and under wet and dry conditions. In addition, the fibers were
also observed under tension and while slack. The cellulose samples, an
Egyptian cotton and a rayon tire cord, were observed only under dry conditions. Two quantities were related over the temperature range 20°-200°C.
The first quantity, the width of the absorption curve at half-height, is referred to as "matrix rigidity" by the author. The three samples, slack
rayon, taut rayon, and slack cotton, showed decreases in line width of a
fraction of a gauss over the temperature range.

It is apparent that the supramolecular morphology is affected little by
temperature over the range observed. The other quantity reported is the

ratio of the narrow to broad peak to peak intensity of the first derivative of
the absorption (x/y in Fig. 3). Although the samples had been previously
dried, the existence of the narrow component indicates the presence of some
water. Since the data are given only to minimum value of 1 and the samples
drop below this value over much of the temperature range observed, little
can be said for this experiment except that it shows the disappearance of
water with increasing temperature.

Pittman and Tripp [9] observed the spectra of a number of celluloses
(dried for about 24 h at 110° C), which differed in crystallinity fraction and
type as determined by x-ray methods [10] . The samples ranged numerically
from 100% to 0% crystallinity and included both cellulose I and cellulose II.
The line widths, spin-spin relaxation times, and second moments of the
samples are given in Table 1 along with the parameter k, which is the product of the spin-spin relaxation time in microseconds and the line width in
gauss. The line width was measured between the maximum and minimum
points on the first derivative curves as obtained from the spectrometer.
The variation among the values of k indicates that the line shapes of the
spectra are different. The value of k for a Gaussian curve would be about
9×10^{-5}. A plot of the second moments of the absorption curves versus x-
ray crystallinity gave a good straight-line fit for all the samples except cotton and purified marine alga Valonia ventricosa. The spin-spin relaxation
times of all the samples, except the Valonia, were in fair agreement. For
all the samples, except cotton and Valonia, the following observations may
be made. First, the variation in the values of k indicates that the absorption curves may be a combination of curves of differing line shapes. Second,
the agreement among the spin-spin relaxation times suggests that, if the
absorption curves are a combination of several basic curves arising from
different groups of protons in the samples, these groups all have the same
spin-spin relaxation times or nearly so. Thirdly, since the samples' second moments fit the following straight line,

$$< \Delta H >^2 = 0.0255 X_t + 14.4,   \tag{8}$$

where $< \Delta H >^2$ is the second moment in gauss^2 and X_t is the percent total
crystallinity of the sample, one may conclude that these spectra are a linear
combination of a crystalline fraction having a second moment of 16.9 G^2
(obtained from Eq. (8) when $X_t = 100$) and an amorphous component with a second moment of 14.4. Since the samples contain both cellulose I and II it
appears that both crystalline types have the same second moments. Nothing
in the above analysis explains the failure of cotton and Valonia to follow the
other samples. However, it may be noted that all the samples except these
have had their original crystalline structure modified to some extent.

TABLE 1

Data from Wide-Line Spectra of Dry Celluloses

Sample	ΔH_c (G)	T_{2c} (μsec)	(ΔH_c^2)[a] (G^2)	$k \times 10^5$	Estimated crystallinity[b] and type
Cotton fabric	11.4	10.5	15.5	11.97	80%-I
Mercerized cotton	12.0	10.3	15.6	12.36	40%-II 10%-I
Hydrocellulose I	10.6	10.5	16.6	11.13	90%-I
Hydrocellulose II	11.6	10.4	16.7	12.06	90%-II
Fortisan	12.4	10.1	16.3	12.52	70%-II
Rayon	12.0	10.6	15.0	12.72	25%-II
Saponified acetate	11.8	10.7	14.4	12.63	0%
Valonia	8.8	11.5	14.7	10.12	100%-I

[a]Corrected for modulation sweep width.

[b]Estimated by method of Patil et al. [10] .

B. Wet Cellulose

Wet cellulose and its associated water have received far more attention than dry cellulose. The reasons for this are manifold. Cellulose is usually found in association with moisture, its affinity for water being so great that dry cotton will immediately absorb water upon exposure to the atmosphere and, indeed, drying cotton completely is no mean task. The amount of water present in cellulose has a profound effect on its physical and chemical properties. Additionally it has been found that the NMR study of the absorbed water may offer clues to the structure of the cellulosic substrate.

Early investigators [11] recognized the possibilities of using NMR in the measurement of water sorbed on solid samples since the water normally showed as a narrow signal easily distinguishable from the wider signal from the solid substrate. Swanson et al. [12] investigated the water-cellulose sorption system for wood pulp, cotton, and rayon. The authors divided the absorption curves into two components in a fashion similar to Fig. 3, the narrow component being assigned to the water and the wide component to the cellulose. The intensity of the narrow component, the quantity in Fig. 3, and the second moment of the narrow component, calculated according to Eq. (7), were plotted against the water content. The quantity x (called D_{max} by Swanson et al.) was shown to increase slowly with water content at low levels with the slope increasing until a value of about 5% moisture was reached, after which a linear relationship was observed. The authors recommended the simplicity of this measurement in the determination of water content, observing also that, with suitable calibration curves, the method was useful both above and below fiber saturation, an advantage not realized in many other methods. Plots of the second moment of the narrow components against moisture content gave a curve which decreased from an initial value with increasing moisture content and finally leveled off to a constant value at about 20 - 25% moisture content. The authors also mention using spin-lattice relaxation techniques but no data are given.

Sasaki et al. [13] obtained approximate values for the spin-spin relaxation time of the water sorbed on cotton and, using the revised theory of Bloemborgen et al. [14] and Kubo and Tomita [15] , concluded that the mechanisms governing spin-spin and spin-lattice relaxations were different. The spin-spin relaxation times ranged from 0.5 to 4.5 msec as the water content ranged from 10% to about 60% (see Section IV).

Tanaka and Yamagata [16] observed the NMR spectrum of water sorbed on cotton and carbon and pointed out that proton signals were obtained from cellulose in the neighborhood of zero relative humidity in contrast to carbon which showed none. It is not clear whether the signal was due to cellulose or water protons or both. These investigators observed the narrowing of the signal as water content increased, as did later workers. The authors concluded from the behavior of the line shapes as the water content increased that the sorption mechanism of cellulose was characteristic of a multiple layer.

The theoretical aspects of the narrowing of the NMR line width of sorbed water have been discussed by Miyake [17] in terms of the theory of Kubo and Tomita [15]. He concluded that the narrowing can be explained in terms of two species of water, one with spin-spin relaxation time T_{12} and the other T_{22}, so that the observed relaxation time T_2 is given by

$$1/T_2 = (1 - x)T_{12} + x/T_{22}, \quad T_{12} < T_{22} \qquad (9)$$

where x is the relative humidity at which the sample has been conditioned.

In addition to the appearance of the narrow component in the NMR signal of cellulose upon the addition of moisture, one also observes that the wide component of the signal begins to narrow as the water content increases and

finally levels off to a constant value. Swanson et al. [12] examined the relationship of the second moments of the spectra of wood, cotton, and rayon
against the water content. These data show a steady decrease in second
moment with increasing water content and a leveling-off in the range of 20
to 30% water. They interpret this as an indication that the proximity or
immobility of the cellulose molecules, or both, has been disturbed by the
water induced swelling of the polymer. Glazkov [18] observed a line width
reduction in cellulose on addition of water and suggested that it could be explained by the existence of sorbed water in two phases: mobile and localized.
Two mechanisms are suggested: (a) the water acts as a plasticizer, increasing the molecular motion, or (b) spin exchange between the cellulose
and water protons occur.

Pittman and Tripp [9] have attempted to explain the line width reduction
in terms of the second suggestion of Glazkov. The first and most difficult
step in such a procedure is to divide the wide component of the wet cellulose
spectrum into two parts. This may be accomplished by matching the spectrum of the dry cellulose to the spectrum of the wet cellulose with the intensity of the dry spectrum adjusted to the maximum value that can be fit
under the wet cellulose spectrum. Such a match is shown in Fig. 4. The
curve i represents the difference between the wet (cw) and the dry (c) spectra. The spectrum c, according to this model, is a measure of those cellulose protons not interchanging spins with water protons, while the spectrum
i represents water and cellulose protons which are interchanging spins.
The curves c and i may then be integrated to obtain the absorption curve
(the curves in Fig. 4 represent the left half of the first derivative curve).
The areas under the respective absorption curves give the relative number

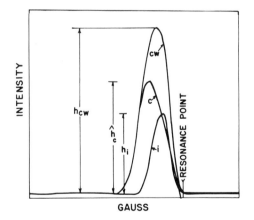

Fig. 4. Resolution of wet cellulose spectrum (cw) into noninterchanging
cellulose protons (c) and interchanging water-cellulose proton (i) components [9].

of noninterchanging cellulose protons, p_c, and interchanging cellulose-water protons, p_i. If the absorption curve of the interchanging protons is normalized to unit area, then the spin-spin relaxation time of the inter-changing protons, T_{2i}, may be obtained from Eq. (3). Table 2 gives the results thus obtained for a number of cellulose samples with about 7% sorbed water. The data have been normalized so $p_c+p_i=1$. The quantity p_i may be divided into interchanging cellulose protons, p_{ic}, and interchanging water protons, p_{iw}, by assuming that the resulting spin-spin relaxation time of the interchanging protons, T_{2i}, arises from the interchange of cel-lulose protons having the spin-spin relaxation time T_{2c} (obtained from the dry cellulose spectrum) and water protons having relaxation time T_{2w} (ob-tained from a separate spectrum of the water sorbed on the cellulose) ob-served with the instrumental parameters set to optimize the observation of the water spectrum. The spin-spin relaxation times will be related by

$$T_{2i} = p_{ic} T_{2c} + p_{iw} T_{2w} \qquad (10)$$

TABLE 2

Calculated Proton Counts and Spin-Spin Relaxation Times

Sample	p_i	p_c	T_{2i} (μsec)	Water concentration[a]
Cotton fabric	0.187	0.813	16.9	0.110
Mercerized cotton	0.055	0.945	31.6	0.153
Hydrocellulose I	0.179	0.821	15.8	0.119
Hydrocellulose II	0.094	0.906	21.3	0.131
Fortisan	0.200	0.800	20.8	0.132
Rayon	0.090	0.910	23.4	0.119
Samponified acetate	0.303	0.697	13.2	0.114
Valonia	0.106	0.894	25.4	0.124

[a]Protons of water/protons of cellulose.

if it is assumed that the interchange is taking place in times much less than the relaxation times involved. Since

$$p_{ic} + p_{iw} = p_i, \tag{11}$$

Eq. (10) and (11) may be solved for p_{ic} and p_{iw}. Knowledge of p_{iw} and the total water in the sorption system enables one to obtain the proportion of noninterchanging water protons, that is, those represented in the narrow component of the wet cellulose spectrum, relative to the number of protons in the wide component which is $p_c + p_{ic} + p_{iw}$. In this way the sorption system can be characterized by four types of protons, viz. , interchanging water and cellulose protons and noninterchanging water and cellulose protons. Table 3 represents the results of such analysis by Pittman and Tripp on a number of celluloses. The data have been renormalized on the basis of the number of protons in each category per cellulose proton. Further evaluation of this model will be necessary before its validity can be determined. However it is clearly capable of providing useful information about the water cellulose sorption system.

TABLE 3

Distribution of Proton Sorption States in Water-Cellulose Systems
(Protons per Total Cellulose Protons)

Sample	Cellulose[a]		Water	
	Inter-changing (n_{ic})	Non-interchanging (n_{fc})	Inter-changing (n_{iw})	Noninter-changing (n_{fw})
Cotton fiber	0.123	0.877	0.078	0.032
Fortisan	0.107	0.893	0.116	0.016
Valonia	0.049	0.951	0.063	0.062
Mercerized cotton	0.017	0.983	0.040	0.113
Saponified acetate	0.258	0.742	0.064	0.050
Hydrocellulose I	0.126	0.874	0.066	0.053
Hydrocellulose II	0.048	0.952	0.051	0.080
Viscose rayon	0.041	0.959	0.054	0.065

[a] $n_{ic} + n_{fc} = 1.$

C. Orientation of Sorbed Water

The water of crystallization in single crystals of hydrates will exhibit an NMR spectrum separated into two or more absorption peaks [19] when the sample is properly oriented in the magnetic field of a spectrometer. The interaction of pairs of protons of moment μ separated by distance r will, when the proton-proton vector makes an angle θ with the magnetic field, produce a splitting ΔH with the form

$$\Delta H = (3\mu/r^3) \ (\cos^2\theta - 1). \tag{12}$$

For sorbed water on parallelized fibers of certain materials, the above equation takes the form

$$\Delta H = K \ (3 \cos^2\theta - 1). \tag{13}$$

Dehl [20] has investigated this phenomenon in oriented rayon fibers and has obtained spectra split into three peaks for both sorbed H_2O and D_2O, although the peaks were not well resolved in the proton spectrum. The central line did not change in intensity as the angle θ was varied; however, the intensity increased with water content of the fibers. The splitting at $\theta=0$ was estimated to be about 0.3 G for the outside lines in the proton spectrum. The deuterium spectrum gave splitting of about 2.5 G for the outside lines, while the central line behaved the same as the proton spectrum as θ and water content were varied. The author concluded that the central lines in both spectra were due to water trapped between the fibrils of the rayon. The spectra for both nuclei obeyed Eq. (12) with different values for K. Two models were examined in the light of these experimental data. Model I assumed that the water molecules rotated about an axis parallel to the fiber axis in such a way that its dipole-dipole axis maintained a constant angle with the axis. Model II proposed that the molecules are tumbling almost randomly but with a slightly higher probability of orientation parallel to the axis than perpendicular to it. By interpreting the splitting of the deuterium spectrum in terms of its nuclear quadrupole moment, Dehl was able to show that only model II was consistent with both the hydrogen and deuterium spectra.

IV. SPIN - LATTICE RELAXATION

The spin-lattice relaxation time T_1 of liquid water is about 3 sec, while T_1 for water sorbed on cellulose even at very high concentrations (>1:1) is of the order of tenths of a second. The sorbed water that can be observed in the narrow signal of the water-cellulose NMR spectrum, however, has two or more spin-lattice relaxation times, imposing difficulties for their measurement and serving, of course, to indicate the probable complexity of the sorption mechanism.

A. Method of Observation

Bloembergen et al. [14] have described a method which involves the observation of the NMR signal as the rf field strength is progressively increased to saturate the spin system. The resulting plot of signal strength vs rf field strength (or its logarithm) will show an increase in signal as field strength is increased until some maximum signal is observed after which the signal will decrease. Samples A and B with spin-lattice relaxation times T_{1A} and T_{1B} and spin-spin relaxation times T_{2A} and T_{2B} are observed over a range rf powers such that saturation occurs. Their signal intensity (normalized to the maximum signal observed before saturation) is then plotted against the log of the rf power H_1, usually observed as the voltage output to the rf coils. In this fashion two saturation curves are obtained which are shifted relative to one another by an amount $H_{1A} - H_{1B}$, where the ratio $(H_{1A}/H_{1B})^2$ is related to the relaxation times by

$$T_{1A}/T_{1B} = (H_{1B}/H_{1A})^2 T_{2B}/T_{2A}. \tag{14}$$

Since the spin-spin relaxation times may be obtained by methods previously described, this technique permits evaluation of the ratio of the spin-lattice relaxation times. If some reference material of known spin-lattice relaxation time is available, then absolute spin-lattice relaxation times may be obtained. Aqueous solutions of paramagnetic salts, e.g., gadolinium chloride, are convenient reference materials since T_1 usually equals T_2.

An alternative technique is to subject the sample to a short pulse of rf power and then observe the signal intensity over various time intervals t_i. The spin-lattice relaxation time may be obtained from the slope of the semilogarithmic plot of signal strength against t. A number of other techniques are available to measure T_1 but so far have found little application to the water sorbed on cellulose.

B. Applications

Sasaki et al. [13] have applied the pulse technique to protons of water sorbed on cellulose over a range (12.5% to 148%) of water contents and temperatures (10° to 100°C). The expected straight-line plot was not obtained, but rather one which could be represented by

$$y_1 - y_2 = y_{11} \exp(-t/T_{11}) + y_{12} \exp(-t/T_{12}). \tag{15}$$

where $y_1 - y_2$ is the difference in signal immediately after the first and second pulses. This suggests that two species of water are participating in the signal spin-lattice relaxation times T_{11} and T_{12}. Plots of the signal immediately after the first pulse y_1 against water content gave a straight-line plot indicating to the authors that all the water protons but none of the cellulose protons were responsible for the signal observed. The proportion of the water associated with relaxation time T_{11} is $y_{11}/(y_{11} + y_{12})$, the remainder being associated with T_{12}. The longer relaxation time, T_{12}, showed an almost linear increase with water content, ranging from about 160 msec to about 250 msec over the water content range observed. T_{11} displays an initial increase with water content, leveling off at about 25 msec when the water content reached about 30%. Semilogarithmic plots of T_{11} and T_{12} against the reciprocal of the absolute temperature were essentially linear. The apparent quantity of water associated with T_{11} decreased when water content or temperature was increased. Such data could be interpreted as indicating that the water with the shorter T_1 was localized, perhaps hydrogen-bonded to the cellulose, while the longer T_1 was characteristic of mobile water not directly adhering to the substrate. They further observed that their data generally agreed with dielectric measurements.

Using the method of progressive saturation, Pittman and Tripp [9] obtained curves and spin-lattice relaxation times from water absorbed on a variety of celluloses at about 7% water content. These data, along with the spin-spin relaxation times and the line width of the narrow component, are presented in Table 4. Figure 5 is a typical saturation curve obtained by these investigators. They suggest that the anomalous behavior at points indicated by the arrows imply that more than one spin-lattice relaxation time is associated with the water.

TABLE 4

Parameters of Narrow Component of Wet Cellulose Spectra

Sample	ΔH_w (G)	T_{2w} (msec)	T_{1w} (μsec)
Cotton fabric	0.23	0.408	11.0
Mercerized cotton	0.30	0.312	4.9
Hydrocellulose I	0.24	0.390	9.8
Hydrocellulose II	0.23	0.408	6.1
Fortisan	0.19	0.594	7.8
Rayon	0.30	0.312	2.5
Saponified acetate	0.37	0.254	2.4
Valonia	0.17	0.552	6.2

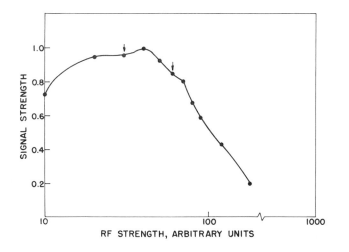

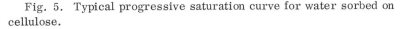

Fig. 5. Typical progressive saturation curve for water sorbed on cellulose.

V. POTENTIAL APPLICATIONS

Apparently only the proton spectrum of cellulose and the proton and deuterium spectra of water sorbed on cellulose have been reported in the literature thus far. The potentialities for work on nuclei other than protons in the cellulose structure and the use of sorbates other than water on a cellulose substrate, as well as the variety of isotopes that might be observed in such sorbates, have barely been touched upon.

There are a number of potential applications related to the sorption of and interaction with cellulose of reagent species and solvents. Of great current interest is the use of nonaqueous systems for the finishing treatment and chemical modification of both textiles and paper. The mode of sorption of agents such as dimethyl sulfoxide, dimethyl formamide, and other substances which exhibit swelling power for cellulose is poorly understood. In cases where nonaqueous solvent procedures are being considered to eliminate or diminish stream pollution, the efficiency of a large number of systems with respect to cellulose interaction deserves investigation, and wide-line NMR would appear to be a prime tool for such studies. Other nuclei whose observation may well be of value are prominently incorporated in cellulose treating agents.

Fluorine is used in modified cottons to impart water repellency oleophobicity. ^{19}F occurs in 100% abundance and has a spin of 1/2 and thus no quadrupole moment to widen its signal in the solid state. The observation of the fluorine signal from modified celluloses should not offer difficulty, and thus its use in the study of such materials should be feasible. The importance of sodium in cellulose technology is well known, and the observation of the ^{23}Na (100% abundant) signal in liquid reagents in the presence of cellulose is undoubtedly of value in the study of swelling and mercerization. The spin of 3/2 of this isotope may complicate its observation in the solid state; however, its importance in the fiber field is usually associated with its ionic form. Phosphorus is used to impart flame retardance to cotton, and the 100% abundance and 1/2 spin of ^{31}P should simplify the interpretation of the spectra of such samples. Other nuclei of technological importance can be listed and the feasibility of their use evaluated in terms of their natural abundance and the presence of a nuclear quadrupole moment.

REFERENCES

[1] E. R. Andrew, Nuclear Magnetic Resonance, 1st. ed., Cambridge Univ. Press, London and New York, 1958.

[2] G. E. Pake, Solid State Phys., 2, 1 (1956).

[3] R. E. Richards, in Determination of Organic Structures by Physical Methods (F. C. Nachod and W. D. Phillips, eds.), Vol. 2, Academic, New York, 1962, pp. 537-562.

[4] F. Bloch, W. W. Hansen, and M. Packard, Phys. Rev., 69, 127 (1946).

[5] F. Bloch, W. W. Hansen, and M. Packard, Phys. Rev., 70, 474 (1946).

[6] J. H. Van Vleck, Phys. Rev., 74, 1168 (1948).

[7] V. Glazkov, Dokl, Akad. Nauk, SSSR, 142, 387 (1962). Translation.

[8] W. Statton, Am. Dyestuff Reptr., 54, 314 (1965).

[9] R. A. Pittman and V. W. Tripp, J. Polymer Sci., A-2, 8, 969 (1970).

[10] N. B. Patil, N. E. Dweltz, and T. Radhakrishnan, Textile Res. J., 32, 460 (1962).

[11] T. M. Shaw, and R. H. Elskin, J. Chem. Phys., 18, 1113 (1950).

[12] T. Swanson, E. O. Stejskal, and H. Tarkow, Tappi, 45, 929 (1962).

[13] M. Sasaki, T. Kawai, A. Hirai, T. Hasig, and A. Odajima, J. Phys. Soc. Japan, 15, 1652 (1960).

[14] E. Bloembergen, E. M. Purcell, and R. V. Pound, Phys. Rev., 73, 679 (1948).

[15] R. Kubo and K. Tomita, J. Phys. Soc. Japan, 9, 888 (1954).

[16] K. Tanaka and K. Yamagata, Bull. Chem. Soc. Japan, 28, 90 (1955).

[17] A. Miyake, J. Chem. Phys., 27, 1425 (1957).

[18] V. Glazkov, Vysokomolckul. Soedin., 5, 120 (1963); Translated by B. J. Hazzard, Polymer Sci., USSR, 4, 736 (1963).

[19] G. E. Pake, J. Chem. Phys., 16, 327 (1948).

[20] R. Dehl, J. Chem. Phys., 48, 831 (1968).

Chapter 8

INFRARED SPECTRA OF CHEMICALLY MODIFIED COTTON CELLULOSE

Robert T. O'Connor

Southern Regional Research Laboratory
New Orleans, Louisiana

I. INTRODUCTION

Although until almost 1950 the infrared absorption spectroscopy of cellulose was practically unknown, once it was introduced its popularity, in
the analytical investigation of physical and optical properties and in the detection, identification, and determination of chemical constituents, increased
very rapidly. Today it is probably one of the most, if not the most, widely
used instrumental method of the cellulose chemist for investigating physical,
optical, and chemical properties.

Earliest applications of infrared absorption spectroscopy to cellulose
were confined to studies of the cellulose molecule; to investigations of
physical, optical, and crystal properties of more or less pure celluloses.
Its uses in such investigations as degree of crystallinity or extent of accessibility, to the type or extent of hydrogen bonding, or to identify or determine relative amounts of crystalline modifications or polymorphic forms,
have been described in Chapter 2.

As long as the analytical spectroscopist confined his efforts to investigations of the more or less pure unmodified cellulose molecule, contributions which could be made to cellulose chemistry were limited to a more
complete understanding of these physical and optical properties of the
molecule. The major uses of infrared absorption, as a tool of analytical
chemistry, have been derived from the ability of this branch of spectroscopy to detect, to identify, and to quantitatively measure specific organic
functional groups. Measurement of such functional groups in the unmodified
cellulose molecule are, of course, severely limited.

The earliest uses of infrared spectroscopy which were concerned with
the investigation of organic functional groups appear to have been in connection with its oxidation and investigations of oxidation and pyrolysis products. The use of infrared absorption spectroscopy in connection with the
oxidation of cellulose and of the products produced thereby are described in
Chapter 5, along with other investigations by thermal analyses.

The major use of infrared absorption spectroscopy as an analytical tool
in cellulose chemistry was prompted by two more or less recent developments in textile chemistry. First was the introduction of blends or admixtures of cotton with other natural fibers, or more often with synthetic or
man-made fibers, to impart to the resulting fabric properties not inherent
in native cotton. The second factor was the chemical modification or resin
treatment of cotton to impart to the fibers properties not innate in the native
cotton fiber, such as chemical modification to impart heat resistance,
flame-proof properties, mildew resistance, rot-proofing, or resistance to the
low-actinic radiation, or to provide durable-press, easy-care, or wash-
and-wear properties. These admixtures of natural and synthetic fibers in
so-called blended fabrics, and these chemical modifications and resin treatments have added to the cellulose molecule practically every organic functional group which can be found in Beilstein. It is not surprising then that

once introduced to the industry, a tool, such as infrared absorption spectroscopy, long-known to organic chemists to be ideal for the detection,
identification, or even quantitative determination of such specific functional
groups, would find considerable use in cellulose research.

This chapter is concerned with the use of infrared absorption spectroscopy as an analytical tool to detect, identify, or quantitatively measure the
type and extent of admixture of other fibers with the cotton fiber and to
similarly detect, identify, and quantitatively measure the extent of chemical finishing whether by chemical modification of the cellulose molecule
or by the use of additives which modify the properties of the resultant
fabric. The discussion will be divided into four main sections. First, the
general principles involved and the analytical techniques used in the application of infrared absorption to investigations of modified cotton will be
reviewed. Applications to cotton blends, with either other natural or with
synthetic fibers, will then be considered. These discussions will be followed
by illustrations of the use of infrared absorption spectroscopy in the analysis of chemically modified cotton, such as by esterification or etherification reactions or other type modifications usually at the site of the hydroxyl
group. The complicated analysis of chemical finishing agents, including
resins used for durable-press properties, softners, optical whiteners,
sizes, etc., will be described. Finally, the major problems as they exist
today in the use of infrared absorption spectroscopy as an analytical tool
for the qualitative or quantitative analysis of chemical finishes will be
discussed.

II. GENERAL PRINCIPLES AND ANALYTICAL TECHNIQUES USED
IN THE APPLICATION OF INFRARED ABSORPTION
SPECTROSCOPY TO CELLULOSE

Infrared spectroscopy, when used as a tool for analytical chemistry, becomes a comparative technique. Thus, one essential requirement in the
establishment of any analytical method by means of infrared absorption
spectroscopy is that the spectra of the materials to be detected, identified,
or quantitatively estimated must be known. There are two general methods.
First, the functional group – frequency correlation technique first proposed
by Julius [1]. In the use of this technique specific functional groups are
correlated with the wavelength or frequency of absorption bands arising
from their vibrations in the infrared region. Correlation charts, or tables,
relating band wavelength or frequency with specific organic groups are
prepared, including, ideally, all the bands which will be encountered in the
specific investigation, i.e., all bands which will be seen in the infrared
spectra of cellulose and chemically modified cellulose. From comparisons
of the bands which appear in the spectrum of an unknown analytical sample,
the presence of specific organic functional groups are detected and the
presence of specific fibers other than cotton in blends or the manner in
which a specific sample has been chemically modified can be deduced.

The second general technique of infrared absorption spectra by the analytical chemist is based on the so-called "fingerprint" technique, often credited to Coblentz [2] who first suggested the compilation of a library of reference infrared absorption spectra. With this technique the infrared absorption spectrum of the unknown analytical sample is compared with a series of spectra of known, identified molecules, much as a fingerprint may be compared with a file of fingerprints to identify an individual. Just as it would obviously be impractical, and costwise prohibitive, to obtain the identification of a person by individually comparing his fingerprints to each in a fingerprint file, it is obvious that the number of organic molecules which an adequate reference library of infrared spectra would contain would make identification by "fingerprint" matching an equally too time-consuming and prohibitively costly means of analysis. Fingerprint identification of individuals is accomplished by means of fingerprint classification. In a similar manner the problem of "fingerprint" matching of infrared absorption spectra is being accomplished. The correlation data between wavelength or frequency and specific organic functional group are being coded, the coded data for tens of thousands of infrared absorption "fingerprints" transferred to magnetic tapes and the matching of infrared spectra accomplished by means of a computer. By these means, identification can be established from "fingerprint" matching with as many as 100,000 spectra of identified molecules, in a matter of a very few minutes. Details of the preparation of these coded spectra, of their use in "fingerprint" matching, and of availability of the coded spectra on magnetic tape from the American Society for Testing and Materials, where this activity has been a major project of Committee E-13 on Absorption Spectroscopy for over a decade, will be found in one of the ASTM publications [3].

Both the Julius' functional group - frequency correlation and the Coblentz' "fingerprint" system of identification can be used, and both should be incorporated into a method of analysis of chemically modified cellulose. Both techniques will be referred to in the following discussions of the applications of infrared absorption spectroscopy to the analysis of cellulose.

Undoubtedly one of the principal factors in the delay of popular use of infrared absorption spectroscopy by the cellulose chemist, or by the textile chemist in general, was the simple fact that no truly satisfactory procedure for obtaining the infrared absorption of textile materials, especailly cellulose, existed. Difficulty in applying procedures used in other applications arose from the fact that no solvents suitable for measuring the infrared absorption of these materials were available, and none has ever been found. Furthermore, the more or less classical method of dealing with insoluble materials, their mixture with heavy mineral mulls, was less than satisfactory. The difficulty of obtaining useable infrared spectra is emphasized by one of the first procedures introduced to the cellulose chemist by Rowen et al. [4]. In this method the cellulose was first acetylated and the deacetylated cellulose acetate dissolved in redistilled acetone in order that a cast film on a glass plate could be obtained. Obviously, this procedure was not conducive to rapid analysis and was severely limited in application, any modified cellulose which could not be further acetylated could not, for example, be analyzed by this technique. It was shortly abandoned by its originators in favor of the unsatisfactory mull technique [5]. Cast films were used successfully for certain types of fibers, such as viscose rayons, which could be cast into a film [6], but these techniques were of no use to the analyst dealing with native cotton cellulose or with any modified cotton cellulose.

O'Connor et al. [7] introduced the use of KBr disks to cellulose analysis and found that the fortuitous almost exact index of refraction match between cotton cellulose and potassium bromide permitted the preparation of a very satisfactory disk without extensive grinding of the sample, and hence even physical or crystalline properties (which would be severely modified or even completely destroyed by such grinding) could be measured satisfactorily by this technique. Later McCall et al. [8] demonstrated that the multiple internal reflectance technique could also be used to measure the infrared absorption spectra of cellulose and that, with specific precautions, the technique could be made quantitative. In special analyses where the investigation of surface effects were of concern, this technique was demonstrated to be capable of yielding data of considerable additional value.

Details of the various techniques which have been introduced to measure the infrared spectra of cellulose have been reviewed in specific detail in a chapter in the new (2nd) edition of the American Association of Textile Chemists and Colorists Manual of Analytical Methods [9], dealing with the applications of infrared absorption techniques. The reader is referred to this publication for further specific details and to the many references contained therein to the original papers introducing various specific techniques. No further details of specific techniques will be described here. However, in the following discussions of the applications of infrared absorption spectra to specific problems, references will be cited to the original sources of the technique whose applications are being described.

In most of the investigations to be described the KBr disk or pellet technique, as first described by O'Connor et al. [7] as a tool for the cellulose chemist, has been used to obtain the spectra of cotton or modified cotton samples, taking advantage of the favorable match in index of refraction to obtain very satisfactory spectra. However, in numerous analyses for impurity or chemical finishes on cotton fabric, solvent extraction or acid hydrolysis has been resorted to to obtain sufficient sensitivity for the specific analysis. In these cases the extracted material, a chemical resin, a softener, whitener, etc., may have an index of refraction considerably different from that of potassium bromide, and the KBr disk technique will not result in satisfactory spectra. However, these extracts and hydrolysates are usually soluble in one of the infrared solvents, such as carbon disulfide, chloroform, carbon tetrachloride, or one of the longer-chain chlorinated hydrocarbons, and very satisfactory spectra can be had by use of the oldest,

and probably still the most popular, technique of the infrared spectroscopist - a direct measurement from the solution.

III. DETECTION, IDENTIFICATION, AND QUANTITATIVE DETERMINATION OF INDIVIDUAL COMPONENT FIBERS IN ADMIXTURES (BLENDS) WITH COTTON CELLULOSE FIBER

From the above discussion it should be apparent that first, infrared absorption spectra are ideal to detect, identify, or quantitatively determine
specific functional groups in admixtures with other groups. Secondly, such
a tool should be a very suitable means for the cellulose chemist involved in
such problems as identifications of fibers admixed with cotton in the ever
growing list of popular blends of fibers in the newer fabrics.

The investigation of fiber blend composition could be looked upon as the
determination of cellulose in such an admixture or it could be considered
as the analysis of a cellulose sample for the fibers, natural or synthetic
mixed into the cotton fibers. Either approach will, of course, result in
essentially the same data. However, as this text is concerned with cotton
cellulose and the analysis of cotton-containing samples, and not with the
determination of cotton, the second view will be used in the following
discussions.

The analysis of cotton-containing blends, or, as more often encountered,
fabrics composed of such blends, has become increasingly difficult as the
number of synthetic fibers which can and are being used in such blends is
increasing at an accelerated pace. Blends of cotton with other natural
fibers have at least the advantage that there are, perhaps, less than a doz-
en such blends, and only three or four of any commercial importance. However, the number of synthetic fibers which can be and which are being
blended with cotton fibers has already reached several score. The American Association of Textile Chemists and Colorists, through its Technical
Committee RA24 on Fiber Analysis, which is divided into two Subcommit-
tees, S-1 Qualitative Identification and S-2 Quantitative Determination, has
inaugurated a major effort to devise analytical procedures which will permit
rapid and reliable techniques for such qualitative and quantitative analyses.
The Committee has recognized the vast numbers of synthetic fibers which
might be considered by starting their efforts with the compilation of a list
of such possible fibers. The Journal of the Association "Textile Chemists
and Colorists" has recently published a list of 189 blends arranged in 19
groups, as obtained by the Committee, of the synthetic fibers which might
be considered in the analysis of fiber blends [10]. Infrared absorption spectra is by no means the only technique to be used in these analyses, but it is
considered as an instrumental technique, especially for the final confirmation of the specific fibers. For this purpose it would appear necessary,
since infrared, as explained above, is a comparison technique, to obtain

the infrared spectra of as many of these fibers as possible. Once a truly representative library of spectra of all synthetic fibers is available, infrared absorption "fingerprint" matching cannot fail to be a useful tool for final confirmation of identity. In addition, to the use of a suitable library of infrared spectra for final confirmation, the use of properly prepared correlation charts of functional groups vs wavelength or frequency of absorption band maxima can furnish several useful initial clues as to the route to follow to obtain satisfactory identification. But the preparation of such a chart also involves obtaining spectra of well-characterized standard samples of as many of the synthetic fibers as it is decided need to be considered as of commercial importance.

The problem is difficult as the composition of blends is not static. New and entirely different blends are continually appearing. In addition to keeping up with the many newly introduced blends is the problem of judging which will ultimately achieve the degree of commercial importance that will justify their inclusion in an analytical scheme of analysis.

In the investigation of cotton-containing blends (the only blends considered in this article) for the identification of other fibers admixed with them, by means of infrared absorption spectra, it is necessary that one become reasonably familiar with the spectra of cotton cellulose. The bands in the spectrum of cotton are those which can be eliminated in the detailed consideration of the spectra of the blends - they are the absorption bands common to all cotton-containing blends.

In Fig. 1 the infrared spectra of cotton cellulose I (native cotton) and cotton cellulose II (mercerized cotton) are produced. These spectra were obtained from KBr disks with a Perkin-Elmer model 621 spectrophotometer affording high resolution throughout the rock salt region 2 to 15 μ . The spectra of cellulose I and II are very similar, although careful inspection of the two spectra in Fig. 1 will reveal subtle differences by means of which previous workers have been able to differentiate one from the other. A list of the bands characteristic of cellulose I and II has been given in Chapter 2 (along with descriptions of the identification of the vibrations giving rise to them). In the investigation of cotton-containing blends, the problem is to recognize bands other than those identified as from the absorption by cotton. Illustrations of how specific natural or synthetic fibers blended with cotton may be identified is attempted in the paragraphs which follow. A tabulation of the bands of cotton along with those of selected natural and synthetic fibers is reproduced in the Appendix.

In Fig. 2 the infrared spectrum of cotton is compared with those of the vegetable fibers jute and flax. As flax consists of some 60 to 65% cellulose, there is considerable similarity among these spectra. Jute can be recognized by bands with maxima at 5.75 μ and at 6.10 μ , the latter being more intense than the weak water band found in the spectra of all celluloses. These bands arise, probably from the lignocellulose nature of jute, from some C=O stretching vibration of the lignone, and are relatively weak as

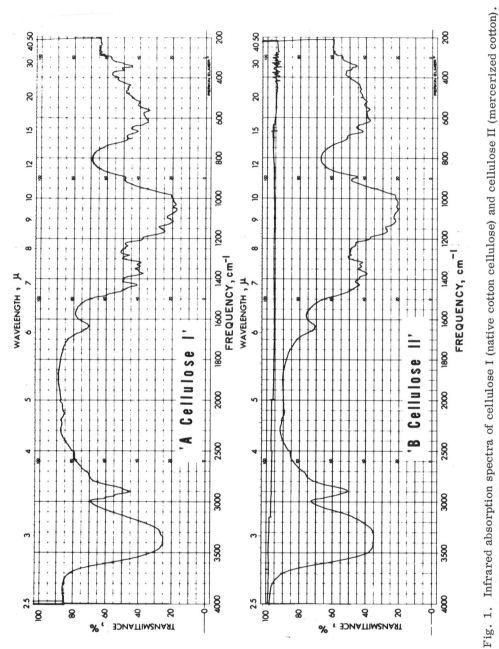

Fig. 1. Infrared absorption spectra of cellulose I (native cotton cellulose) and cellulose II (mercerized cotton).

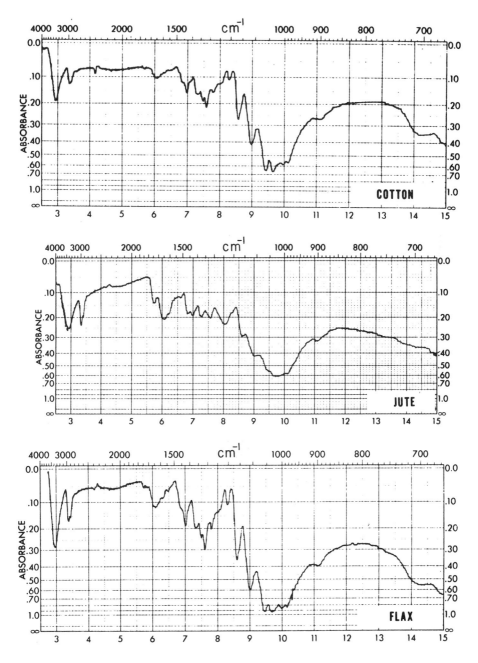

Fig. 2. Infrared absorption spectra of cotton, jute, and flax.
(Courtesy of Wilks Sci. Co.)

this component is a minor constituent of the jute. The C-O component of
the carbonyl in jute is shown by the appearance of a band at 8.00μ . This
band is particularly useful in differentiating between jute and flax. Flax,
which may contain over 80% cellulose, is more difficult to differentiate from
cotton, the most distinguishing feature being the band with maximum at
7.60μ , which is, relatively, more intense in flax than in cotton cellulose,
presumably from contributions of minor constituents.

As shown in Fig. 2, the "fingerprint" technique is probably the most reliable technique to differentiate cotton cellulose from the bast fibers, jute
and flax. While this technique would not be reliable for detection of small
amounts of one of these fibers in the presence of either of the others, with
some experience, the analytical spectroscopist could reliably identify the
pure fibers or the major components in mixtures. The sensitivity of the
infrared technique is sufficient to detect, identify, and determine the components to the levels at which the minor components will be found in such
blends, i.e., 15 - 35%.

In Fig. 3 the spectrum of cotton cellulose is compared to two natural
protein fibers, silk and wool. The amino acid bands, arising from the protein components, the fibroin and the sericin of silk and the keratin of wool,
provide a ready and dependable means for differentiating either of these
natural fibers from cotton cellulose. The amide I band at 6.18μ and the
amide II band at 6.67μ can be used to both identify silk or wool and to determine its content in admixture with cotton cellulose. Silk and wool can be
differentiated from each other with more difficulty. Silk exhibits a weak
but definite band with maximum at 5.75μ , arising from a C=O stretching,
presumably of the fatty acids which are minor components of this fiber.
The spectra of wool and silk differ at the 8.0-μ region, silk exhibiting a
single asymmetric band with maximum at 8.18μ , while wool reveals two
bands, one at 7.85μ , the second at 8.20μ .

In Fig. 4, two groups of the large number of synthetic fibers, the poly-
amides and the polyesters, are represented by nylon and Dacron. These
fibers are readily differentiated from cotton cellulose and from each other
by the very strong amide I and amide II, C=O stretching and N-H deformation, in the spectrum of nylon at 6.13 and 6.52μ and by the very strong
absorption arising from the C=O stretching of the carbonyl groups in the
polyethylene-terephthalate polymer, Dacron, which is observed at 5.80μ .
Nylon exhibits a band at about this wavelength; but it can be readily differentiated as it has a maximum at shorter wavelength, and it is relatively
weak. These bands can be used to identify nylon or Dacron in the presence
of cotton cellulose or to determine the amounts of these two synthetic fibers
in admixtures with cotton, as in cotton-synthetic fiber blends.

In Fig. 5 are shown the spectra of cotton and Dacron and five levels of
Dacron blended with cotton. The strong characteristic band in the spectrum
of Dacron with maximum of 5.80μ can be used to identify the presence of
this fiber in the cotton even at a concentration of 10%, and can be used to

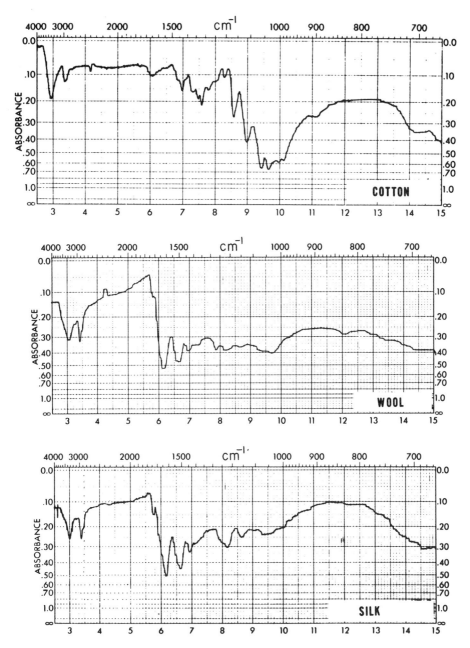

Fig. 3. Infrared absorption spectra of cotton, wool, and silk. (Courtesy of Wilks Sci. Co.)

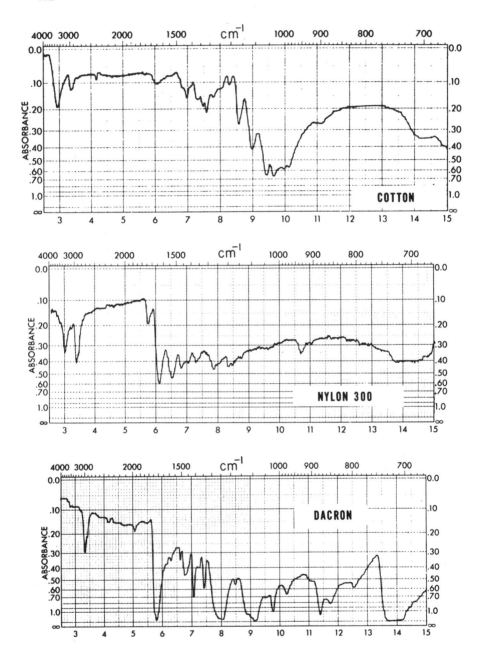

Fig. 4. Infrared absorption spectra of cotton, nylon, and Dacron. (Courtesy of Wilks Sci. Co.)

determine the amount of Dacron in the specific blend, as shown in Fig. 6. In this figure, the intensity (absorbance) of the 5.80 μ maximum is plotted against the known Dacron concentration for a series of Dacron-cotton cellulose blends. Once such a Beer-Lambert relationship is established for a specific instrument, the Dacron content of any Dacron-cotton blend can be established rapidly, with satisfactory precision and accuracy, and to reasonably low sensitivity levels.

In Figs. 7 and 8 are shown the spectra for a series of nylon-cotton blends and a quantitative relationship (Beer-Lambert curve) of the intensities of the 6.52μ band against the known nylon content. The 6.52μ band is selected over the 6.13 μ, which is a bit more intense, in order to avoid any overlapping by the water band of cotton with maximum at 6.10μ in the spectrum of cotton. Again the 6.52μ band provides a method for confirming the presence of nylon in the blend with cotton and for measuring its concentration rapidly, precisely, accurately, and with reasonable sensitivity.

In Fig. 9, the spectra of cotton and two polyacrylic synthetic fibers, Orlon and Acrilan, are compared. These two synthetic fibers can be most readily identified by the bands at 4.40 and 4.48μ in the spectra of Acrilan and Orlon, respectively, arising from a $C\equiv N$, nitrile, stretching vibration. Acrilan can be differentiated from Orlon by the strong band with maximum at 5.70 μ, a $C=O$ stretching of the ester group. Identity of this group is confirmed by the strong band at 8.08 μ, arising from a C-O stretching of the ester moiety. Orlon can be distinguished by the strong intense $C=O$ stretching band with maximum at 5.80 μ.

The spectra of synthetic fibers of the polyvinyl group are compared with cotton in Fig. 10. The $C\equiv N$, nitrile, stretching at 4.43 μ identifies Dynel in the presence of, and differentiates it from, Saran. Dynel is differentiated from cotton by the complete absence of strong absorption between 9.3 and 10.3 μ, a region where C-O stretching vibrations and O-H bending modes produce a series of incompletely resolved bands in the spectrum of cotton. Both Dynel and Saran are characterized also by the appearance of very strong bands at about 7.0 μ, and this band can also be used to differentiate Dynel from Saran, as it appears in the spectrum of Dynel below 7.0 μ, at 6.92 μ, while in the spectrum of Saran it is found above 7.0 μ, at 7.09 μ, as shown in Fig. 10.

While all of these group frequency correlations are helpful to the analytical spectroscopist in attempting to determine components of blends, direct comparison of the spectrum of the "unknown" blend with the spectra of likely or suspected components is, probably, the most reliable technique for confirming their presence. With some familiarity with the spectra of various natural and synthetic fibers and with some experience in examination of the spectra of blends, infrared spectroscopy can provide a satisfactory technique for identifying components in blends and frequently can be used to determine the concentrations of individual components rapidly, reliably, and with reasonable sensitivity.

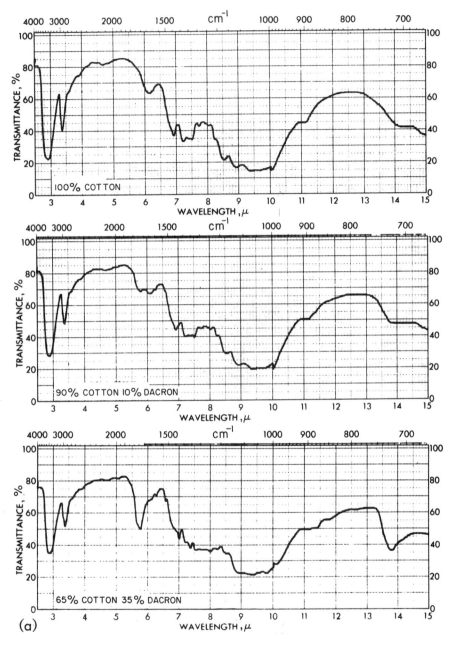

Fig. 5a. Infrared absorption spectra of blends of cotton and Dacron.

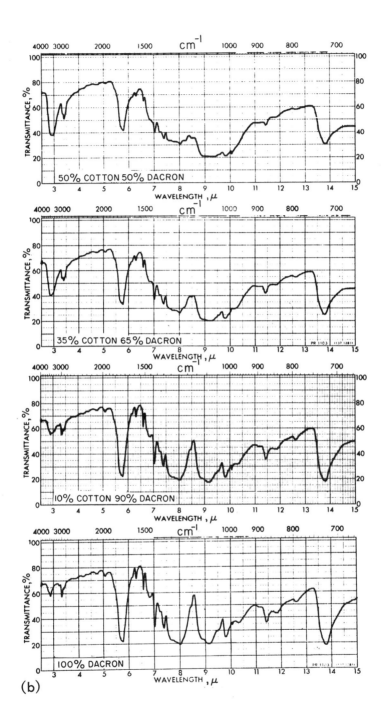

(b)

Fig. 5b. Infrared absorption spectra of blends of cotton and Dacron.

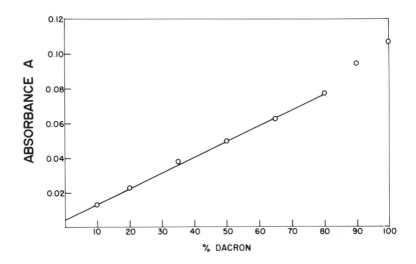

Fig. 6. Beer-Lambert law relationship – calibration curve for determination of Dacron in cotton–Dacron blends by measurement of infrared
absorption maximum at 5. 80 μ.

IV. ANALYSES OF CHEMICALLY MODIFIED COTTON CELLULOSE

While studies of the infrared absorption spectrum of cotton cellulose to
investigate physical and crystalline properties and changes which occur in
these properties under specifically controlled conditions have been of interest to many cellulose chemists, as indicated by the number of papers which
have been published during the past decade, to the spectroscopist such investigations are limited solely to observations and measurements of C-H
and O-H stretching and bending modes.

The chemical modification of cotton cellulose to obtain properties not
native to cellulose, such as flameproofing, durable press, mildew resistance, resistance to low-wavelength radiation, rot resistance, and many
others, is accomplished by the introduction of new and different organic
functional groups, mainly by reactions with the hydroxyl groups of the cellulose molecule. In this manner the cellulose chemist has introduced a host
of new and different organic functional groups, each designed to impart some
additional property to the cotton. Investigations of such modified cottons
consist mainly of identification of the modifying functional groups, measuring their concentration, and correlations of these data with specific textile
properties imparted to the final fabric product. As infrared absorption is
the ideal tool for both the identification and for the quantitative estimation
of organic functional groups, it becomes a logical tool available, ready, and
almost waiting to assist in such investigations.

Considering the cellulose molecule as containing polyhydroxyl groups as the most characteristic and reactive sites, the most obvious manner in which such a molecule could be modified is by esterification. Esterification with acetic acid (or more accurately acetic anhydride or acetic chloride) to obtain acetylated or partially acetylated PA cotton is one of the oldest methods of modification. PA cottons have been used in about every published description of the use of infrared absorption involving investigations of chemically modified cotton. The spectra of PA cotton so well illustrates the potential use of infrared absorption measurements in connection with such investigations that we are using this example again. In Fig. 11 the infrared spectrum of cotton is compared with that of a PA cotton. The esterified cotton can be recognized by the appearance in its spectrum of the intense band at 5.68μ, arising from a stretching vibration of the C=O group of the ester. This band can also be conveniently used to estimate the extent of acetylation. In Fig. 12 the intensity of the maximum of this band (absorbance) is plotted against the chemically determined acetyl content of a series of PA cottons. A linear relationship is obtained to a concentration of about 10% acetyl content. Above this value further dilution of the sample is required to maintain a base-line corrected absorbance of not more than approximately 0.12. With this technique the acetyl content of any PA cotton can be obtained rapidly with satisfactory precision and accuracy.

Investigation of the rather simple reaction to obtain PA cotton by means of infrared absorption spectroscopy has, however, generated some interesting questions, the complete answer and understanding of which may be of considerable importance to subsequent investigations of both the structure of unmodified cotton and of chemically modified cottons. The first question, of course, is why do we detect such a sharp departure in linearity in the Beer-Lambert curve of Fig. 12 at such relatively low acetyl content?

A second interesting problem arose within our laboratory in connection with the use of the technique, as outlined, to obtain rapid estimations of acetyl content on a more or less routine basis. These values, as reported from the infrared analyses, appeared for a period of time to be completely satisfactory, checking with random standards and chemical analyses for acetyl content. However, subsequently, a specific set of analyses was shown to be in considerable variance with the infrared absorption determined values. A lengthy investigation revealed that all of these PA cottons had been obtained by acetylation with acetyl chloride rather than by use of the more common acetic anhydride technique. The problem of erroneous analyses is readily "explained" by obtaining a series of PA cottons acetylated by the acetyl chloride technique. A Beer-Lambert curve, very similar to that shown in Fig. 12, is obtained, but the curve has an appreciably different slope. Obviously, if maximum intensities of the C=O stretching vibration are to be used for analysis of acetyl content, the infrared spectroscopist would have to have some knowledge of how the PA cotton was prepared in order to make the correct selection of which calibration curve to use-- and this is information which, unfortunately, is often not available.

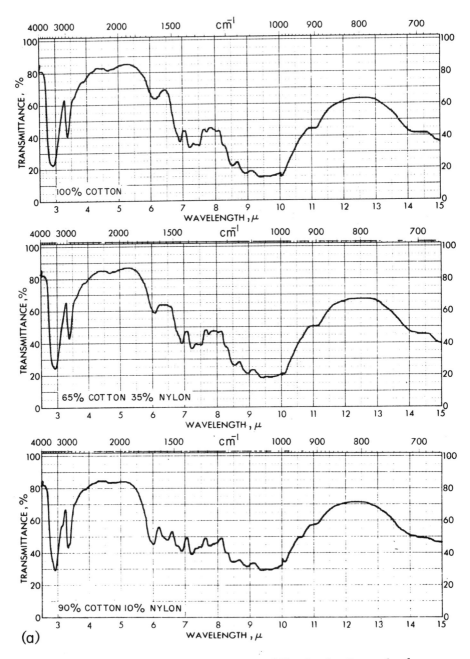

Fig. 7a. Infrared absorption spectra of blends of cotton and nylon.

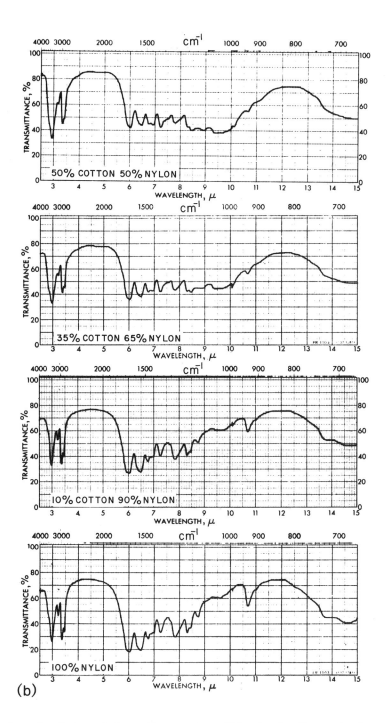

(b)

Fig. 7b. Infrared absorption spectra of blends of cotton and nylon.

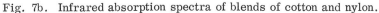

Important observations which either compounded the confusion or somewhat clarified it, depending upon the view taken, were made by Hurtubise
[11] in a paper entitled "The Analytical and Structural Aspects of the Infrared Spectroscopy of Cellulose Acetate." Dr. Hurtubise investigated the infrared spectra of heterogeneously acetylated wood pulps and cotton linters
and showed that Beer-Lambert curves of the intensity of the 1750 cm^{-1}
(5.72 μ) band gave different slopes for acetates made from spruce and from
western hemlock pulps. These curves as published by Hurtubise are reproduced in Fig. 13. They indicate a tendency to depart from linearity, but at
somewhat higher levels than found for cotton cellulose. The interesting
comparison is that spruce and western hemlock after acetylation, give
curves with two different slopes, very closely resembling the phenomena
previously reported for cotton cellulose acetylated by means of acetic anhydride and by acetyl chloride!

Hurtubise made several important observations regarding the infrared
spectra of the wood pulps. Higgins 12 had previously reported differences
in optical densities (absorbances) of acetates made from a long-fibered
bleached kraft pulp and those from a high α-cellulose pulp, and suggested
that these anomalies may arise mainly from differences in the capacity of
different acetates to be dispersed in the halide. Higgins also commented on

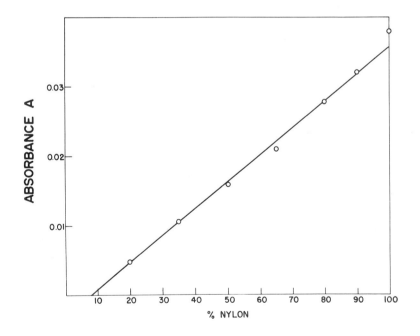

Fig. 8. Beer-Lambert law relationship - calibration curve for determination of nylon in cotton-nylon blends by measurement of infrared absorption maximum at 6.52μ.

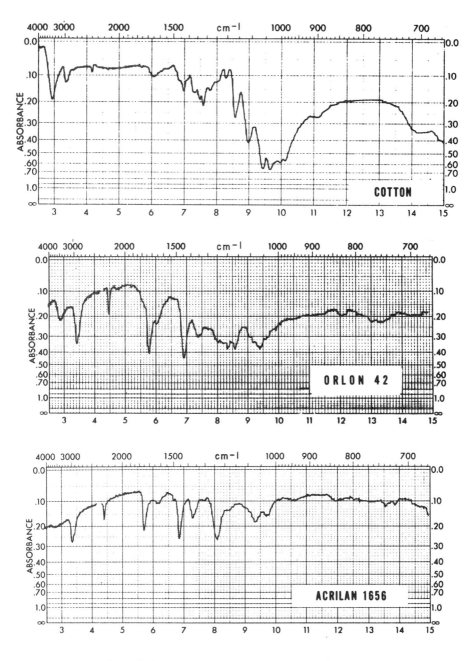

Fig. 9. Infrared absorption spectra of cotton, Orlon, and Acrilan. (Courtesy of Wilks Sci.Co.)

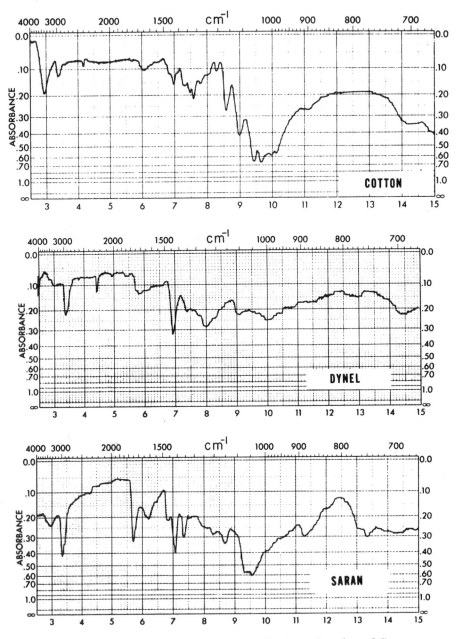

Fig. 10. Infrared absorption spectra of cotton, Dynel, and Saran.
(Courtesy of Wilks Sci. Co.)

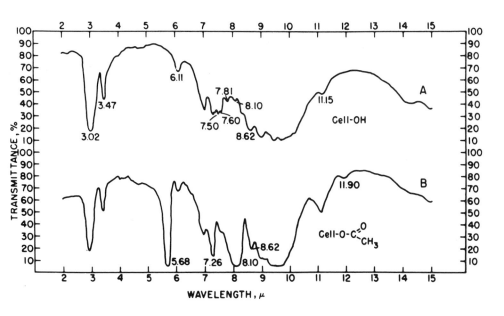

Fig. 11.   Infrared absorption spectra of cotton and of PA cotton.

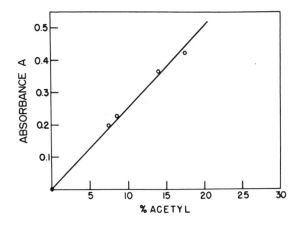

Fig. 12.   Beer–Lambert law relationship – calibration curve for determination of acetyl content of PA cotton at $5.68\,\mu$.

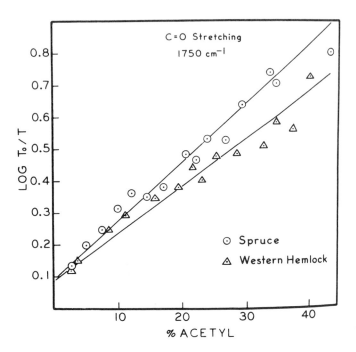

Fig. 13. Beer-Lambert law relationships - calibration curves for determination of acetyl content of partially acetylated western hemlock and spruce pulps at 1750 cm^{-1}. (Courtesy of Tappi.)

the scatter of the results around the Beer-Lambert curves for plots of intensities (absorbances) vs acetyl content.

Arguing that particle size would have much less effect on integrated intensities than on band heights, Hurtubise plotted these integrated areas under the specific bands and, as shown in Fig. 14 from his paper, the slopes of the integrated intensity vs acetyl content are the same for acetates made from the spruce and the western hemlock wood pulps. Thus, the method of total integration under the area of the curve provides an analytical method, based on infrared absorption spectroscopy, which can be used for acetates from different wood pulps and cotton linters.

Hurtubise proposed another "universal" technique, consisting of plotting the ratio of the absorbance of the C=O stretching band at 1750 cm^{-1} to the absorbance of the O-H stretching band at 3400 cm^{-1} against the acetyl content. This ratio was called the "infrared acetyl index" and was shown to result in a single linear relationship for all wood pulps and cotton linters. Details for obtaining the "infrared acetyl index" and for using it to determine the degree of substitution for any pulp acetate are described in the paper by Hurtubise [11].

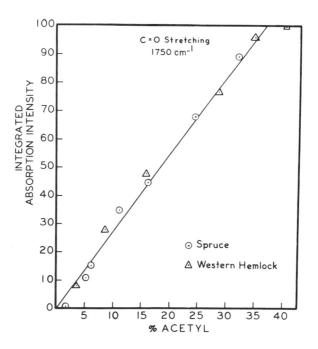

Fig. 14. Calibration curve for determination of acetyl content of partially acetylated western hemlock and spruce pulps from integrated absorption at 750 cm^{-1}. (Courtesy of Tappi.)

While scattering of experimental values about the calibration line undoubtedly arises, in part at least, from particle size, we do not believe that this effect can account for the differences in slopes of the working curves. Higgins' data were obtained from KCl spectra, where the match in index of refraction between the matrix KCl and cellulose is not ideal. In the experiments as obtained from KBr data the effect of particle size should be considerably less as there is an almost perfect match between the refractive index of KBr and of cotton cellulose. A series of partially acetylated cottons, prepared by acetic anhydride and by acetyl chloride methods, was ground excessively in a vibratory ball mill. (This mill can grind strips of heavy army duck fabric to a fine material, resembling talcum powder.) Calibration curves were plotted from infrared data obtained from the two series of samples both before and after grinding. (The unground fabric was merely cut in a Wiley mill to pass a 20-mesh screen.) The finely ground samples gave data which showed, probably, a somewhat decrease in scatter of the points about the calibration line, although scatter has not been much of a problem with cotton cellulose, as seen in Fig. 12. However, the data for the acetic anhydride acetylated cellulose and for the acetyl chloride prepared acetates were as far apart as they had been in the unground samples. It is difficult to accept the fact that excessive grinding would not reduce this difference in slope of the calibration curves if it is to be attributed solely to scattering arising from particle size or to distribution of sample in the KBr matrix.

Another experiment also led to some doubt as to whether particle size and distribution could account entirely for the difference in the slopes of the two working curves. A series of cotton cellulose partially acetylated to various levels was measured in the infrared spectrum and the data plotted as in Fig. 12. The highest acetylated sample, approximately 30%, was then deacetylated and a series, all orginating from the single 30% acetylated sample, was prepared. A plot of infrared absorption intensities of this series of samples against their known acetyl content resulted in a calibration curve with a slope appreciably different from that of the original series of partially acetylated cottons. These results are in agreement with the facts that a specific partially acetylated cotton, acetylated to a specific acetyl content, will differ in physical properties from the same cellulose, with the same acetyl content, but obtained by deacetylation from a higher acetylated sample.

With our present knowledge, we can identify partial acetylation and measure it quantitatively with good accuracy either by the total integration or by the "infrared acetyl index" of Hurtubise. The importance of the differences in slopes of calibration curves obtained from band maxima is that this may be a clue to the position or to the site of the chemical modification. An infrared method for ascertaining reliably and rapidly the site of a chemical modification has not been achieved. If this could be accomplished, it would be another and probably the greatest contribution that infrared absorption spectroscopy could make to cellulose chemistry.

Esterification reactions with cotton are not, of course, confined to PA cotton. Esters have been described that have been obtained from long-chain fatty acids and from unsaturated and aromatic acids. The infrared spectra of PA cotton is compared with the spectra of some of these types of esterification products in Fig. 15 to illustrate how various esterification products can be differentiated by means of infrared absorption spectra. The esterification can be identified and is probably best measured quantitatively by the C=O stretching band at about 5.8μ. A wide band arising from a C-O stretching of the ester group will appear in PA cotton at just about 8.10μ, while in esterification with saturated acids of long-chain length, it is found at longer wavelengths. In Fig. 15B the spectrum of a cotton reacted with stearic acid reveals this band with a maximum at about 8.50μ. Esterification with aromatic acids can, of course, be readily recognized by appearance in their spectra of the phenyl bands, as also shown in Fig. 15D, E. The esterification is clearly identified by the strong 5.80μ C=O stretching vibration. The presence of an aromatic group is confirmed by appearance in these spectra of the C=C stretching vibrations of the aromatic rings in the

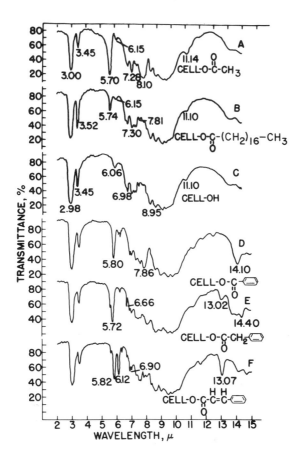

Fig. 15. Infrared absorption of esterified cotton cellulose: A. cellulose acetate; B. cellulose stearate; C. untreated cotton cellulose; D. cellulose benzoate; E. cellulose phenylacetate; F. cellulose cinnamate.

6.0μ region. The phenyl groups give rise to two C=C stretching vibrations at about 6.2 and 6.7μ, with the longer wavelength band the more intense. However, conjugation with the benzene ring, as in benzoic acid, gives rise to a third C=C stretching vibration at about 6.3μ and a marked decrease in the intensity of the 6.7μ band, so that the band at 6.2μ is the most intense. Thus, as shown in Fig. 15, cellulose benzoate and cellulose phenylacetate can be readily differentiated, and both are readily distinguished from PA cotton.

Esterification with unsaturated aliphatic acids can be detected by appearance in their spectra of a rather sharp, narrow band about 6.1μ, arising from C=C stretching in a linear molecule. This sharp band is clearly

distinguishable from the broad shallow band at about 6.15μ , arising from
absorbed water and attributed to the H-O-H stretching mode, as illustrated
in Fig. 15F by the spectrum of a cotton esterified with cinnamic acid.

A second general type of reaction involving hydroxyl groups of the cellulose molecule is etherification. This reaction has been illustrated by partial
cyanoethylation, the reaction of cellulose with acrylonitrile in the presence
of sodium hydroxide and water. The infrared absorption of a cyanoethylated
cotton is shown in Fig. 16. The C≡N stretching of the nitrile group at
4.45μ readily identifies the type of treatment, and measurement of the intensity of this band for a series of partially cyanoethylated cottons provides
the data shown in Fig. 17. It will be noted that the individual points fall
very well on a straight line; scattering from the KBr disks is no problem.
From a calibration curve of this type, the extent of cyanoethylation in any
partially cyanoethylated cotton can be obtained readily, rapidly, and with
high accuracy. This infrared method has the advantage over several
chemical methods, particularly the estimation by means of nitrogen content,
as only nitrile nitrogen is measured and the analysis is independent of the
presence, in the particular fiber or fabric, of any other type of nitrogen,
especially of amino nitrogen.

The number of possible functional groups that can be introduced into the
cotton molecule by direct reaction with the hydroxyl group is, of course,
limited. However, the cellulose chemist has been able to overcome this
limitation by replacement or multiple reaction. The hydroxyl group is replaced with a more reactive group which will in turn permit replacement of
the group desired. Reactions of this type offer another challenge to the analytical spectroscopist, as the problem consists of not only identifying the

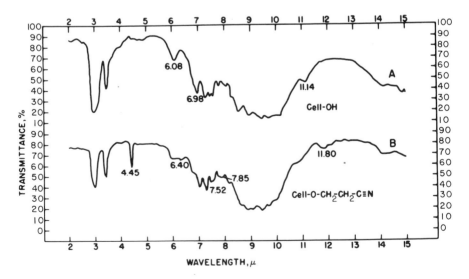

Fig. 16.   Infrared absorption spectra of cotton and CN cotton.

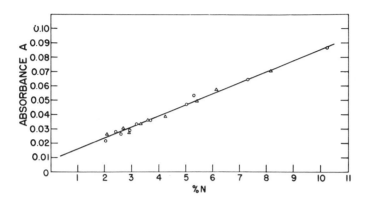

Fig. 17. Beer-Lambert law relationship - calibration curve for determination of cyanoethyl content of CN cotton at 4.45 μ.

presence of the groups to be substituted but often of verifying the complete
replacement of the intermediate groups used to affect its introduction into
the cellulose molecule.

The well-known (and probably among the oldest) reagents for reactions
with cellulose are the sulfonyl chlorides, methyl sulfonyl chloride, and p-
toluene sulfonyl chloride. Klein and Snowden [13] used these reagents to
introduce reactive groups into the cellulose molecule, to produce mesylated
and tosylated cottons, which were subsequently used as intermediates for
making new cellulose derivatives in which the cellulosic hydroxyl group is
first replaced by either the mesyl or the tosyl group, and these groups replaced by other desirable functional groups.

Mesylated and tosylated cotton can be readily identified from unreacted
cotton and from each other by their infrared absorption spectra, Fig. 18.
Both the spectra of mesylated and tosylated cotton reveal prominent bands
at about 7.32 and 8.50 μ, arising from the covalent sulfonate group, $-O-$
SO_2-R. They also exhibit bands in the 12 to 13μ region of the spectrum
which both distinguish them from unmodified cotton cellulose and differentiate them from one another. These bands arise from a vibration of the
sulfur group, as they are exhibited in the spectra of both mesylated and
tosylated cotton. The bands at 12.10μ in the spectrum of mesylated cotton
and at 12.30μ in the spectrum of tosylated cotton are not observed either
in the spectrum of unmodified cotton or in the spectra of these two reagents.
They must, therefore, arise from the C-O-S linkage formed during the
modification reaction. Appearance of such bands, which appear neither in
the spectrum of unmodified cotton nor in the spectra of the reagents, occasionally offers the analytical spectroscopist an opportunity to establish
some indication of the type of linkage formed during the chemical modification of the cotton. The bands at 13.25μ in the spectrum of mesylated cotton, and at 12.62μ in the spectrum of tosylated cotton most probably arise

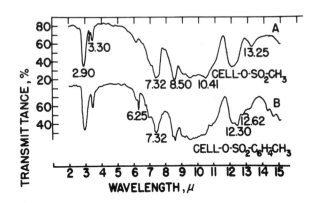

Fig. 18. Infrared absorption spectra of mesylated and tosylated cotton cellulose.

from a C–S stretching. Both of these bands are observed in the spectra of the reagents. The spectrum of tosylated cotton exhibits the C=C skeletal in-plane vibrations about the aromatic ring, characteristic of the spectra of aromatic compounds with maxima at about 6.25 and 6.70μ. These bands distinguish tosylated from mesylated cellulose. The spectrum of mesylated cotton reveals two bands at about 3.25 and 3.40μ. The longer wavelength band is found in the spectrum of unmodified cotton and in that of tosylated cotton and is attributed to C–H stretching. Any C–H stretching of the aromatic ring C–H groups are apparently unresolved in the spectrum of tosylated cotton as no additional bands which could be assigned to an aromatic C–H stretching are observed in the spectrum of this modified cotton or in the spectrum of this reagent. The spectrum of mesylated cotton, however, reveals a second band with maximum at 3.25μ, very probably arising from a C–H stretching of the methyl group adjacent to the sulfonate. The intensity of this band increases with degree of mesylation and can be used to identify the process of mesylation, to differentiate it from tosylation, and to quanti-tatively measure the degree of mesylation. Once identified, either mesy-lation or tosylation can be quantitatively determined by use of the sulfonate bands at 7.3 and 8.5μ. Alternatively the bands in the 12 to 13μ region can be used for quantitative measurements, as illustrated in Fig. 19. Choice would be dictated by the specific problem. The sulfonate bands at 7.3 and 8.5μ would be more sensitive in the determination of very small amounts of mesylation or tosylation. However, the slopes of the Beer-Lambert calibration curves are considerably steeper for the longer wavelength bands in the 12 to 13μ region, and use of these bands would be more sensitive if the problem is to detect small differences in degree of substitution between samples.

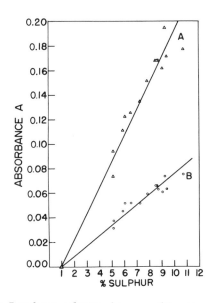

Fig. 19. Beer-Lambert relationships - calibration curves for determination of tosyl content of tosylated cotton cellulose: A. at 12.35 μ ; B. at 6.25 μ.

The following examples of the use of infrared absorption spectra as a tool to follow replacement reactions and multiple modification reactions are from the original paper describing them [14]. Klein and Snowden [13] reacted cotton cellulose with the sulfonyl chlorides and used the cellulose sulfonates as intermediates for making new cellulose derivatives in which the hydroxyl groups were replaced with functional groups which would not react readily with the cellulose.

Fig. 20A is the infrared spectrum of mesylated cotton cellulose which has been subsequently reacted with the sodium salt of p-toluene sulfonamide. Mesylated cellulose is detected by the bands at about 7.3, 8.5, and 12.1 μ . The p-toluene sulfonamide reagent exhibits similar bands at 7.6, 8.7, and 12.4 μ. The spectrum of the reaction product (Fig. 20A) reveals both a decrease in the intensity of these bands as the mesyl groups are replaced and a shift to longer wavelengths as the bands of the reagent are replacing them. The bands observed at about 7.32, 8.60, and 12.27μ are probably unresolved combinations of the vibrations of the mesylated cotton and the reagent. Introduction of the phenyl group into the cellulose molecule can be verified by the appearance of the aromatic C=C stretching band at 6.24 μ and the aromatic ring band (deformation of C-H groups about the phenyl ring) at 14.17 μ. Either of these bands could be used to measure the extent of the reaction.

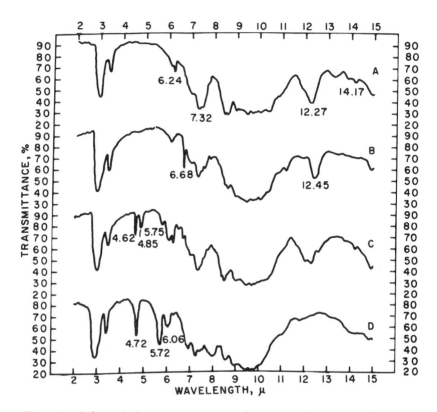

Fig. 20. Infrared absorption spectra of cotton cellulose modified by replacement reactions: A. mesylated cotton subsequently modified with sodium
salt of p-toluene sulfonamide; B. mesylated cotton subsequently modified
with sodium salt of p-methylthiophenol; C. tosylated cotton subsequently
modified with potassium thiocyanate; D. cotton cellulose acetylated and then
mesylated and the mesyl group subsequently replaced with sodium azide in
N, N-dimethylformamide.

The product whose spectrum is shown in Fig. 20B was obtained by replacement of the mesyl group of mesylated cellulose with the sodium salt of
p-methylthiophenol. This spectrum is differentiated by the strong, sharp
band with maximum at 6.68 μ. This band is observed also in the spectrum
of the reagent. Its presence can be used to verify the reaction and its intensity to measure the extent of the replacement.

A tosylated cellulose was treated with potassium thiocyanate to produce
the replacement product whose spectrum is reproduced in Fig. 20C. Two
bands at 4.62 and 4.85 μ, assigned to C≡N stretching vibration, identify the
type of replacement and could be used to measure its extent.

The spectrum in Fig. 20D is an example of a multimodified cellulose. This sample of cotton cellulose was modified by three successive reagents. The sample was first acetylated, then mesylated, and the mesyl group then replaced by treatment with sodium azide in N,N-dimethyl formamide. The acetyl group is identified by the usual carbonyl band at 5.72μ. Complete replacement of the mesyl group is shown by the absence of any trace of the 3.3 and 12.1μ bands. The azide can be identified and quantitatively measured by the band at 4.72μ arising from an $N\equiv N$ vibration.

These spectra illustrate how infrared absorption can be used to follow the presence or fate of each reagent in multimodified or in replacement reactions. Incomplete replacement can be verified by traces of the initial reagent, and the extent of replacement can be followed quantitatively by consideration of the intensities of the bands of the replacing reagents.

Aminized cotton celluloses can also be used as intermediates for further modification reactions, as originally demonstrated by Reeves and Guthrie [15]. In Fig. 21 the infrared spectra of aminized cotton and three cottons further modified with the aminized sample as the starting point are reproduced. In curve B of this figure the original aminized cotton (Fig. 21A) was treated with acetophenone. This replacement reaction can be readily identified and quantitatively measured by means of the intense 5.96μ band of the C=O stretching of the phenone.

The aminized cotton was modified by reaction with nitromethane, and the spectrum of the reaction product is shown in Fig. 21C. This modification is readily identified and again could be easily measured quantitatively by means of the characteristic band at 6.47μ arising from the N=O stretching vibration. The third sample of aminized cellulose was treated with chlorophenol. The resulting product can be identified by means of the 6.09 and/ or 6.78μ band of its infrared spectrum, Fig. 21D, indicating a phenyl ring with no conjugation.

V. ANALYSES OF RESIN-TREATED COTTON CELLULOSE

With a method for identification and quantitative estimation available and in use on a more or less routine basis for cottons modified by chemical reactions such as esterification or etherification, interest, quite naturally, arose as to whether similar identification and determination could be made of the reagents used in resin treatments of fibers or fabrics to produce wash-and-wear, crease-resistance, or durable-press properties.

When a fiber or fabric is treated to produce these properties, the reagent is usually considered to act to form crosslinks between layers of the cellulose polymer. The reagents which have been widely used are mostly derivatives of urea or melamine, as shown in Fig. 22. These are not, perhaps, a class of compounds an infrared spectroscopist would select to demonstrate the applications of infrared absorption spectroscopy as a tool for qualitative or quantitative analysis. The spectra of most of these materials

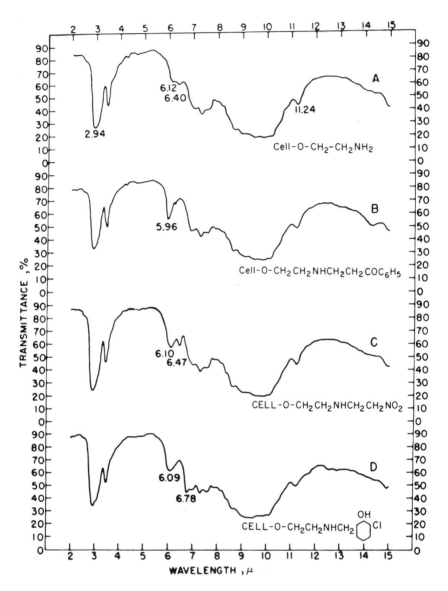

Fig. 21. Infrared absorption of multimodified cotton cellulose: A. aminized cotton cellulose; B. aminized cotton further modified with acetophenone; C. aminized cotton further modified with nitromethane; D. aminized cotton further modified with chlorophenol.

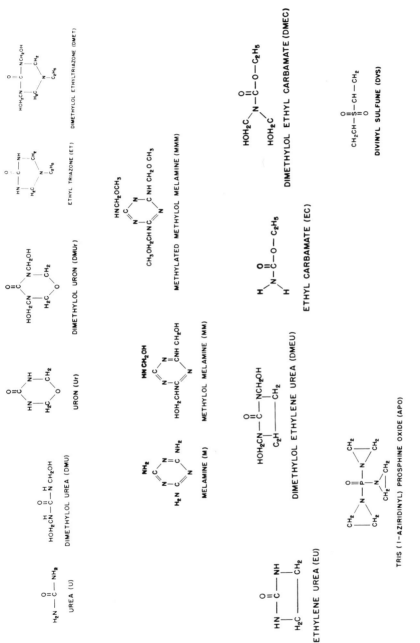

Fig. 22. Structures of organic reagents used in resin treatment of cotton cellulose to produce wash-and-wear, crease-resistant, or durable-press cotton fabrics.

are rather nondescript, with no particular strong identifying bands. Further-more, the cellulose chemist is working to produce a sufficient degree of the desirable property to be imparted to the cotton with a minimum of reagent. First the new property can be imparted only at a cost, a decrease in one or more of the desirable properties of cotton. The modification must be made in a manner so as to keep these losses very small, negligible if possible. Since loss of desirable properties follows the extent of treatment, the cel-lulose chemist attempts to impart the new property at the lowest add-on level of resin which will give satisfactory crease-recovery, or durable-press performance. Secondly, the cost of the resin must be considered, and for economical reasons the use of the modifying reagent must be kept to a minimum. Thus, the infrared spectroscopist has both the problem of iden-tifying a class of materials whose spectra do not exhibit sharp and intense bands and of identifying these materials at relatively low levels. To the problems encountered in the identification of chemically modified cotton, that is esterification, etherification, etc., there is superimposed the prob-lem of sensitivity.

Note from the formulas for the various types of resins, Fig. 22, used in resin treatment of cotton, first, that most of them contain the C=O group adjacent to the N-H group, the so-called amide grouping, which has been investigated extensively in infrared spectroscopy in connection with studies of amino acids and proteins. Secondly, most of these reagents contain rings. Both of these observations are important as spectra of compounds contain-ing the amide group exhibit bands characteristic of this grouping which are very sensitive to the particular environment in which it is found. The wave-length position of the maximum can be expected to differ for the various structures shown. Secondly, the ring vibrations, particularly the C-H de-formations, are also very characteristic of the particular ring. These bands, therefore, might be expected to offer opportunity for identification of the specific resins by the Julius functional group – frequency technique. A list of the bands from the spectra of several resins is shown in the Appendix. Location of the exact position of the amide I and amide II bands in the 1500 and 1600 cm^{-1} region and observation of the position of one or more of the ring vibrations in the 700 to 800 cm^{-1} region should provide a satisfactory identification of a specific resin.

However, upon examination the KBr disk spectra of many resin-treated cotton cellulose samples revealed no differences from the spectrum of un-modified cotton. Obviously the level of resin treatment is not sufficiently high to permit direct observation within the limits of the sensitivity of such spectra. For this reason, several special techniques or spectroscopic "tricks" were employed. In particular, these special techniques consisted of use of differential spectra, application of linear scale expansion, and a prechemical treatment consisting of an acid hydrolysis of the resin-treated sample.

If an equivalent amount of cellulose is placed in the reference beam of the double-beam spectrophotometer and the resin-treated cotton in the analysis beam, the resulting spectrum should be that of the modifying resin only, with the contribution of the cellulose electronically canceled. In practice, it is difficult to prepare KBr disks containing the same amount of KBr and unmodified cellulose and keep them both at the same moisture content. Nevertheless, by compensating to a large degree for the absorption of cellulose, while keeping the quantity of KBr equal and by equilibrating to avoid gross differences in moisture content, the presence of bands which would otherwise be masked or obscured can be detected. Examples are shown in Fig. 23, from the original report of McCall and co-workers [16].

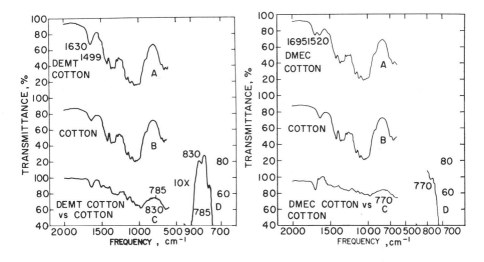

Fig. 23. Differential and linear-scale expanded infrared absorption spectra of (left) DMET cotton and (right) DMEC cotton. Left: A. DMET cotton; B. unmodified cotton; C. differential spectrum of DMET cotton; D. linearally expanded differential spectrum of DMET cotton. Right: A. DMEC cotton; B. unmodified cotton; C. differential spectrum of DMEC cotton; D. linearally expanded differential spectrum of DMEC cotton.

As can be seen in this figure, even when the spectrum of cotton has been balanced to permit the bands of the resins to be seen, several of the characteristic ring vibrations in the 700 to 800 cm^{-1} region are so weak as to be almost unidentifiable. At this point linear scale expansion, available now with most double-beam absorption spectrophotometers, is used. Linear scale expansion simply means that instead of automatically plotting on a transmittance scale from 0 to 100%, the scale is expanded in the region where the band is sought, so that, in this region, the scale may be, for example, from 60 to 90%. Use of linear scale expansion is illustrated in the inserts in Fig. 23. Here the weak, almost invisible bands of DMET cotton at 830 and 785 cm^{-1} and of DMEC cotton at 770 cm^{-1} in the differential

spectra, become, upon linear scale expansion, sufficiently intense to permit detection and identification of the triazone and carbamate resins,
respectively.

The third technique consists of an acid hydrolysis. This hydrolysis may
be preceded by a series of solvent extractions to remove optical whiteners,
softeners, etc., which may interfere with the identification of the resin.
The American Association of Textile Chemists and Colorists is now studying,
in Committee RA 45 - "Finish Identification," the use of infrared absorption
spectroscopy to identify chemical finishing agents such as softeners, whiteners, and soil release agents after separation by a series of extractions with
organic solvents. If these attempts are successful, infrared absorption
spectra will be destined to play an even more important role in the analysis
of cotton finishing agents.

The acid hydrolysis technique will, very likely, be the last step in the
qualitative scheme for separating various chemical finishing agents prior
to identifying them by infrared absorption spectra. Acid hydrolysis is the
most sensitive technique for the detection, identification, or quantitative
estimation of a resin finish by means of infrared absorption spectroscopy.

Spectra of typical acid hydrolysis products are illustrated in Fig. 24,
from the original publication of Miles et al. [17]. The spectra of dimethylol
urea (DMU) and of dimethylol ethyleneurea (DMEU) are compared to the
spectra of these reagents after acid hydrolysis and to the spectra of the acid
hydrolysate from a cotton treated with each of these reagents.

Similarly, in Fig. 25, the spectra of the acid hydrolyzed products from
dimethylolethyltriazone (DMET) and from trisaziridinyl phosphine oxide
(APO) are compared, respectively, with the hydrolysate products of cotton
fabric treated with these reagents. Note the easy identification of the resins
from the spectra of their cotton hydrolysates by direct comparison with the
spectra of acid hydrolyzed pure resins. Note also the considerable differences which sometimes occur during the acid hydrolysis of these resins, and
the difficulty of identification unless the spectrum of the hydrolysate from
the cotton fabric is compared to that of the resins after acid hydrolysis.

For the identification and measurement of degree of resin treatment, some
confusion existed regarding the merits of KBr disks, differential spectra,
the use of linear-scale expansion, or the technique of combining the infrared measurements with acid hydrolysis. For these reasons a description
of a combined method was published, designed to achieve identification and
reasonable estimation of the amount of resin treatment as simply, rapidly,
and accurately, as the particular sample would permit [18]. This combined
method merely advocates first the measurement of the KBr infrared absorption spectrum of the unknown resin-treated sample. If the identification and
quantitative estimation can be made from this spectrum, no further modification is recommended. If, however, such a spectrum does not reveal the
resin, the sample can be remeasured using a differential technique. Again,
if the problem is now solved, no further analysis is required. However, if

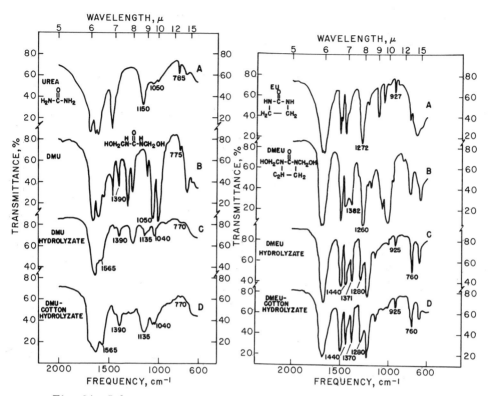

Fig. 24. Infrared absorption spectra of hydrolysate products of resin-treated cotton cellulose. Left: A. Urea; B. DMU; C. DMU after hydrolysis in situ; D. DMU after acid-hydrolysis from cotton fabric. Right: A. EU; B. DMEU; C. DMEU after hydrolysis in situ; D. DMEU after acid hydrolysis from cotton fabric.

bands still cannot be identified, it is now a simple step to use linear scale expansion, without any further sample preparation. If all these techniques fail, recourse must be had to the acid hydrolysis treatment. It is not intended by this order in the combined procedure to relegate the acid hydrolysis technique to the position of being a method of last resort. It might, under specific conditions be the preferred technique. Thus, it has particular merit if the specific investigation involves knowledge of softeners, whiteners, etc. The scheme as outlined in the so-called combined procedure was designed to permit the identification of the resin as simply and rapidly as possible, and the acid hydrolysis technique, involving a prechemical treatment, is not the most rapid nor the most simple.

It has been demonstrated that the combined KBr - differential - linear scale expansion - acid hydrolysis method will permit detection and reasonable qualitative estimation of any resin that has been used at a level to achieve satisfactory crease-recovery or durable-press properties. Obviously, it is possible to treat a fabric with a level of resin which cannot be detected by means of infrared absorption spectroscopy. Committee RA-45

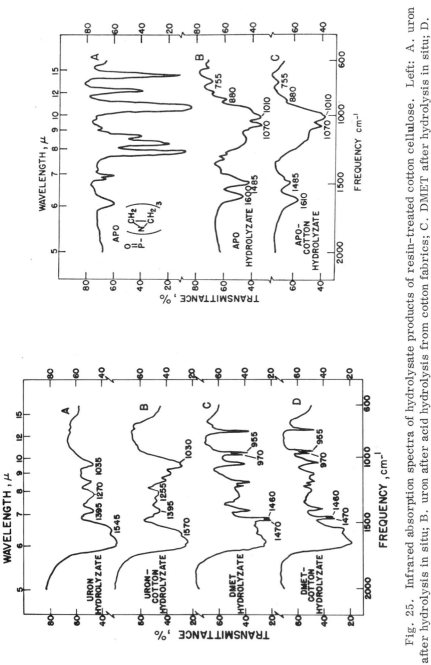

Fig. 25. Infrared absorption spectra of hydrolysate products of resin-treated cotton cellulose. Left: A. uron after hydrolysis in situ; B. uron after acid hydrolysis from cotton fabrics; C. DMET after hydrolysis in situ; D. DMET after acid hydrolysis from cotton fabrics. Right: A. APO; B. APO after hydrolysis in situ; C. APO after acid hydrolysis from cotton fabrics.

of the AATCC has adopted, in its collaborative testing, the criteria that, if the sample does not have satisfactory durable-press properties, the identification of the resin is to be abandoned. Thus, a lower limit at which the analytical spectroscopist can be reasonably expected to work is established. With this lower limit, the combined method should permit both identification and reasonable quantitative estimation of any resin treatment.

VI. SPECIALIZED ADVANTAGES OF MULTIPLE INTERNAL REFLECTANCE METHODS OF INFRARED

Within the last few years, a new technique has been introduced to infrared spectroscopists. This is the so-called attenuated total reflectance technique. Measurement of absorption by any of the techniques thus far referred to involves, in reality, measurement of loss in transmission. The transmission of the sample is compared to some standard, usually the solvent in which it is dissolved or the matrix with which it is mixed. This loss of transmission, measured under conditions where there are no losses due to reflectances, or where, at least hopefully, such losses due to any reflectance will be equal in each beam of the spectrophotometer, must be due to the radiation absorbed by the analytical sample. We can reverse this technique and measure the loss in reflectance under conditions where there are either no transmission losses or where they are equalized in both beams. It is interesting to note that this technique was the first, and for many years the only method for measuring the ultraviolet absorption of cotton cellulose [19, 20]. Use of attenuated total reflectance in the infrared spectra of cotton, however, was unsuccessful as the intensity was not sufficient. However, the introduction of multiple internal reflectance (MIR), or as it was earlier called frustrated multiple internal reflectance (FMIR), provides a very satisfactory technique, which, as has been illustrated [8] , can offer particular advantages in specific investigations. Multiple internal reflectance or frustrated multiple internal reflectance consists, as the names imply, of a technique for trapping the radiation, as in a prism. Several reflections are required before the radiation finally escapes from the prism (during which it is said to be frustrated) and is measured in the detection portion of the spectrophotometer.

Krentz [21] has recently published a comprehensive report on the applications of multiple internal or frustrated multiple internal reflectance to artificial leather and textile products, including cotton cellulose, oxidized cellulose, and urea derivatives of cellulose. Krentz includes an excellent review of the principles of multiple reflection as a means of obtaining infrared absorption spectra and considerable useful information regarding laboratory techniques to obtain satisfactory infrared absorption spectra of macromolecular substances. He presents several illustrations of its successful application, examples designed mainly to demonstrate the advantages of reflectance over transmission techniques for obtaining infrared absorption spectra.

MIR is not, however, a technique which is a panacea for all the problems of infrared spectrophotometry within the textile area. We still believe, for example, that for the problem, illustrated earlier, of identifying PA cotton and of quantitatively determining the extent of partial acetylation, the KBr disk technique is superior, particularly in quantitative accuracy. MIR does, however, offer particular advantages in specific types of problems. First, it can be used to obtain the infrared absorption spectrum of a small sample of fabric without in any way damaging or modifying the sample. Second, as it can be used to make measurements directly from a piece of fabric, it is the most rapid technique. The absorption spectrum of a fabric can be obtained by MIR in less time than is required to prepare a KBr disk. A third advantage involves the properties of reflectance. The absorption bands from multiple internal reflectance spectra at the longer wavelengths (shorter frequencies) will appear to be relatively more intense than those at shorter wavelengths, compared to the spectra of the same samples by means of transmission techniques. It will be recalled that, in the identification of resin-treated cotton cellulose, one of the problems was to obtain spectra with sufficient intensity to identify the positions of the ring bands, the C-H deformations about the various rings which are characteristic of various resins, and that linear scale expansion was resorted to in examples shown to afford this identification (Fig. 23). It appears that MIR spectra might, in specific instances, enable the identification to be made of the ring bands in the 700-800 cm^{-1} region without use of linear scale expansion.

An additional advantage of MIR spectra in specific applications again involves a characteristic property by this type of measurement. The reflectance is obtained only of the surface of the analytical sample. A painted wooden panel, for example, will by reflectance techniques yield a spectrum about identical to that of the paint in solution, with no bands characteristic of the wood. Thus, this technique can be used to investigate surface effects. This specific application is illustrated in Fig. 26. Fig. 26A is the KBr disk spectrum of an unmodified cotton and B is the KBr disk spectrum of a polyvinyl chloride reagent which was used to modify it. Spectrum C in Fig. 26 is that of the cotton treated with polyvinyl chloride resin obtained by the MIR technique. Fig. 26D is the spectrum of the identical treated cotton, but measurement by the MIR technique has been made of the reverse side of the fabric. Note the close resemblance between spectrum A, the unmodified cotton, and D, one side of the treated cotton; and between B, spectrum of the pure reagent, and C, spectrum of the reverse side of the treated cotton fabric.

The MIR technique is, as illustrated, a technique which will enable the spectroscopist to tell the cellulose chemist more about the migration of certain resins to the surface of fabrics, to inform him as to when he has accomplished a desired surface treatment without "bleeding" through the fabric, or alternately when he has achieved a uniform treatment throughout the fabric.

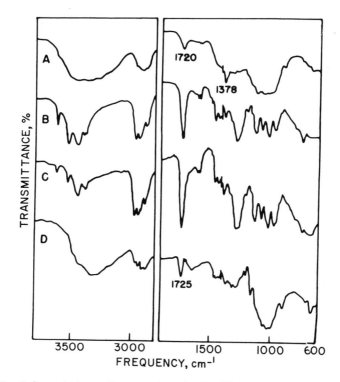

Fig. 26. Infrared absorption spectra obtained by MIR technique: A. unmodified cotton cellulose; B. polyvinyl chloride reagent; C. cotton fabric treated with polyvinyl chloride - front side of fabric sample; D. cotton fabric treated with polyvinyl chloride - reverse side of fabric sample.

Another technique which is quite new is the direct pressing of fibers, as described by Knight and his co-workers [22, 23]. The technique consists of pressing, in a press identical to that used for pressing KBr pellets, a thin layer of essentially parallel fibers of the short staple fibers of either a natural or synthetic material. The technique is, as will be recognized by x-ray diffractionists, similar to techniques used for obtaining x-ray diffraction powder patterns of fibers. It has not, probably, been introduced long enough to permit adequate evaluation. It might, hopefully, provide a method by which fibers, such as native cotton cellulose from which films cannot be generated, can be deuterated for infrared absorption investigations. A critical point will be to maintain the thin layer, 3 to 5 fibers thick, in the fiber press (as described by Knight and his co-workers) in the presence of D_2O, that is, in the presence of water, in either the vapor or liquid state.

VII. CONCLUSIONS

This review has illustrated how infrared absorption spectroscopy can be used to aid the cellulose chemist in the identification of finishing agents. However, there is today no specific analytical procedure, no standard method for the identification or determination of chemical finishing agents on cotton or on any other natural or synthetic fiber. In the opinion of this writer, if the research group within any textile organization has, as a high priority assignment, the problem of identifying and measuring the extent of chemical finishing on a specific fabric, a complete identification and reasonably accurate quantitative estimation could, and very probably would, be obtained. But the procedure would not be that of an analytical laboratory. The identification of each finishing agent and the estimation of the amount of each finish would not be dictated by a demonstrated reproducible and accurate analytical method. The attack would be, rather, in the nature of a minor research project, the identification and quantitative estimations made by a series of trials, reliability being attained by repetitive attacks which appeared most successful.

The present status of the analytical chemist to perform a complete analysis of a chemically finished fabric can be itemized by a series of quotations:

(1) "For a good many years, the identification of finishing agents on textile fabrics has been a matter of trial and error. In the case of inorganic materials, it is true, an ashing or a Kjedahl digestion may be performed, followed by a regular qualitative inorganic analysis of the resulting material; this is satisfactory and needs no further description." (See also, however, Chapter 1 of this volume.) "Organic materials, on the other hand, had to be guessed at, tested for, and if the first guess did not happen to be correct, further guesses and trials were made. This is obviously inefficient in principle and if several agents are present, one or more may be missed entirely."

(2) "The widespread use of the newer finishing agents has brought with it the problem of identification. This problem is complicated by the fact that many of these finishing materials do not possess the same chemical and physical properties before their application that they possess on the finished textile fabric." "The detection and identification of applied textile finishes has been, and will continue to be, a more and more difficult task as the number and types of commercial finishes increase."

(3) "Certain other finishing agents have not been included" (in an attempted systematic procedure of analysis for chemical finishing agents) "because (1) authentic samples could not be obtained, (2) because they have come onto the market too recently to be included in the investigation which resulted in this scheme, or (3) because they have not yet been definitely established as commercially acceptable finishes. It is hoped that these finishes may be fitted into the scheme of analysis as soon as possible."

Present-day analysts concerned with the complete identification and reliable estimation of chemical finishing agents will recognize in these quotations a summary of the major problems with which they are faced. The rather astonishing fact regarding these quotations is that they are from reports of John H. Skinkle [24] and of Ray Krammes and Charles Maresh [25], published seventeen and twenty-four years ago, respectively!

Analytical chemists in the field of textile chemistry have not kept up with the increase in number and types of commercial finishes, as predicted by Krammes and Maresh, despite the rapid advancements in analytical methods, including spectroscopic, microscopic, and other instrumental techniques. The greatest advance during the past two or three years has been a growing recognition of the problem, which is now probably well defined:

(1) There is a great, almost endless, variety of chemically identifiable materials which are being used in the chemical finishing of cotton fabric. Included in such a list will appear resins used to obtain durable-press or wash-and-wear characteristics and chemicals used to produce water repellency, fire resistance, resistance to mildew, or protection from actinic radiation. Then there are the endless lists of lubricants, plasticizers, softeners, antibacterial agents, deodorants, fungicides, germicides, and rot-resistant agents, and so on and so on.

(2) Not only is the analyst faced with the problem of detecting, identifying, and measuring such an endless number of chemical species, but he cannot have a prepared list even of possibilities. The techniques within the cotton fabric industry are changing so rapidly that by the time any list of finishing agents could be prepared with any claim for completeness, several of the chemicals would have been discarded and several new agents (of which the analyst might well not even be aware) will have replaced them.

(3) There is the problem as to how a chemical finish is to be identified. The analyst can, certainly with sufficient effort, identify a specific agent by its generic name. However, in specific analyses, trade names of the chemicals used are often highly desirable. Identification by trade name is, however, often difficult as it is not always possible to differentiate between two or more trade names merely by establishment of generic identity. It is often difficult for the analyst to obtain authentic samples of a specific trade name chemical, a factor which augments the difficulties of detection and identification as most analytical methods, particularly instrumental methods, are based on comparison techniques.

(4) The identification of many finishes by comparison techniques becomes more difficult by the fact that the chemical, physical, including optical, properties of many of these finishes undergo considerable modification during the chemical finishing process. It is not, usually, satisfactory to attempt to identify finishing agents, removed from a fabric, by direct comparison with a sample of the reagent from a supply bottle. The comparison must be made between the unknown finish with similar measurements of

known finishes removed from the fabric in a manner identical to the procedure used to obtain the unknown fraction.

Despite these apparent difficulties, it must not be conceded that a completely coordinated method for the analysis of a cotton fiber for the detection, identification, and measurement of all chemical finishing agents which it contains, which will be rapid, reproducible, and accurate, cannot be devised. While the problem has increased in magnitude, we have today more tools for attack than the analyst has ever had at his command.

The contributions of Skinkle and of Krammes and Maresh were to establish procedures for the successive removal of finishing agents by a series of extracting solvents, generally in order of increasing polarity. Their proposals might be compared, in general principle, to the qualitative scheme of inorganic analysis, familiar to all college chemistry majors.

The American Association of Chemists and Colorists, through its Committee RA-45 on Finish Identification, has embarked on a two-pronged attack on the problem to establish a means for the identification of finishing agents by a well-established, completely reproducible and reliable procedure [26]. The qualitative extraction schemes of first Skinkle and later of Krammes and Maresh are essentially identical and similar to the method recommended in the American Association of Chemists and Colorists annual Technical Manual to separate various finishing agents into specific categories [27]. A survey by Committee RA-45, however, reveals that there are as many variations in the use of this scheme, probably, as there are laboratories and analytical chemists using it. One of the first problems of the Committee then is to study the extraction procedure, to establish the most advantageous order of extraction solvents, and to attempt to reach agreement as to a standard uniform method of extraction. Once separated into qualitative groups, the chemical finishing agents within each group will have to be identified.

Infrared absorption spectroscopy, as illustrated in this chapter, will be one of the major tools, but not necessarily the only tool, for these identifications. However, advantageous use of infrared absorption spectroscopy, as indicated, will be as a comparison technique, using either established organic group -- band frequency correlation tables or complete "fingerprint" matching. Thus the second task of Committee RA-45 will be to compile and publish an adequate library of infrared absorption spectra of chemical finishing agents. Several attempts to achieve this goal on a limited basis, usually restricted to a single type or a closely related group of chemical finishing agents, have been published. The Northern Piedmont Section of AATCC has prepared a compilation entitled "Infrared Spectroscopy - An Aid in the Identification of Textile Surfactants" [28], which contains both several infrared spectra of surfactants and a number of tables of organic functional group - infrared absorption band frequency correlations. Hummell [29] has published a book entitled "Identification and Analysis of Surface Active Agents by Infrared and Chemical Method." An ideal model of the type of publication need for the identification of chemical finishes by means of infrared absorption spectra may be found in the recent publication of a new edition of "Infrared Spectroscopy - Its Use in the Coating Industry" by the Chicago Society for Paint Technology [30].

To achieve a standard, uniform analytical procedure for the identification of chemical finishes which is rapid, reliable, and accurate, and to establish a model library of infrared absorption spectra, comparable to that of the Coatings Industry, will not be easy. We will still face the problems itemized above, especially the almost endless growing list of chemical finishing agents, and with this ever growing list the problems which perplexed Skinkle and Krammes and Maresh two decades ago. How will we establish which proposed finishing agents should be included in a scheme of analysis? How can we identify and measure, by a procedure established today, a finishing agent which will be introduced next week or next month?

To achieve these goals will require effort - a great effort. But more than effort it will require cooperation. Unless a small nucleus of dedicated workers, with backing from their organizations, can be found, there will be little, if any, chance for success. With such a nucleus, despite the difficulties, the envisioned analytical procedure can be made a routine operation. The attainment of such a goal must be preceded by a true desire for it. If such a desire is not now awakened, then in the latter part of this century, another two decades from now, some future writer can again itemize the needs for a method to identify chemical finishes and compare them, again with appropriate exclamation points, with the needs as published way back in 1970.

REFERENCES

[1] W. H. Julius, Verhandel. Koninkl. Ned. Akad. Wetenschap., Amsterdam, Sec. 1, I, 49 pp. 1892.

[2] W. W. Coblentz, Investigations of Infrared Spectra. Washington, D.C. 1905. Carnegie Inst. of Washington, Publication No. 35.

[3] R. T. O'Connor, Chairman, Recommended Practices for Identification of Material by Absorption Spectroscopy Using the Wyandotte-ASTM (Kuentzel) Punched Card Index, Manual on Recommended Practices in Spectrophotometry, Sponsored by ASTM Committee E-13, pp. 29-38.

[4] J. W. Rowan, C. M. Hunt, and E. K. Plyler, J. Res. Natl. Bur. Std., 39, 133-140 (1947).

[5] F. H. Forziati, W. K. Stone, J. W. Rowen, and W. D. Appel, J. Res. Natl. Bur. Std., 45, 109-113 (1950).

[6] H. J. Marrinan and J. Mann, J. Appl. Chem. (London), 4, 204-211 (1954).

[7] R. T. O'Connor, E. F. DuPre, and E. R. McCall, Anal. Chem., 29, 998-1005 (1957).

[8] E. R. McCall, S. H. Miles, and R. T. O'Connor, Am. Dyestuff Reptr., 55 (No. 11, 31-35), 400-404 (1966).

[9] R. T. O'Connor, in Absorption Spectroscopy, Analytical Methods for Textile Laboratory (J. W. Weaver, ed.), 2nd ed., Chap. 10, Am. Assoc. Textile Chemists and Colorists, Research Triangle Park, N.C., 1968.

[10] Am. Association of Textile Chemists and Colorists Questionnaire, Which Fiber Blends Should Be Included? Textile Chemists and Colorists, 1 (No. 19, 53-54), (1969).

[11] F. G. Hurtubise, Tappi, 45, 460-465 (1962).

[12] H. G. Higgins, Australian J. Chem., 10, 496-501 (1957).

[13] E. Klein and J. E. Snowden, Ind. Eng. Chem., 50, 80-2 (1958).

[14] R. T. O'Connor, E. R. McCall, and D. Mitcham, Am. Dyestuff Reptr., 49 (No. 7, 35-43), 214-222 (1960).

[15] W. A. Reeves and J. D. Guthrie, Textile Res. J., 23, 522-527 (1953).

[16] E. R. McCall, S. H. Miles, V. W. Tripp, and R. T. O'Connor, Appl. Spectry., 18, No. 3, 81-84 (1964).

[17] S. H. Miles, E. R. McCall, V. W. Tripp, and R. T. O'Connor, Am. Dyestuff Reptr., 53 (No. 12, 23-27), 440-444 (1964).

[18] E. R. McCall, S. H. Miles, and R. T. O'Connor, Am. Dyestuff Reptr., 56 (No. 2, 13-17), 35-39 (1967).

[19] G. Champetier and R. Morton, Bull. Soc. Chim. France, Ser. 5, 10, 102-106 (1952).

[20] J. D. Dean, C. M. Fleming, and R. T. O'Connor, Textile Res. J., 22, 609-616 (1952).

[21] V. Krentz, Melliand Textilber., 50 (5) 557-563 (1969) (in German).

[22] J. A. Knight, M. P. Smoak, R. A. Porter, and W. E. Kirkland, Textile Res. J., 37, 924-927 (1967).

[23] J. A. Knight, H. L. Hicks, and K. W. Stephens, Textile Res. J., 39, 324-328 (1969).

[24] J. H. Skinkle, Am. Dyestuff Reptr., 35 (No. 14, Sept. 23, 1946), 449-452 (1946).

[25] R. Krammes and C. Maresh, Am. Dyestuff Reptr., 42 (No. 11, May 25, 1953), 317-327 (1953).

[26] R. T. O'Connor, Chairman, Report of Technical Committee RA-45, Finish Identification, Vol. 45, Tech. Manual of the American Association of Textile Chemists and Colorists, Research Triangle Park, N.C., 1969, p. 22.

[27] AATCC Test Method 94-1969 Finishes in Textiles: Identification, Vol. 45, Tech. Manual of the American Association of Textile Chemists and Colorists, Research Triangle Park, N.C., 1969, pp. 71-83.

[28] J. E. Nettles, Textile Chem. Color., 1 (No. 20, 49-59), 430-440 (1969).

[29] D. Hummel, Identification and Analysis of Surface Active Agents by Infrared and Chemical Methods. Text and Spectra Volumes. Translated by E. M. Walkow, Wiley-Interscience, New York.

[30] Infrared Spectroscopy Committee of the Chicago Society for Paint Technology, Infrared Spectroscopy - Its Use in the Coating Industry, Federation of Societies for Paint Technology, Philadelphia, Pa., 1969.

APPENDIX

Infrared Absorption Bands of Particular Significance in the
Investigation of Modified Cottons and Cotton Blends

Unmodified cotton spectra, approx. position of band maxima (μ)	Vibrating group most probably giving rise to observed absorption band	Modified cotton spectra, approx. position of band maxima (μ)	Specific use in application of infrared absorption spectra to cotton and modified cotton	Blend component other than cotton spectra, approx. position of band maxima (μ)
2.8–3.2	Free O—H and bonded O—H ...stretching		Investigations of extent and type of hydrogen bonding [1]	
	N—H stretching	3.0	Appears as a broadening of the 2.8-3.2 μ O—H stretching band in the spectra of aminocellulose	
	C—H stretching of CH_3 group adjacent to SO_3 group	3.30	Identification of mesylation, especially to differentiate from tosylation	
3.4	C—H stretching of C—H and CH_2 groups		Characteristic of all cottons	

Vibration			Remarks
C—H stretching of CH$_3$ group	3.45-3.55		Indicates methyl or ethyl cellulose
C≡N stretching		4.40	Characteristic of Acrilan
C≡N stretching		4.43	Differentiate Dynel from Saran
C≡N stretching	4.45		Identification and quantitative measurement of cyanoethylation
C≡N stretching		4.48	Characteristic of Orlon
C≡N stretching	4.62 and 4.85		Characteristic of cotton modified by replacement reaction with thiocyanate
N≡N stretching	4.72		Identification and quantitative measurement of cotton modified by replacement reaction with azide
C=O stretching (anhydride)	5.57		Coupled with deformation bands in 13μ region, verifies that esterification with anhydride has been accomplished without breaking anhydride ring, e.g., aconitic anhydride

Unmodified cotton spectra, approx. position of band maxima (μ)	Vibrating group most probably giving rise to observed absorption band	Modified cotton spectra, approx. position of band maxima (μ)	Specific use in application of infrared absorption spectra to cotton and modified cotton	Blend component other than cotton spectra, approx. position of band maxima (μ)
	C=O stretching	5.58	In absence of evidence of anhydride ring, may indicate position of substituent	
	C=O stretching		Differentiate Acrilan from Orlon	5.70
	C=O stretching (ester)	5.75	Identification and quantitative measurement of esterification	
	C=O stretching		Identify jute	5.75
	C=O stretching		Characteristic of silk and wool	5.75
	C=O stretching		Characteristic of Dacron	5.80
	C=O stretching		Differentiate Orlon from Acrilan	5.80

C=O stretching (acid)	5.85	If this band appears only after a HC1 wash it can be used to differentiate esterification and etherification with an acid. Absence of this band after esterification with a dicarboxylic acid may indicate crosslinkage
C=O stretching, N–H bending – amide I	5.88	Identify dimethylol ethylcarbamate
C=O stretching, N–H bending – amide I	5.95	Identify dimethylol ethyleneurea
C=O stretching	5.98	May indicate esterification with a polybasic acid
C=O stretching (amide)	6.00–6.06	Indicates modification to form an amide, e.g., carbamoylethylation with acrylamide; replacement of mesyl with N,N-dimethylformamide; replacement of cyanoethyl with hydroxylamine (see 6.30μ)
C=O stretching, N–H bending – amide I	6.01	Identify dimethylol urea
C=O stretching, N–H bending – amide I	6.06	Identify dimethylol uron

Unmodified cotton spectra, approx. position of band maxima (μ)	Vibrating group most probably giving rise to observed absorption band	Modified cotton spectra, approx. position of band maxima (μ)	Specific use in application of infrared absorption spectra to cotton and modified cotton	Blend component other than cotton spectra, approx. position of band maxima (μ)
	C=C stretching (aliphatic)	6.10	Indicates modification with aliphatic unsaturated compound or aromatic compound containing unsaturated side chain	
	C=O stretching, N-H bending – amide I	6.10	Identify dimethylol ethyltriazone	
	C=O stretching		Identify jute	6.10
6.13	Absorbed H₂O		Characteristic of all cottons	
	C=O stretching, N-H bending – amide I		Characteristic of nylon	6.13
	C=O stretching, N-H bending – amide I	6.15	Identify dimethylol perhydropyrimidinone	

Wavelength (μ)	Assignment	Remarks
6.13, 6.18	C=O stretching, N-H bending – amide I	Characteristic of silk and wool
6.25	C=C stretching (aromatic ring)	Evidence of acylation with aromatic compound. Means for differentiation between tosylation and mesylation (see 6.70 μ)
6.30	N-H deformation (amide)	See 6.00–6.06 μ
6.2–6.4	COO⁻ ion stretching mode	Evidence for etherification with an acid and resulting salt formation
6.30	N-H deformation	Probably indicates amino cellulose or 2-aminoethyl cellulose (see 3.0 μ)
6.32	C=C stretching of conjugated aromatic ring	Evidence for acylation with benzoic-type reagent
6.41	C=O stretching, N-H bending – amide I	Identify methylol melamine
6.47	C=O stretching, N-H bending – amide II	Identify dimethylol urea

Unmodified cotton spectra, approx. position of band maxima (μ)	Vibrating group most probably giving rise to observed absorption band	Modified cotton spectra, approx. position of band maxima (μ)	Specific use in application of infrared absorption spectra to cotton and modified cotton	Blend component other than cotton spectra, approx. position of band maxima (μ)
	C=O stretching, N–H bending – amide II		Characteristic of nylon	6.52
	C=O stretching, N–H bending – amide II	6.58	Identify dimethylol ethylcarbamate	
	C=O stretching, N–H bending – amide II	6.62	Identify dimethylol uron	
	C=O stretching, N–H bending – amide II	6.64	Identify dimethylol perhydropyrimidinone	
	C=O stretching, N–H bending – amide II		Characteristic of silk and wool	6.67
	C=O stretching, N–H bending – amide II	6.67	Identify dimethylol ethyltriazone	

Wavelength (μ)	Assignment	Remarks
6.7	C=C stretching (aromatic ring)	See 6.25 μ
6.76	C=O stretching, N-H bending – amide II	Identify dimethylol ethyleneurea
6.85	C=O stretching, N-H bending – amide II	Identify methylol melamine
6.92	C-H bending	Differentiate Dynel and cotton from Saran
6.96	CH_2 wagging	Measure of degree of crystallinity
7.26	C-H deformation	Splitting of this band into two components indicates esterification with isobutyl
7.09	C-H bending	Differentiate Saran from cotton and Dynel
7.3	S=O stretching of sulfonate, $-O-SO_2-R$ group	Evidence of mesylation or tosylation (see 8.5 μ)
7.46	O-H in-plane deformation	Characteristic of all cottons

Unmodified cotton spectra, approx. position of band maxima (μ)	Vibrating group most probably giving rise to observed absorption band	Modified cotton spectra, approx. position of band maxima (μ)	Specific use in application of infrared absorption spectra to cotton and modified cotton	Blend component other than cotton spectra, approx. position of band maxima (μ)
7.58	O–H deformation and/or CH_2 wagging		Characteristic of all cottons	
	O–H deformation		Characteristic of flax	7.60
7.78	O–H deformation and/or CH_2 wagging		Characteristic of all cottons	
	P=O stretching	8.0	Evidence for modification with phosphoryl — not observed in spectra of any modified cottons examined — see text	
	C–H bending		Differentiate wool from silk	7.85
8.02	O–H deformation and/or CH_2 wagging		Characteristic of all cottons	

Wavelength (μ)	Assignment	Remarks
8.00	C–O stretching	Differentiate jute from flax
8.08	C–O stretching	Differentiate Acrilan from Orlon
8.18	C–O stretching	Differentiate silk from wool
8.20	C–O stretching	Differentiate wool from silk
8.5	S=O stretching of sulfonate, $-O-SO_2-R$ group	See 7.3 μ
8.61	Unassigned	Characteristic of all cottons
8.95	Unassigned	Characteristic of all cottons
9.45	Unassigned	Characteristic of all cottons
9.5–9.7	P-O-C stretching	Evidence for modification with phosphoryl (see 8.0 μ)
9.71	Unassigned	Characteristic of all cottons
10.62	C–H deformation	Evidence for methyl cellulose
11.20	CH_2 wagging	Measure of degree of crystallinity

Unmodified cotton spectra, approx. position of band maxima (μ)	Vibrating group most probably giving rise to observed absorption band	Modified cotton spectra, approx. position of band maxima (μ)	Specific use in application of infrared absorption spectra to cotton and modified cotton	Blend component other than cotton spectra, approx. position of band maxima (μ)
	C–H deformation	10.85 and 11.30	Evidence for ethyl cellulose	
	C–H deformation	11.82	Evidence for esterification with isobutyric acid	
	C–H deformation about ring	12.05	Identify dimethylol ethyltriazone	
	C–O–S stretching	12.1–12.3	Identification and quantitative measurement of mesylation or tosylation	
	C–H deformation about ring	12.27	Identify tris–azidinal phorpine oxide	
	C–H deformation about ring	12.35	Identify methylol melamine	

C–H deformation about ring	12.42	Identify dimethylol uron
C–S (tosyl) stretching	12.62	Identification of tosylation
C–H deformation about ring	12.82	Identify dimethylol ethyltriazone
C–H deformation	12.88	Evidence for esterification with capric acid
C–H deformation about ring	12.90	Identify dimethylol ethylcarbamate
$PhCH_2C=O$	13.0–13.1	Identification and quantitative measurement of esterification with phenylacetic acid type reagent
C–H deformation about ring	13.16	Identify dimethylol ethyleneurea
C–S (mesyl) stretching	13.25	Identification of mesylation
C–H bending about anhydride ring	13.0 and 13.6	Verifies esterification with anhydride (see 5.57 μ)

Unmodified cotton spectra, approx. position of band maxima (μ)	Vibrating group most probably giving rise to observed absorption band	Modified cotton spectra, approx. position of band maxima (μ)	Specific use in application of infrared absorption spectra to cotton and modified cotton	Blend component other than cotton spectra, approx. position of band maxima (μ)
	C-H deformation about ring	13.33	Identify dimethylol ethyltriazone	
	C-H deformation about ring	13.33	Identify dimethylol perhydropyrimidinone	
	C-H bending about aromatic ring	14.1-14.4	Identification and quantitative measurement of modification with aromatic compound